STUDENT'S SOLUTIONS MANUAL

BEVERLY FUSFIELD

MATHEMATICAL REASONING FOR ELEMENTARY TEACHERS

SIXTH EDITION

Calvin A. Long
Washington State University

Duane W. DeTemple
Washington State University

Richard S. Millman
Georgia Institute of Technology

Addison-Wesley
is an imprint of

The author and publisher of this book have used their best efforts in preparing this book. These efforts include the development, research, and testing of the theories and programs to determine their effectiveness. The author and publisher make no warranty of any kind, expressed or implied, with regard to these programs or the documentation contained in this book. The author and publisher shall not be liable in any event for incidental or consequential damages in connection with, or arising out of, the furnishing, performance, or use of these programs.

Reproduced by Pearson Addison-Wesley from electronic files supplied by the author.

Publishing as Pearson Addison-Wesley, 75 Arlington Street, Boston, MA 02116.

ISBN-13: 978-0-321-69386-0
ISBN-10: 0-321-69386-8

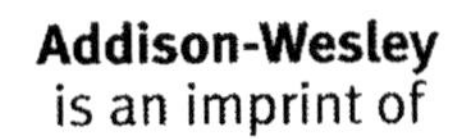

www.pearsonhighered.com

Contents

Chapter 1 Thinking Critically

Problem Set 1.1

1. (a) Using guess and check:
Guess 14 bikes, 13 trikes. The number of wheels is $14 \times 2 + 13 \times 3 = 67$.
Too many wheels, too many trikes.
Guess again: 17 bikes, 10 trikes. The number of wheels is $17 \times 2 + 10 \times 3 = 64$.
Still too many wheels.
Guess again: 21 bikes, 6 trikes. The number of wheels is: $21 \times 2 + 6 \times 3 = 60$.
O.K.

4. (a) The only possible sums of 3 digits totaling 19 are:
$4 + 7 + 8 = 19$
$4 + 6 + 9 = 19$.
Since 4 is used twice, it must be in the middle.

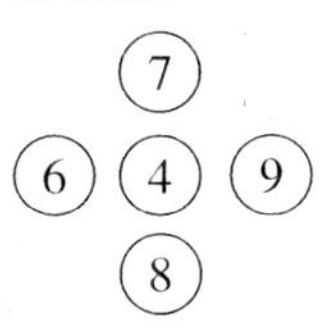

5. Use the make an orderly list strategy.

Dimes	Nickels	Pennies	Total Value
4	0	5	45¢
4	1	4	49¢

4 dimes are too many; try 3 dimes.

Dimes	Nickels	Pennies	Total Value
3	1	5	40¢
3	2	4	44¢
3	3	3	48¢

Xin has 3 dimes, 3 nickels, and 3 pennies.

6. Work backwards from 52.
Add 8: $8 + 52 = 60$.
Divide by 5: $60 \div 5 = 12$.
Guess and check or a carefully structured guess and check strategy, as in 1(b), also work well.

9. (a) Answers will vary. Two possibilities are

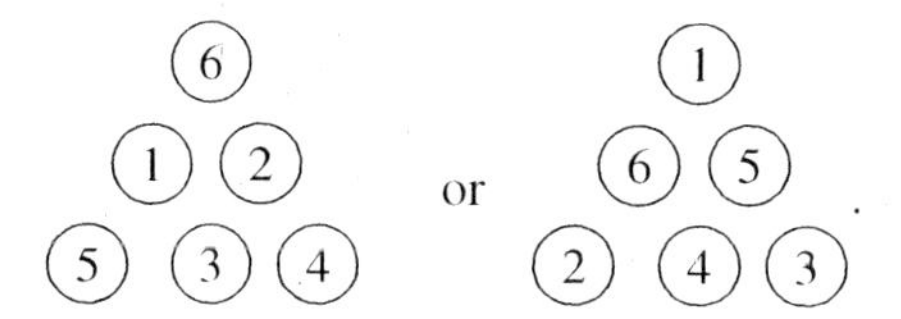

Another possibility is given in part (c).

11. Parts (a), (b), and (c) have more than one solution. You can place an arbitrary number in the upper left circle and then complete the rest of the circles.

(a)
7 15 8
11 20
4 16 12

(b)
7 12 5
10 21
3 19 16

(c)
1 7 6
9 16
8 18 10

(d) Note that in each of (a), (b), and (c), the sum of the top and bottom numbers given is the sum of the left and right numbers; i.e., $11 + 20 = 15 + 16$, $10 + 21 = 12 + 19$, and $9 + 16 = 7 + 18$. For such a problem to have a solution, this must always be the case. Thus, there is no solution.

12. (a) Since $2 + 3 = 5$, $3 + 5 = 8$, $5 + 8 = 13$, and $8 + 13 = 21$, the sequence is 1, 2, 3, 5, 8, 13, 21.

(c) This can be solved using the guess and check strategy. A more formal solution is as follows: Suppose the second number is n. Then the third number is $3 + n$, and the fourth number is 13. Therefore, $n + (3 + n) = 13$, so $n = 5$. The sequence is 3, 5, 8, 13, 21, 34, 55.

(e) Suppose the second number is n. Then the sequence is 2, n, $2 + n$, $2 + 2n$, $4 + 3n$, 11. Therefore $(2 + 2n) + (4 + 3n) = 11$, so $6 + 5n = 11$ and $n = 1$. The sequence is 2, 1, 3, 4, 7, 11.

14. (b) Each result is 5. For example, in the top row, $6 + 8 - 9 = 5$.

			5
6	9	8	5
3	5	7	5
2	1	4	5
5	5	5	5

Problem Set 1.2

1. No, because $5 \times 10 + 13 = 63$, not 48.

2. Using the guess and check and making an orderly list of strategies, we construct the following table.

Guess for Lisa's number	Number that results
1	$7 \cdot 1 - 4 = 3$
2	$7 \cdot 2 - 4 = 10$
3	$7 \cdot 3 - 4 = 17$

Lisa's number is 3.

4. (a) Yes. Yes. Juan's rule could have been either one of these. The result obtained by the first rule is $5n - 3$. The result obtained by the second rule is
$5(n - 1) + 2 = 5n - 5 + 2 = 5n - 3$, which is the same as the first rule.
Without algebra, you can reason that multiplying one less than a number by five gives a result that is 5 less than the original number multiplied by 5. Then adding 2 gives a number which is 3 less than the original number multiplied by 5.

5. (a) Add: $7 + 1 = 8$, $1 + 4 = 5$, $4 + 7 = 11$.

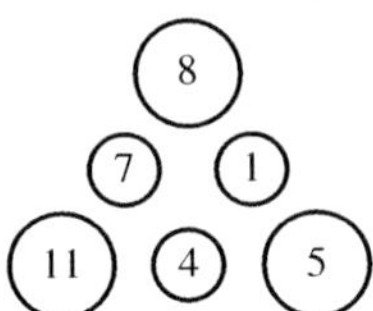

(b) Note that $11 + 1 = 12$, $19 - 11 = 8$, and $8 + 1 = 9$.

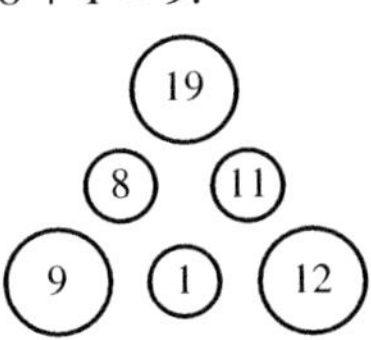

(c) The sum of the three new numbers must be $(7+11+13)/2 = 15.5$. Note that $15.5 - 7 = 8.5$, $15.5 - 11 = 4.5$, and $15.5 - 13 = 2.5$.

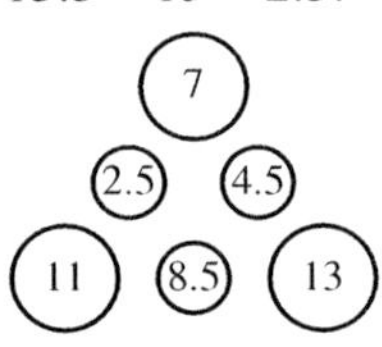

(d) The sum of the three new numbers must be $\dfrac{-2+7+11}{2} = 8$. Note that
$8 - (-2) = 10$, $8 - 7 = 1$, and $8 - 11 = -3$.

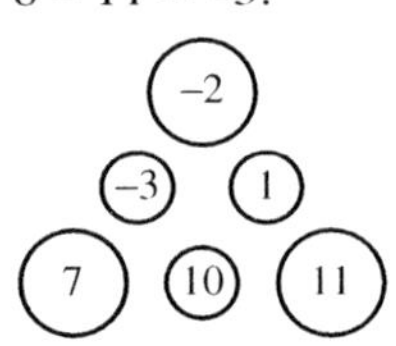

7. There are 49 ways, listed below in order of number of pennies used. For example, Q + D + 15P means a quarter, a dime, and 15 pennies.

2Q
Q + 2D + N
Q + D + 3N
Q + 5N
5D
4D + 2N
3D + 4N
2D + 6N
D + 8N
10N
Q + 2D + 5P
Q + D + 2N + 5P
Q + 4N + 5P
4D + N + 5P
3D + 3N + 5P
2D + 5N + 5P
D + 7N + 5P
9N + 5P
Q + D + N + 10P
Q + 3N + 10P
4D + 10P
3D + 2N + 10P
2D + 4N + 10P
D + 6N + 10P
8N + 10P
Q + D + 15P
Q + 2N + 15P
3D + N + 15P
2D + 3N + 15P
D + 5N + 15P
7N + 15P
Q + N + 20P
3D + 20P
2D + 2N + 20P
D + 4N + 20P
6N + 20P
Q + 25P
2D + N + 25P
D + 3N + 25P
5N + 25P
2D + 30P
D + 2N + 30P
4N + 30P
D + N + 35P
3N + 35P
D + 40P
2N + 40P
N + 45P
50P

There are 49 different ways to make change.

11. Use the strategy of making an orderly list. Assume that the pearls and the bags are identical, so that, for example, 21, 1, 3 is considered the same as 21, 3, 1. Then we need list only the possibilities in which the number of pearls in bag 1 is at least as great as the number of pearls in bag 2, and the number of pearls in bag 2 is at least as great as the number of pearls in bag 3. We have the following:

Bag 1	Bag 2	Bag 3
23	1	1
21	3	1
19	5	1
19	3	3
17	7	1
17	3	5
15	9	1
15	7	3
15	5	5
13	11	1
13	9	3
13	7	5
11	11	3
11	9	5
11	7	7
9	9	7

13. Make an orderly list.

Number pair	Sum
4, 24	28
6, 16	22
8, 12	20

15. A diagram of the situation will show that Bob has to make 9 cuts to get 10 2-foot sections. Since each cut takes one minute, it will take Bob 9 minutes to do this.

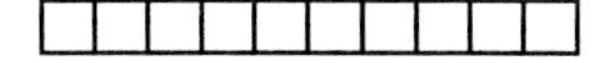

18. Though it may seem unusual, the simplest approach to this problem is that of drawing a diagram as in Example 1.5. Make the analogy between the race in Example 1.5 and the political race in the present problem. Draw a line with equally spaced points with the distance between consecutive points representing 1000 votes. Then place A (for Albright), B, C, D, and E on the line according to the statement of the problem.

B A E C D

Place A at an arbitrary point on the line and B two units to A's left since Albright finished 2000 votes ahead of Badgett, and so on. The completed diagram is as shown and the order of finishing from first to last is Dawkins, Chalmers, Ertl, Albright, and Badgett.

Problem Set 1.3

1. **(a)** 2, 5, 8, 11, 14, 17, 20. Each succeeding term is 3 more than the preceding term.

(c) 1, 1, 3, 3, 6, 6, 10, 10, 15, 15. The sequence consists of the triangular numbers repeated twice. (See page 37.)

(e) 2, 6, 18, 54, 162, 486, 1458. Each term is three times the preceding term.

3. **(a)** Each term is 2 more than its predecessor. Since $35 = 5 + 30$ and $30 \div 2 = 15$, we conclude that 15 2s must be added to 5 to get 35. Thus, there are 15 terms after the first one, or 16 terms in all.

4. **(a)** Yes, the sequence is arithmetic because the difference of consecutive terms is the constant -3.

(b) The difference between the first and last terms is $-52 - 8 = -60$. Thus, the number of terms is one more than $-60 \div (-3) = 20$, or 21 terms.

5. **(a)** The middle term on the left side of the nth equation is n. The next two lines are:
$1 + 2 + 3 + 4 + 5 + 4 + 3 + 2 + 1 = 25$
$1 + 2 + 3 + 4 + 5 + 6 + 5 + 4 + 3 + 2 + 1 = 36.$

8. Notice the pattern: Except for the last column of the table, when you move right one square, the units digit increases by 1, and when you move down one square, the tens digit increases by 1. Hence, we fill in the arrays and determine the desired last number as shown.

(a)

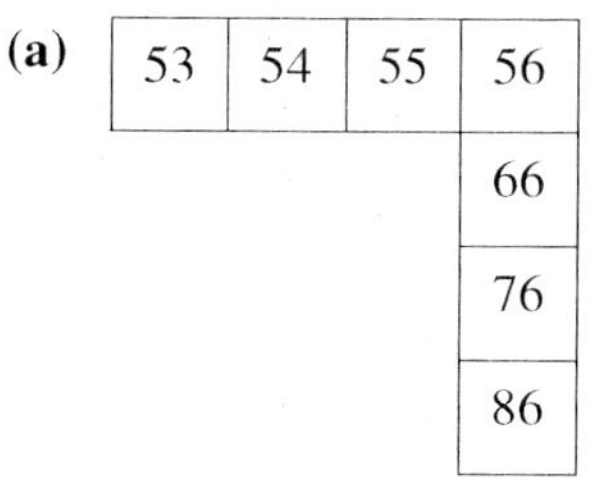

The desired number is 86.

(d) 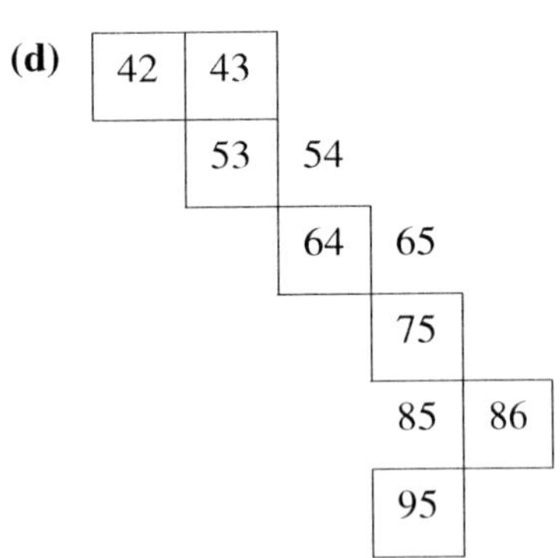

The desired number is 42.

10. (a) The left side of the equation has an additional square number added or subtracted in each step. Additions and subtractions alternate.
$1-4+9-16+25=15$
$1-4+9-16+25-36=-21$

(b) The sequence 1, 3, 6, 10, ..., is the sequence of triangular numbers,
$t_n=\dfrac{n(n+1)}{2}$ (See section 1.4 in the text.)
Since the right sides of the equation alternate between positive and negative, we expect the seventh equation to have $t_7=\dfrac{7\cdot 8}{2}=28$ on the right, and the eighth equation should have $-t_8=-\dfrac{8\cdot 9}{2}=-36$.
The equations are:
$1-4+9-16+25-36+49=28$
$1-4+9-16+25-36+49-64=-36$

(c) n even: $1-4+9-\cdots-n^2=-\dfrac{n(n+1)}{2}$

n odd: $1-4+9-\cdots+n^2=\dfrac{n(n+1)}{2}$

15. (a) As shown in the diagram in the text, there are 6 chords.

(c) Each dot is connected to 99 other dots, so 99 chords end at each dot. By multiplying 100×99, we count each chord twice (since a chord has 2 endpoints), so the number of chords is $\dfrac{100\times 99}{2}=4950$.

19. (a) For 4: $\sqrt{1\cdot 1\cdot 3\cdot 6\cdot 10\cdot 5}=\sqrt{900}=30$.
For 15: $\sqrt{6\cdot 5\cdot 10\cdot 20\cdot 35\cdot 21}$
$=\sqrt{4,410,000}=2100$.
For 35: $\sqrt{21\cdot 15\cdot 20\cdot 35\cdot 70\cdot 56}$
$=\sqrt{864,360,000}=29,400$.

Problem Set 1.4

1. (a) Each pattern has one more column of dots than the previous one. The next three patterns are shown:

● ● ● ● ● ● ● ● ● ● ● ● ● ● ●
● ● ● ● ● ● ● ● ● ● ● ● ● ● ●

(d) The nth even number.

2. Let n = Toni's number. Toni doubles it to get $2n$, and then adds 11, ending with $2n + 11$. Thus, $2n + 11 = 39$ or $2n = 28$, so $n = 14$.

6. (a) Each additional table increases the number of seats by 4. Thus, the number seated at n tables has the form $a+4n$, for some a. Since 6 are seated at the first table (when $n = 1$), we see that $a = 2$. Thus, $2+4n$ people can be seated an n tables.

(b) To seat 24 people, n must be the smallest integer for which $2+4n$ is at least 24. This occurs at $n = 6$, and leave two empty places.

7. (a) Person A shakes with B and C. Since B and C have already shaken hands with A, only one shake remains, B with C. The total is $2 + 1 = 3$.

(c) The logic is the same as in part (b). The first person shakes hands with the other 199, the second with 198, the third with 197, etc. until we get to the penultimate person who has one new hand to shake. The total is
$199+198+197+196+\ldots+2+1$
$=\dfrac{(200)(199)}{2}=19,990$, where we have used Gauss's Insight.

13. (a) In each figure, dots are added to the upper left, upper right, and lower right sides to complete the next larger pentagon.

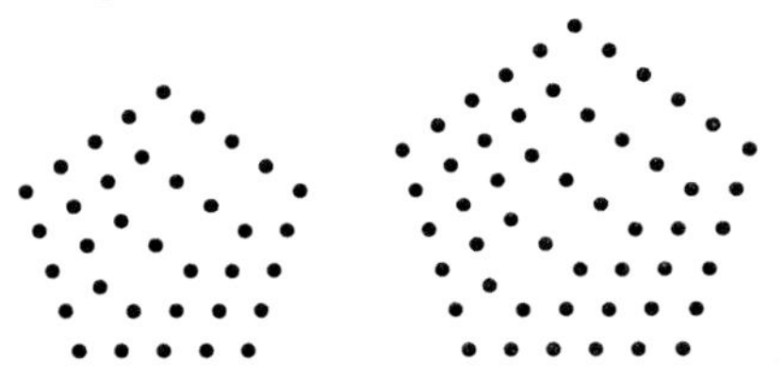

(b) 1, 5, 12, 22, 35, 51, ...

(c) $1 + 4 + 7 + 10 + 13 = 35$
$1 + 4 + 7 + 10 + 13 + 16 = 51$

(d) 10th term = $1 + 3(9) = 28$

(e) Use Gauss's Insight:

$$\begin{array}{l} s = 1 + 4 + 7 + \cdots + 28 \\ s = 28 + 25 + 22 + \cdots + 1 \\ \hline 2s = 29 + 29 + 29 + \cdots + 29 \end{array}$$

$$\text{Sum} = \frac{(10)(29)}{2} = 145$$

(f) nth term = $1 + 3(n - 1) = 3n - 2$

(g) Using Gauss's Insight and the result from part (f), there are n terms of $(3n - 1)$. The sum is $\frac{n(3n-1)}{2}$. Therefore,

$$p_n = \frac{n(3n-1)}{2}.$$

16. **(a)** The entries in the second row are $x + 8$ and 9, so $(x + 8) + 9 = 23$, or equivalently, $x + 17 = 23$. Subtracting 17 from both sides gives $x = 6$. The entries in the second row are then 14 and 9, respectively.

17. Suppose that x denotes the value in the lower small circle. Then the entries in the other small circles are $17 - x$ and $26 - x$, giving the equation $(17 - x) + (26 - x) = 11$. This simplifies to $43 - 2x = 11$, or $2x = 43 - 11 = 32$. Therefore, $x = 16$, and the entries in the other two circles are $17 - 16 = 1$ and $26 - 16 = 10$. Alternatively, one can work clockwise to see that the upper left small circle is $17 - x$ and therefore the remaining small circle value is $11 - (17 - x) = x - 6$. Then $(x - 6) + x = 26$, or $2x - 6 = 26$. As before, $x = 16$.

18. Let x be the unknown value in the lowermost circle. Working clockwise, the entries in the other small circles are $13 - x$, $13 + x$, $18 - x$, and $x - 1$. Thus, $x + (x - 1) = 23$, or equivalently, $2x = 24$ and $x = 12$. In clockwise order from the bottom, the entries are then 12, 1, 25, 6, and 11.

19. **(b)** Let x, y, z, w be integers in the circles:

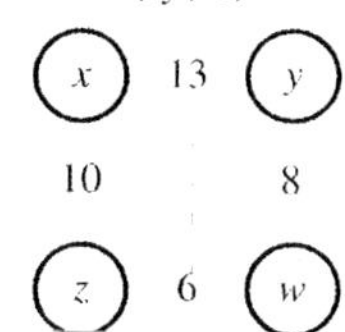

The conditions of the problem are $x + y = 13$, $x + z = 10$, $y + w = 8$, and $w + z = 6$. Subtracting the second equation from the first gives $y - z = 3$. Subtracting the fourth equation from the third gives $y - z = 2$. It is impossible for $y - z = 3$ and $y - z = 2$. Thus, there are no solutions.

20. **(a)**

n	1	2	3	4	5	6
n^2	1	4	9	16	25	36
$(n+1)^2$	4	9	16	25	36	49
difference	3	5	7	9	11	13

24. Let L and W denote the length and width of the rectangle, respectively, and S the length of the sides of the square. Since the rectangle is 3 times as long as it is wide, $L = 3W$. Therefore, the perimeter of the rectangle is $2L + 2W = 6W + 2W = 8W$, and its area is $LW = 3W^2$. The perimeter of the square is $4S$, and its area is S^2. We know both the rectangle and the square have the same perimeter, so $8W = 4S$, or $2W = S$. Also, the area of the square is 4 square feet larger than the area of the rectangle, so $S^2 = 3W^2 + 4$. By substitution, $(2W)^2 = 3W^2 + 4$, or $4W^2 = 3W^2 + 4$. Therefore, $W^2 = 4$, and the positive width of the rectangle is $W = 2$. Its length is $L = 3W = 6$. The square has sides of length $S = 2W = 4$.

Problem Set 1.5

1. The second player can always add a sufficient number of tallies to reach a multiple of 5 at each step. This will force the first player to go over 30.

3. **(a)** Work backward. Before the last jump, Josh had \$16, since $16 \times 2 = 32$. Before the second jump, Josh had $\frac{1}{2}(16+32) = \$24$. Before the first jump, Josh had $\frac{1}{2}(24+32) = \$28$. He started with \$28.

6. Coats: Since Joe was wearing Moe's coat, Hiram must have been wearing Joe's coat. Therefore, Moe was wearing Hiram's coat. Hats: Since Joe was wearing Hiram's hat, Moe must have been wearing Joe's hat. Therefore, Hiram was wearing Moe's hat. To summarize, Moe was wearing Hiram's coat and Joes' hat. Hiram was wearing Joe's coat and Moe's hat.

7. **(a)** Since the number is greater than 20, less than 35, and divisible by 5, it must be either 25 or 30. Since the sum of the digits is 7, it must be 25. Not all information was needed—for example, we did not use the first and second clues.

8.

	Choc. Malt	Straw. Shake	Banana Split	Walnut Cone
Aaron	X	X	X	O
Boyd	X	X	O	X
Carol	O	X	X	X
Donna	X	O	X	X

Aaron had the walnut cone. Boyd had the banana split. Carol had the chocolate malt. Donna had the strawberry shake.

12. **(a)** 3. Use the Pigeonhole Principle—in this case, the children are the "pigeons" and genders are the "holes."

14. If the difference $a - b$ is divisible by 10, that means that $a - b$ is a multiple of 10. In any set of 11 natural numbers, at least two of the numbers must have the same units digit which implies that their difference is a multiple of 10.

16.

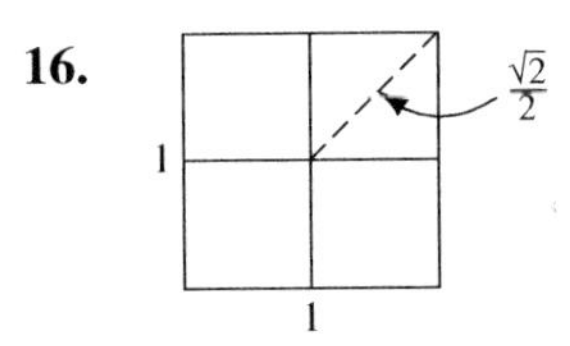

If five points are chosen in a square with diagonal of length $\sqrt{2}$, then, by the Pigeonhole Principle, at least two of the points must be in or on the boundary of one of the four smaller squares shown. The farthest these two points can be from each other is $\sqrt{2}/2$ units, if they are on opposite corners of the small square.

18. If the cups of marbles are arranged as described, each cup will be part of three different groups of three adjacent cups. The sum of all marbles in all groups of three adjacent cups is $3 \cdot (10 \cdot 11/2) = 165$, since each cup of marbles is counted three times. With the marble count of 165 and 10 possible groups of three adjacent cups, by the Pigeonhole Principle, at least one group of three adjacent cups must have 17 or more marbles, since $165 \div 10 = 16.5 > 16$.

20. The number of people at the party with no friends is none, exactly one, or 2 or more.
Case (i): If everyone has at least one friend, then each of the 20 people at the party has between 1 and 19 friends, inclusive. By the Pigeonhole Principle, at least two of them have the same number of friends.
Case (ii): If exactly one person has no friends, then each of the other 19 people has 1 to 18 friends at the party. By the Pigeonhole Principle, at least two of them have the same number of friends.
Case (iii): If 2 or more people have no friends, then they have the same number of friends at the party.

Problem Set 1.6

1. **(a)**
$$9 \times 9 = 81$$
$$79 \times 9 = 711$$
$$679 \times 9 = 6111$$
$$5679 \times 9 = 51{,}111$$

3. **(a)**
$$1 \times 8 + 1 = 9$$
$$12 \times 8 + 2 = 98$$
$$123 \times 8 + 3 = 987$$

6. **(a)** The values at each given point are zero.

10. (a) There are $F_5 = 5$ arrangements of five logs, supporting the generalization.

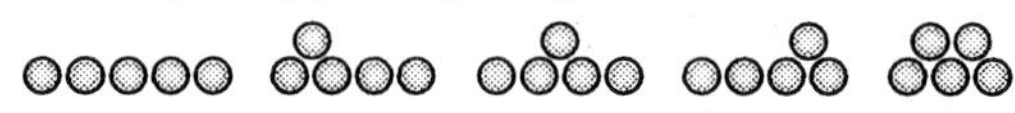

However, there are nine arrangements of six logs, instead of 8 as suggested by the Fibonacci pattern.

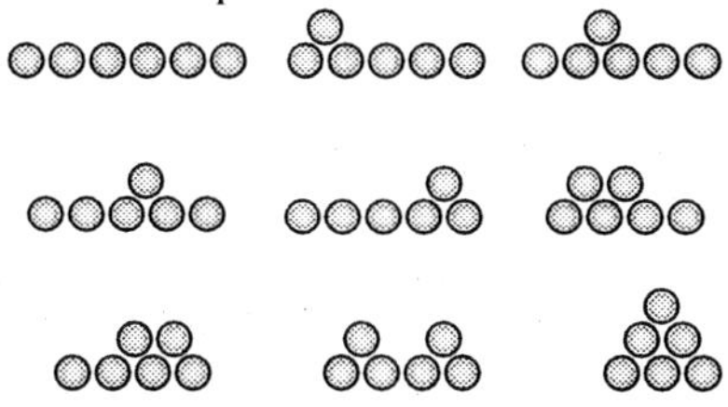

12. (a) Two pennies must be moved.

13. (a) The 1 is represented by the dot in the middle of the bottom row. The 4 is represented by the 4 adjacent dots, the 7 by the next layer of dots, and so on.

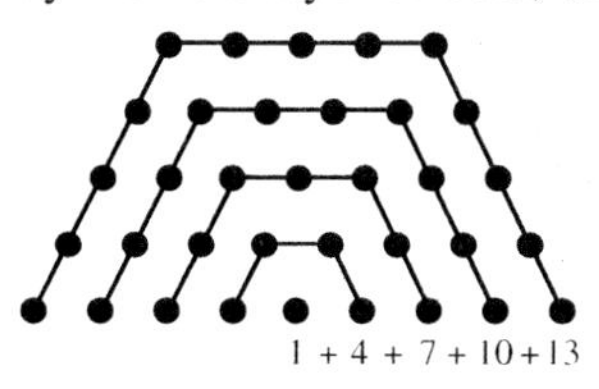

15. (a) The starting position and 15 moves to interchange the frogs are shown in a table:

○ ○ ○ _ ● ● ●
○ ○ ○ ● _ ● ●
○ ○ _ ● ○ ● ●
○ _ ○ ● ○ ● ●
○ ● ○ _ ○ ● ●
○ ● ○ ● ○ _ ●
○ ● ○ ● ○ ● _
○ ● ○ ● _ ● ○
○ ● _ ● ○ ● ○
_ ● ○ ● ○ ● ○
● _ ○ ● ○ ● ○
● ● ○ _ ○ ● ○
● ● ○ ● ○ _ ○
● ● ○ ● _ ○ ○
● ● _ ● ○ ○ ○
● ● ● _ ○ ○ ○

16. (a) Additional examples, such as those shown in the table below, support the conclusion that all such numbers are divisible by 30.

17. If n is a multiple of three, then $n = 3s$ for some whole number s and $n^2 = (3s)^2 = 9s^2$, which shows that n^2 is a multiple of 9.

18. Assume that n is even. Then $n = 2s$ for some whole number s. But then $n^2 = 4s^2 = 2(2s^2)$, which implies that n^2 is even. But this is false since we are given that n^2 is odd. Therefore, the assumption that n is even must be false. Thus, n is odd, as we were to prove.

Chapter 1 Review Exercises

1. *Strategy 1*: Use trial and error and make a chart similar to the one below:

Number of 8′ boards	Number of 10′ boards	Total number of feet
45	45	810 (too small)
40	50	820 (too small)
35	55	830 (too small)
30	60	840 (too small)
28	62	844 (o.k.)

Therefore, there were 28 eight-foot boards.

Strategy 2: Use reasoning. If all boards were 8 feet long, the total length would be $90 \cdot 8 = 720$. He has $844 - 720 = 124$ "extra" feet. Since each 10 foot board has 2 "extra" feet, the number of 10 foot boards is $124 \div 2 = 62$. Therefore, there were 28 8-foot boards.

Table for problem 16(a)

n	1	2	3	4	5	6	7	8	9	10
$n^5 - n$	0	30	240	1020	3120	7770	16800	32760	59040	99990
$(n^5 - n)/30$	0	1	8	34	104	259	560	1092	1968	3333

2. **(a)** Answers will vary. Two solutions are

$$\begin{array}{r} 379 \\ 462 \\ +158 \\ \hline 999 \end{array} \qquad \begin{array}{r} 179 \\ 368 \\ +452 \\ \hline 999 \end{array}$$

(b) Yes. The digits in each column can be arranged in any order. (Other answers, such as 198 + 267 + 534, are also possible.)

(c) No. The hundreds column digit must sum to 8 to allow for a carry from the tens column. If the digit 1 is not in the hundreds column, the smallest that this sum can be is 2 + 3 + 4 = 9. Note that if 2, 3, and 4 are used, it will be impossible to construct a sum that does not require carrying from the tens digit. Thus, the digit 1 must be in the hundreds column.

3. There are 9 ways to produce 21¢ in change. They are listed in order of number of pennies used:

4N + P	3N + 6P	D + 11P
D + 2N + P	D + N + 6P	N + 16P
2D + P	2N + 11P	21P

4. $5 \times 4 \times 3 = 60$

5. The flower bed plus the walkway has a total area of $12 \times 14 = 168$ square feet, and the flower bed alone has an area of $8 \times 10 = 80$ square feet. The area of the walkway is $168 - 80 = 88$ square feet.

6. Karen's number is 9. Work backward. Add 7 to the result, 11, to obtain 18. Divide by 2 to obtain 9.

7. **(a)** Multiply by 5, then subtract 2.

(b) Answers will vary. A good strategy is to give Chanty consecutive whole numbers starting with 0.

8. **(a)** Multiply each term by 2 to get the next term: 3, 6, 12, $\underline{24}$, $\underline{48}$, $\underline{96}$.

(b) Since $16 \div 4 = 2^2$, multiply each term by 2 to get the next term: 4, $\underline{8}$, 16, $\underline{32}$, $\underline{64}$, $\underline{128}$.

(c) Since $216 \div 1 = 6^3$, multiply each term by 6 to get the next term: 1, $\underline{6}$, $\underline{36}$, 216, $\underline{1296}$, $\underline{7776}$.

(d) Since $1250 \div 2 = 5^4$, multiply each term by 5 to get the next term: 2, $\underline{10}$, $\underline{50}$, $\underline{250}$, 1250, $\underline{6250}$.

(e) Since $7 \div 7 = 1^5$, multiply each term by 1 to get the next term: 7, $\underline{7}$, $\underline{7}$, $\underline{7}$, $\underline{7}$, 7.

9. We make a table to show all possibilities and use the clues to delete those that are impossible and, hence, those that are certain. The steps in the argument are numbered and the numbers in the table below indicate the corresponding conclusions at each step.

(1) By (b), Kimberly is neither the teacher nor the writer.

(2) By (e), Terry is not the writer. Therefore, Otis is the writer.

(3) By (f), neither Otis nor Terry is the painter. So, Kimberly is the painter.

(4) By (g), Otis is not the doctor.

(5) By (d), since the doctor hired the painter (Kimberly) and the doctor is not Otis, Terry is the doctor and Kimberly is not the doctor.

(6) Since Kimberly is not the teacher and, by (a), the doctor (Terry) had lunch with the teacher, Otis is the teacher and Terry is not. Also, since Otis has just two jobs, it follows that he is neither the engineer nor the lawyer.

(7) By (c), the painter (Kimberly) is related to the engineer. Therefore, Kimberly is not the engineer and so Terry is.

(8) Since Terry is the doctor and engineer he is not the lawyer. Thus, finally, Kimberly is the lawyer and the table now shows who holds what jobs.

	Doctor	Engineer	Teacher	Lawyer	Writer	Painter
Kimberly	no (5)	no (7)	no (1)	yes (8)	no (1)	yes (3)
Terry	yes (5)	yes (7)	no (6)	no (8)	no (2)	no (3)
Otis	no (4)	no (6)	yes (6)	no (6)	yes (2)	no (3)

10. **(a)** In the nth equation, we add the "next" n even numbers:

$$14 + 16 + 18 + 20 = 4^3 + 4$$
$$22 + 24 + 26 + 28 + 30 = 5^3 + 5$$
$$32 + 34 + 36 + 38 + 40 + 42 = 6^3 + 6$$

(b) $92 + 94 + 96 + 98 + 100 + 102 + 104 + 106 + 108 + 110 = 10^3 + 10$

11. **(a)** Each term is 3 more than its predecessor. Since 79 is 72 more than 7, and $72 \div 3 = 24$, there are 24 terms after the first term—for a total of 25 terms.

(b) Use Gauss's Insight:

$$\begin{aligned} s &= 7+10+13+\cdots+79 \\ s &= 79+76+73+\cdots+7 \\ \hline 2s &= 86+86+86+\cdots+86 \end{aligned}$$

$$s = \frac{25 \cdot 86}{2} = 1075$$

12. **(a)** 11th term: 3, 6, 12, 24, 48, 96, 192, 384, 768, 1536, 3072
This can also be determined by observing that $3072 = 2^{10} \cdot 3$.

(b) By adding the terms listed above, $S = 6141$.

(c) Duly observed.

(d)
$$\begin{aligned} 2S &= \quad 6+12+24+\cdots+3072+6144 \\ -S &= -3-6-12-24-\cdots-3072 \\ \hline S &= -3+0+0+0+\cdots+0+6144 \\ S &= -3 + 6144 = 6141 \end{aligned}$$

13. $5 + 15 + 45 + 135 + 405 + 1215 + 3645 + 10{,}935 + 32{,}805 + 98{,}415 + 295{,}245 = 442{,}865$.
Alternately, use the method of problem 12(d):

$$\begin{aligned} 3S &= \quad 15+45+\cdots+295{,}245+885{,}735 \\ -S &= -5-15-45-\cdots-295{,}245 \\ \hline 2S &= -5+0+0+\cdots+0+885{,}735 \\ 2S &= -5 + 885{,}735 = 885{,}730 \\ S &= 885{,}730 \div 2 = 442{,}865 \end{aligned}$$

14. Complete the chart below and then generalize from the results.

Number of chords	Number of regions	Number of intersections	Number of segments
0	1 = 0 + 1	0	0
1	2 = 1 + 1	0	1
2	4 = 3 + 1	1	4
3	7 = 6 + 1	3	9
4	11 = 10 + 1	6	16
5	16 = 15 + 1	10	25
6	22 = 21 + 1	15	36
⋮	⋮	⋮	⋮
n	$\frac{n(n+1)}{2}+1$	$\frac{n(n-1)}{2}$	n^2

(a) $\dfrac{n(n+1)}{2}+1$

(b) $\dfrac{n(n-1)}{2}$

(c) Each chord is divided into n segments, for a total of n^2 small segments.

15. The product of the entries in the squares equals the product of the entries in the circles. This is true of any hexagon of six entries with two entries along each side as for the hexagon shown. Indeed, the result is true for *any* hexagon of entries however large so long as there are an even number of entries per side. For example, try various placements of the hexagon shown. Incidentally, since the two products are equal, the product of *all* the entries on the boundary of the hexagon is a perfect square!

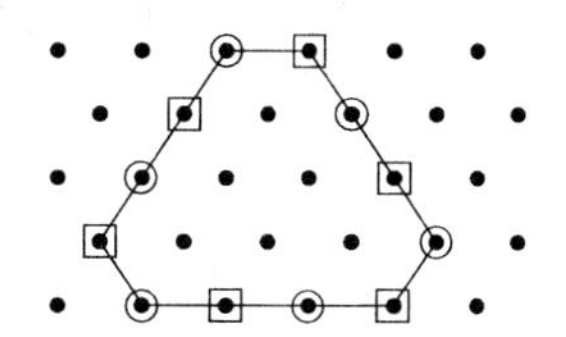

16. **(a)** $1 + 1 \cdot 2 = 3$

(b) $1 + 2 \cdot 2 + 1 \cdot 2^2 = 9$

(c) $1 + 3 \cdot 2 + 3 \cdot 2^2 + 1 \cdot 2^3 = 27$

(d) The left side of the nth equation is obtained by multiplying the entries in the nth row of Pascal's triangle by successive powers of 2, and the right side is 3^n. This suggests that

$$P_0 + P_1 \cdot 2^1 + P_2 \cdot 2^2 + \cdots + P_n \cdot 2^n = 3^n,$$

where P_k is the kth element in the nth row of Pascal's triangle.

(e) $1 + 1 \cdot 3 = 4$
$1 + 2 \cdot 3 + 1 \cdot 3^2 = 16$
$1 + 3 \cdot 3 + 3 \cdot 3^2 + 1 \cdot 3^3 = 64$

(f) Part (e) suggests that
$P_0 + P_1 \cdot 3^1 + P_2 \cdot 3^2 + \cdots + P_n \cdot 3^n = 4^n$.
Taken together, the results suggest that
$P_0 + P_1 \cdot r^1 + P_2 \cdot r^2 + \cdots + P_n \cdot r^n$
$= (r+1)^n$. To check further, one might try another example:
$1 + 4 \cdot 5^1 + 6 \cdot 5^2 + 4 \cdot 5^3 + 1 \cdot 5^4$
$= 1296 = 6^4$.
The result is true in general, but one cannot be sure of this without a mathematical proof.

17. Let x be Bernie's weight. Bernie's weight is also represented by $90 + \frac{x}{2}$. Then, by the condition of the problem, $x = 90 + \frac{x}{2}$, which yields $2x = 180 + x$ or $x = 180$ pounds.

18. (a) A one-car train uses 6 toothpicks to form the hexagon. Adding a square + hexagon combination requires an additional 8 toothpicks, so the trains with 1, 3, 5, 7, ... cars use 6, 6 + 8, 6 + 8 + 8, 6 + 8 + 8 + 8, ... toothpicks. In general, a train with $2m + 1$ cars will require $6 + 8m$ toothpicks, where $m = 0, 1, 2, 3, \ldots$. A two-car train uses 9 toothpicks, so trains with 2, 4, 6, 8, ... cars use 9, 9 + 8, 9 + 8 + 8, 9 + 8 + 8 + 8, ... toothpicks. In general, a train with $2m + 2$ cars uses $9 + 8m$ toothpicks for $m = 0, 1, 2, 3, \ldots$.

(b) Since $9 + 8m$ is always an odd number, a train with 102 toothpicks has an odd number of cars, say $2m + 1$. Then $6 + 8m = 102$, or $8m = 96$. Therefore, $m = 12$, and there are $2(12) + 1 = 25$ cars in the train.

19. (a) Guessing will come up with 8 for one of the values and −1 for the other.

(b) Let x and y be the numbers. Then $x + y = 7$ and $x - y = 9$. Adding the equations gives $2x = 16$, so $x = 8$. Substituting this value into either equation gives $y = -1$. The solution checks.

20. (a) There are four "pigeonholes" (suits), so draw 5 cards.

(b) If only 8 cards are drawn, there could be 2 of each suit. Therefore, draw 9 cards.

(c) If one drew 48 cards, one might get everything except the aces. Therefore, to be absolutely sure of getting two aces, one must draw 50 cards.

21. 17, since it is possible that the first 16 books chosen are 4 each from the 4 types of books.

22. (a)
$67 \times 67 = 4489$
$667 \times 667 = 444,889$
$6667 \times 6667 = 44,448,889$

(b) $6,666,667 \times 6,666,667$
= 44,444,448,888,889. The patterns observed suggest that:
- the number of 4s is one more than the number of 6s in one of the factors.
- the number of 8s is the same as the number of 6s.
- the eights are followed by a single 9.

However, without knowing *why* the pattern holds, or doing the actual calculation, one cannot be completely sure that the guess is correct.

23. (a)
$1 \times 142,857 = 142,857$
$2 \times 142,857 = 285,714$
$3 \times 142,857 = 428,571$
$4 \times 142,857 = 571,428$
$5 \times 142,857 = 714,285$

(b) All of the above answers are obtained by starting at an appropriate place in the following circle.

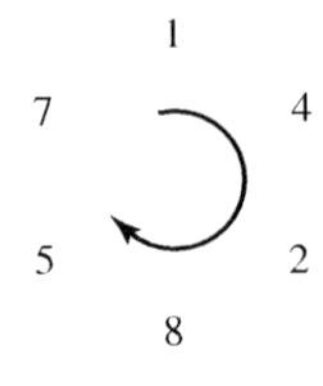

The only remaining place to start is at 8 so we guess that
6 × 142,857 = 857,142. This checks.

(c) The answer to 7 × 142,857 is not clear since there is no unused starting digit in the cycle. In fact, 7 × 142,857 = 999,999.

(d) Apparent patterns may be misleading; they may eventually break down.

24.

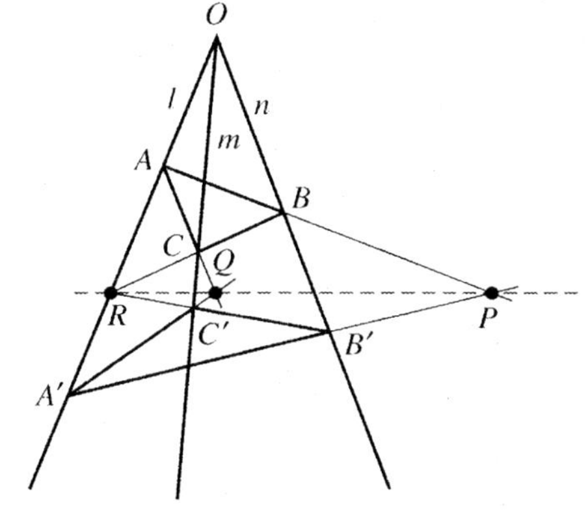

P, Q, and R are collinear for every placement of $\triangle ABD$ and $\triangle A'B'D'$.

25. If n is odd then $n = 2s + 1$ for some whole number s.

$$n^2 = (2s+1)^2 = 4s^2 + 4s + 1$$
$$= 4(s^2 + s) + 1 = 4\left[2 \cdot \frac{s(s+1)}{2}\right] + 1$$
$$= 8\left[\frac{s(s+1)}{2}\right] + 1$$

Since one of any two consecutive whole numbers must be even, $\frac{s(s+1)}{2}$ must be a whole number, say q. Thus $n^2 = 8q + 1$.

26. No, 12 is a multiple of 6 but the sum of its digits is 3, which is not divisible by 6.

Chapter 1 Test

1. The pigeons are most evenly distributed if one hole has 2 pigeons and one has 3 pigeons. This minimizes the number of pigeons in the hole with the most pigeons. The answer is 3 pigeons.

2. Note the pattern: As the "missing digit" on the left side of the equation moves to the right, the 2 on the right side of the equation moves to the right.

$2345679 \times 9 = 21,111,111$
$1345679 \times 9 = 12,111,111$
$1245679 \times 9 = 11,211,111$
$1235679 \times 9 = 11,121,111$
$1234679 \times 9 = 11,112,111$
$1234579 \times 9 = 11,111,211$
$1234569 \times 9 = 11,111,121$
$1234568 \times 9 = 11,111,112$

3. (a) The 100th term in the progression is $2 \times 100 - 1 = 199$.

(b) Let s denote the sum of the first 100 terms. Then, using Gauss's Insight,
$s = 1 + 3 + 5 + \cdots + 199$
and also $s = 199 + 197 + 195 + \cdots + 1$
Therefore,
$2s = 200 + 200 + 200 + \cdots + 200$
$2s = 100 \times 200$
Thus, $s = \frac{100 \times 200}{2} = 10,000$

4. (a) $2 + 5 + 8 + 11 + 8 + 5 + 2 = 41$
$= 3^2 + 2 \cdot 4^2$
$2 + 5 + 8 + 11 + 14 + 11 + 8 + 5 + 2 = 66$
$= 4^2 + 2 \cdot 5^2$

(b) The middle number on the left side is $2 + 9 \cdot 3 = 29$.
$2 + 5 + 8 + \cdots + 26 + 29 + 26 + \cdots$
$+ 8 + 5 + 2 = 281 = 9^2 + 2 \cdot 10^2$

5. 10 days. (He reaches a *maximum* height of 3 feet on the first day, 4 feet on the second day, and so on.)

6. We know that the sum of the elements of the nth row of Pascal's triangle (starting with the top row as the 0th row) is 2^n. Therefore, the desired form is
$S = 1 + 2 + 4 + 8 + 16 + 32 + 64 + 128$
$+ 256 + 512 = 1023$

7. Work backward. Before meeting the third guard, he had $2 \cdot (1 + 2) = 6$ apples. Before meeting the second guard, he had $2 \cdot (6 + 2) = 16$ apples. Before meeting the first guard, he had $2 \cdot (16 + 2) = 36$ apples. He originally stole 36 apples.

8. (a) Notice that each equation has 3 parts. The first part is a sum of consecutive numbers. It starts with the number after the last number in the previous equation and each equation sums 2 more numbers than the previous sum. Since the fourth equation summed 7 numbers and its last number was $16 = 4^2$, the fifth equation will sum 9 numbers and will start with $17 = 4^2 + 1$. The second part of each equation is the sum of consecutive cubes. The first number is the last number of the previous equation.

(*continued on next page*)

(continued)

$17 + 18 + 19 + 20 + 21 + 22 + 23 + 24 + 25 = 64 + 125 = 225 - 36$

$26 + 27 + 28 + 29 + 30 + 31 + 32 + 33 + 34 + 35 + 36 = 125 + 216 = 441 - 100$

(b) Continuing the same pattern, the tenth equation should be

$$82+83+\ldots+100 = 729+1000 = 3025-1296$$

(c) In general, the nth equation should be

$$[(n-1)^2+1]+[(n-1)^2+2]+\ldots+n^2 = (n-1)^3+n^3 = \left[\frac{n(n+1)}{2}\right]^2 - \left[\frac{(n-2)(n-1)}{2}\right]^2.$$

9. (a) Answers will vary. Each sum must be

$$\frac{2+7+12+27+32+37+52+57+62}{3} = 96$$

One possibility is shown.

57	2	37
12	32	52
27	62	7

(b) Answers will vary. A magic subtraction square can be obtained by interchanging the numbers at the ends of each diagonal in (a). One possibility is shown.

7	2	27
12	32	52
37	62	57

10. It appears that the sum always equals the quotient of the two numbers in the denominator of the last fraction being added on. Thus we would guess that

$$S_n = \frac{1}{n(n+1)} + \frac{n-1}{n} = \frac{1}{n(n+1)} + \frac{(n-1)(n+1)}{n(n+1)} = \frac{1+n^2-n+n-1}{n(n+1)} = \frac{n^2}{n(n+1)} = \frac{n}{n+1}$$

as expected. Thus,

$$S_n = \frac{1}{1\cdot 2} + \frac{1}{2\cdot 3} + \cdots + \frac{1}{n(n+1)} = \frac{n}{n+1}.$$

Chapter 2 Sets and Whole Numbers

Problem Set 2.1

1. **(a)** {Arizona, California, Idaho, Oregon, Utah}

2. **(a)** {l, i, s, t, h, e, m, n, a, o, y, c}

3. **(a)** {7, 8, 9, 10, 11, 12,13}

(c) {4, 8, 12, 16, 20}

4. Answers will vary.

(a) $\{x \in U \mid 11 \le x \le 14\}$ or $\{x \in U \mid 10 < x < 15\}$

(c) $\{x \in U \mid x = 4n \text{ and } 1 \le n \le 5\}$

5. Answers will vary.

(a) $\{x \in N \mid x \text{ is even and } x > 12\}$ or $\{x \in N \mid x = 2n \text{ for } n \in N \text{ and } n > 6\}$

7.

(a) $B \cup C = \{\text{a, b, c, h}\}$

(c) $B \cap C = \{\text{a, b}\}$

(e) $\overline{A} = \{\text{f, g, h}\}$

8. **(a)** $M = \{45, 90, 135, 180, 225, 270, 315, \ldots\}$

(b) $L \cap M = \{90, 180, 270, \ldots\}$
This can be described as the set of natural numbers that are divisible by both 6 and 45, or as the set of natural numbers that are divisible by 90.

(c) 90

10. **(a)** $A \cap B \cap C$ is the set of elements common to all three sets.

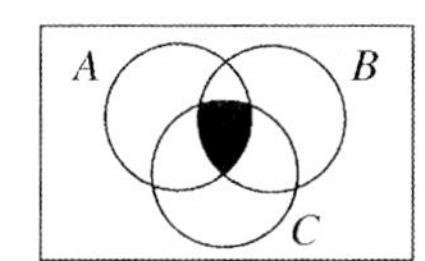

(c) $(A \cap B) \cup C$ is the set of elements in C or in both A and B.

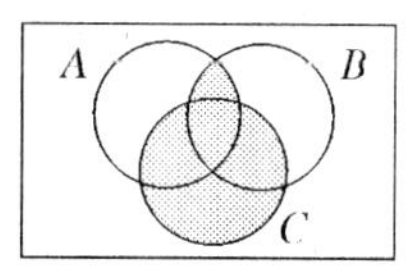

(e) $A \cup B \cup C$ is the set of all elements in A, B, and/or C.

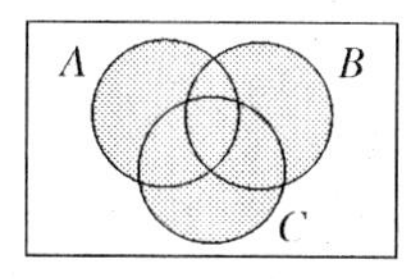

11. **(a)** A and B must be disjoint sets contained in C.

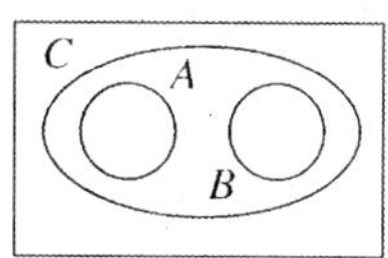

12. No. It is possible that there are elements of A that are also elements of B but not C, or C but not B. For example, let $A = \{1, 2\}$, $B = \{2, 3\}$, $C = \{3\}$. $A \cup B = A \cup C = \{1, 2, 3\}$, but $B \ne C$.

13. **(a)** Since
$A \cap B = \{6, 12, 18\}$,
$\overline{A \cap B} = \{1, 2, 3, 4, 5, 7, 8, 9, 10, 11, 13, 14, 15, 16, 17, 19, 20\}$
Since
$\overline{A} = \{1, 3, 5, 7, 9, 11, 13, 15, 17, 19\}$ and
$\overline{B} = \{1, 2, 4, 5, 7, 8, 10, 11, 13, 14, 16, 17, 19, 20\}$,
$\overline{A} \cup \overline{B} = \{1, 2, 3, 4, 5, 7, 8, 9, 10, 11, 13, 14, 15, 16, 17, 19, 20\}$
Since
$A \cup B = \{2, 3, 4, 6, 8, 9, 10, 12, 14, 15, 16, 18, 20\}$
$\overline{A \cup B} = \{1, 5, 7, 11, 13, 17, 19\}$.
Since
$\overline{A} = \{1, 3, 5, 7, 9, 11, 13, 15, 17, 19\}$ and
$\overline{B} = \{1, 2, 4, 5, 7, 8, 10, 11, 13, 14, 16, 17, 19, 20\}$,
$\overline{A} \cap \overline{B} = \{1, 5, 7, 11, 13, 17, 19\}$.

14. (a) $R \cap C$ is the set of red circles. Draw two red circles—one large and one small.

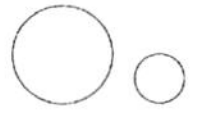

(c) $T \cup H$ is the set of shapes that are either triangles or hexagons. Draw two large triangles, two large hexagons, two small triangles, and two small hexagons—a red and a blue of each.

(e) $B \cap \overline{C}$ is the set of blue shapes that are not circles. That is, the set of four elements consisting of the large and small blue hexagons and the large and small blue triangles.

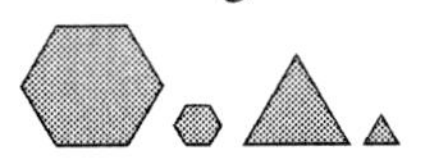

15. (a) $L \cap T$

(c) $S \cup T$

16. (a) Answers will vary. One possibility is B = set of students taking piano lessons, C = set of students learning a musical instrument.

19. (a) 8 regions

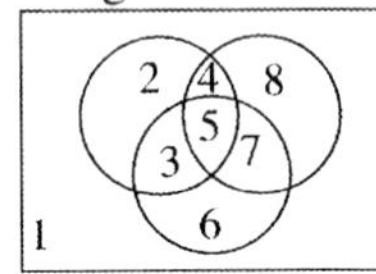

(c) Verify that $A \cap \overline{B} \cap \overline{C} \cap D$ has no region by observing that the region for $A \cap D$ is entirely contained in $B \cup C$. Likewise, the region for $B \cap C$ is entirely contained in $A \cup D$. The other missing region is $\overline{A} \cap B \cap C \cap \overline{D}$.

20. (a) These Venn diagrams show that $\overline{A \cap B} = \overline{A} \cup \overline{B}$:

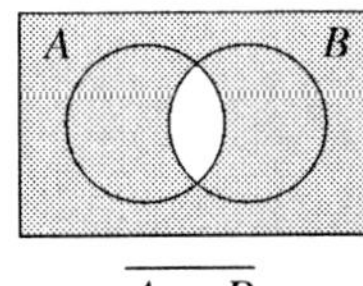

$\overline{A \cap B}$

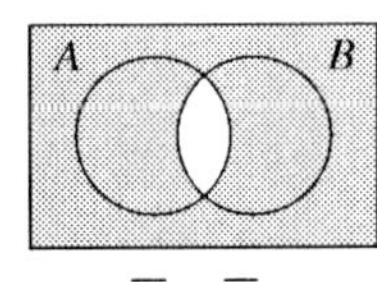

$\overline{A} \cup \overline{B}$

21. (a) There are 6 choices for shape, 2 choices for size, and 2 choices for color, so the number of pieces is $6 \times 2 \times 2 = 24$.

23. (a) There are eight subsets: ∅, {P}, {N}, {D}, {P, N}, {P, D}, {N, D}, {P, N, D}.

24. Note that in this diagram, O refers to the region outside both circles.

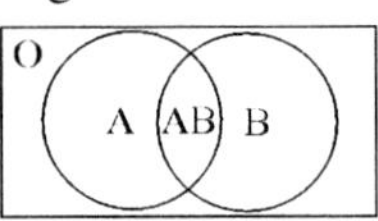

26. Answers will vary. Consult a thesaurus.

Problem Set 2.2

1. (a) 13: ordinal
first: ordinal

2. (a) Equivalent, since there are five letters in the set {A, B, M, N, P}

4. (a) $n(A) = 7$ because $A = \{21, 22, 23, 24, 25, 26, 27\}$.

(c) The solutions for $(x-1)(x-9) = 0$ are $x = 1$ and $x = 9$, so $C = \{1, 9\}$ and $n(C) = 2$.

6. (a) The correspondence $0 \leftrightarrow 1, 1 \leftrightarrow 2, 2 \leftrightarrow 3, \ldots, w \leftrightarrow w+1, \ldots$ shows $W \sim N$.

7. (a) Finite. The number of grains of sand is large, but finite.

8. (a) Answers will vary. For example, $Q_1 \leftrightarrow Q_2$, $Q_3 \leftrightarrow Q_4$, and so on. (Note that P need not be the center—it can be any fixed point inside the small circle.)

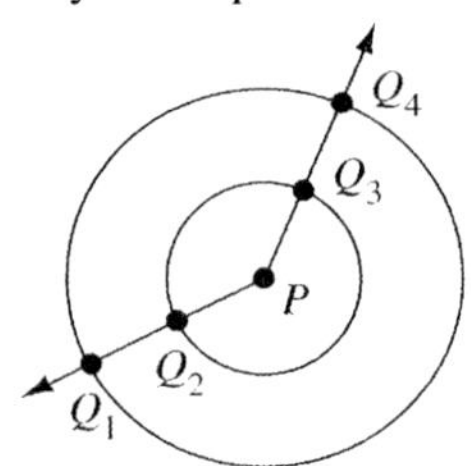

(c) Answers will vary. For example, $Q_1 \leftrightarrow Q_2$, etc.

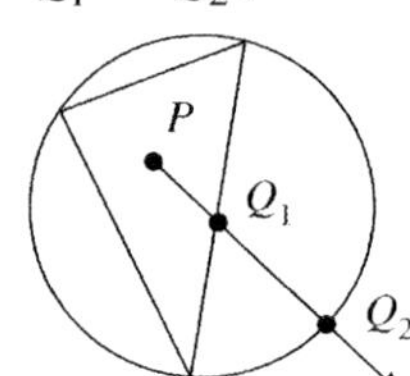

9. **(a)** True. A set B cannot have fewer elements than its subset A.

(c) True. The union $A \cup B$ does not include any more elements than just A, so the elements of B must already be elements of A.

10. **(a)** $n(A \cap B) \leq n(A)$. The set $A \cap B$ contains only the elements of A that are also elements of B. That is, $A \cap B \subseteq A$. Thus, $A \cap B$ cannot have more elements than A.

12. **(a)** $1000 \div 6 = 166.666\ldots$, so the largest element of S is $166 \cdot 6 = 996$. Therefore, $n(S) = 166$.

13. Use the fact that the eight regions pictured below are mutually disjoint sets and apply the strategy of working backwards, that is, start with $n(A \cap B \cap C) = 7$. Next use $n(A \cap B)$, $n(B \cap C)$, and $n(A \cap C)$ to find the values 10, 5, and 8, respectively. Then use $n(A)$, $n(B)$, and $n(C)$ to find the values 15, 28, and 10, respectively, and finally $n(U)$ to find the value 17.

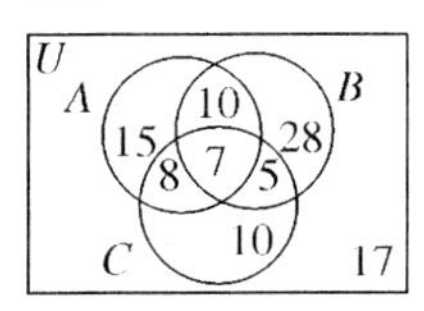

15. **(a)** 1 4 9 16 25

16. Let A, H, and P be the sets of students taking anthropology, history, and psychology, respectively. Then $n(A) = 40$, $n(H) = 11$, $n(P) = 12$, $n(A \cap H \cap P) = 3$, $n(A \cap H) = 6$, and $n(A \cap P) = 6$. *Note*: If one chooses to solve this problem by using a Venn diagram, there are several possibilities, since the number of students taking both history and psychology cannot be determined. However, the answers to parts (a), (b), and (c) are uniquely determined.

(a)
$$\begin{aligned} &n(A - (A \cap H) \cup (A \cap P)) \\ &= n(A) - n(A \cap H) - n(A \cap P) \\ &\qquad + n(A \cap H \cap P) \\ &= 40 - 6 - 6 + 3 = 31 \end{aligned}$$

Note: One might argue that the answer is zero, since all of the students in this situation are also taking mathematics.

(b)
$$\begin{aligned} n(A \cup H) &= n(A) + n(H) - n(A \cap H) \\ &= 40 + 11 - 6 = 45 \end{aligned}$$

(c)
$$\begin{aligned} &n((H \cap A) - \overline{P}) \\ &= n(H \cap A) - n(H \cap A \cap P) \\ &= 6 - 3 = 3 \end{aligned}$$

23. **(a)** Six ways: yrg, rgy, gyr, ygr, gry, ryg

24. **(a)**

Row 0	1				
Row 1	1	1			
Row 2	1	2	1		
Row 3	1	3	3	1	
Row 4	1	4	6	4	1

25. Since we are given that $k < l$ and $l < m$, we can choose sets K, L, and M satisfying $K \subset L \subset M$ and $n(K) = k$, $n(L) = l$, and $n(M) = m$. By the transitive property of set inclusion (see Section 2.1, or just look at a Venn diagram) we know that $K \subset M$ and so $k < m$.

29. **(a)** The diagram below gives $36 + x = 40$, so $x = 4$. (This result can also be obtained using guess and check.) 4 students have been to all three countries.

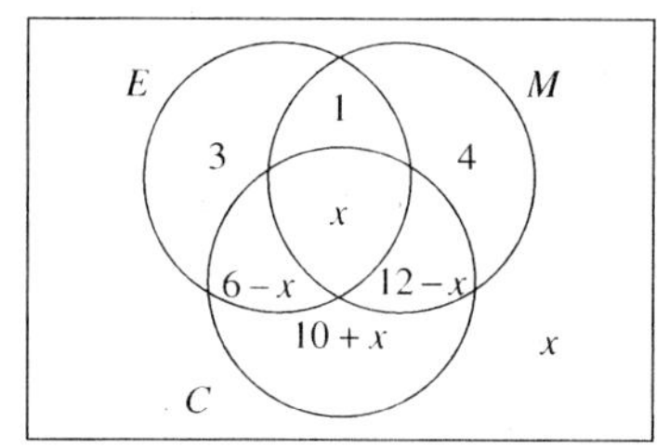

(b) $10 + 4 = 14$ students have been only to Canada.

30. **(a)** All three friends will meet at the mall every $3 \times 4 \times 5 = 60$th day, so there will be 6 days when all three get together. $(365 \div 60 = 6$ with a remainder of 5.)

Problem Set 2.3

1. **(a)** **(i)** $A \cup B = \{$apple, berry, peach, lemon, lime$\}$ so $n(A \cup B) = 5$.

(ii) $A \cup C = \{\text{apple, berry, peach, lemon, prune}\}$ so $n(A \cup C) = 5$.

(iii) $B \cup C = \{\text{lemon, lime, berry, prune}\}$ so $n(B \cup C) = 4$ since lemon is a member of both sets.

(b) (ii) and (iii) because the two sets in each case are not disjoint, that is, they have at least one member in common.

2. (a) B may contain 4, 5, 6, 7, or 8 elements. If B were to contain more than 8 elements, then $n(A \cup B)$ would be greater than 8 which contradicts the fact that $n(A \cup B) = 8$. Similarly, if B were to contain less than 4 elements, $n(A \cup B)$ would be less than 8.

(b) If $(A \cap B) = \varnothing$, then $n(A) + n(B) = n(A \cup B)$, so $n(B) = 4$.

3. (a)

4 + 6

4. (a)

(c)

(e)

7. (a) Closed (since the sum of two positive multiples of 5 is a larger multiple of 5)

(c) Closed (since 0 + 0 = 0)

(f) Closed (since the sum of two multiples of 3 is a multiple of 3: $3a + 3b = 3(a + b)$)

8. (a) Commutative property of addition

(c) Additive-identity property of zero

9. (a) $(1+20)+(2+19)+(3+18)+\cdots+(10+11)$
$= 21+21+21+\cdots+21$ (for 10 terms)
$= (10)(21) = 210$

(b) Associative and commutative properties

10. (a)

(c)

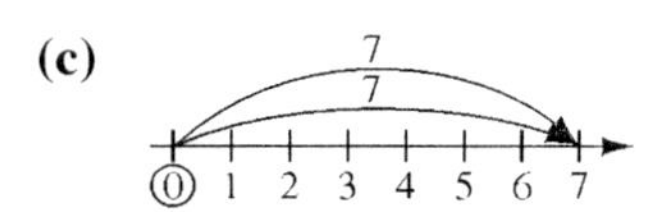

11. (a) $5 + 7 = 12$ $\quad 12 - 7 = 5$
$7 + 5 = 12$ $\quad 12 - 5 = 7$

(b) $4 + 8 = 12$ $\quad 12 - 8 = 4$
$8 + 4 = 12$ $\quad 12 - 4 = 8$

12. (a)

5 + 9 = 14

14 − 9 = 5

9 + 5 = 14

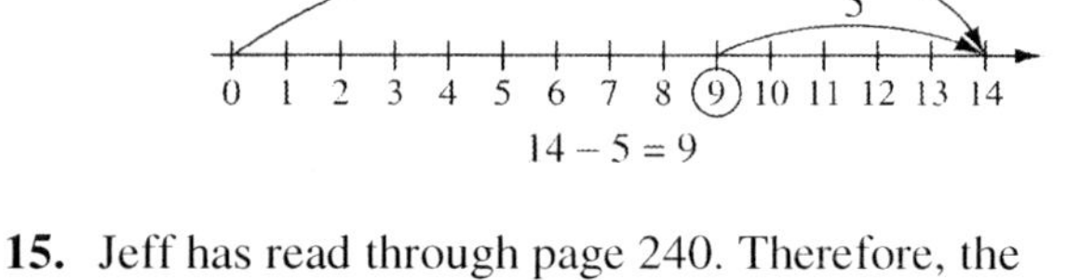

14 − 5 = 9

15. Jeff has read through page 240. Therefore, the number of pages is 257 − 240 = 17 pages or 257 − 241 + 1 = 17 pages.

16. Use the guess and check method. Some answers will vary.

(a) $(8 - 5) - (2 - 1) = 2$

(c) $((8 - 5) - 2) - 1 = 0$

17. (a) First fill in the squares by noting that $3 - 1 = 2$, $4 - 2 = 2$, and $7 - 2 = 5$. Then complete the circles.

5	2	(7)
1	2	(3)
(6)	(4)	(7)

18. Write all possible combinations of 3 different numbers whose sum is the indicated number in the triangle. Then, place numbers which occur in more than one sum at the vertices as shown below.

(a) $1 + 3 + 6$
$2 + 3 + 5$
$4 + 5 + 1$

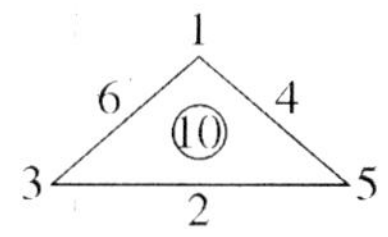

1
6 (10) 4
3 2 5

28.

$A \cup B$ = A + B − $A \cap B$

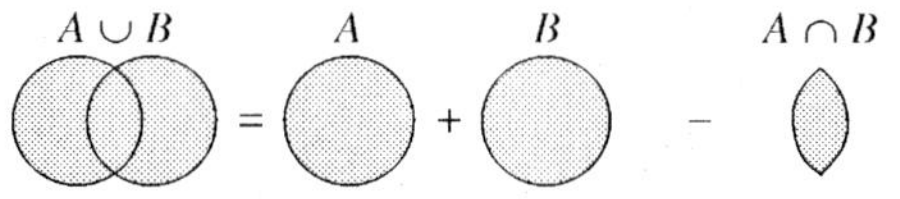

$n(A \cup B) = n(A) + n(B) - n(A \cap B)$

Note that elements of $A \cap B$ are counted twice when calculating $n(A) + n(B)$. We compensate by subtracting $n(A \cap B)$, giving $n(A \cup B)$.

32. $\{0\}$, since $0 - 0 = 0$.

34. **(a)** Use the formula $t_n = \dfrac{n(n+1)}{2}$.

n	1	2	3	4	5	6
t_n	1	3	6	10	15	21
n	7	8	9	10	11	12
t_n	28	36	45	55	66	78
n	13	14	15			
t_n	91	105	120			

(b) $11 = 10 + 1$
$12 = 6 + 6$
$13 = 10 + 3$
$14 = 10 + 3 + 1$
$15 = 15$
$16 = 15 + 1$
$17 = 15 + 1 + 1$
$18 = 15 + 3$
$19 = 10 + 6 + 3$
$20 = 10 + 10$
$21 = 21$
$22 = 21 + 1$
$23 = 10 + 10 + 3$
$24 = 21 + 3$
$25 = 15 + 10$

36. First find the top row and the left column. For example, the fourth entry in the left column must be $6 - 2 = 4$, and then the third entry in the top row is $5 - 4 = 1$, and so on. The completed table is shown at the bottom of the page.

38. **(a)** $0 = 5 - (1 + 4)$, $1 = 5 - 4$, $2 = (1 + 5) - 4$, $3 = 4 - 1$, $4 = 5 - 1$, $5 = 1 + 4$, $6 = 1 + 5$

42. Statement B does not correspond to the number sentence $15 - 8 = \square$. The other problems illustrate take-away (A), comparison (C), and missing addend (D).

Table for problem 36.

+	5	4	1	6	9	2	0	8	7	3
3	8	7	4	9	12	5	3	11	10	6
9	14	13	10	15	18	11	9	17	16	12
6	11	10	7	12	15	8	6	14	13	9
4	9	8	5	10	13	6	4	12	11	7
0	5	4	1	6	9	2	0	8	7	3
7	12	11	8	13	16	9	7	15	14	10
5	10	9	6	11	14	7	5	13	12	8
2	7	6	3	8	11	4	2	10	9	5
1	6	5	2	7	10	3	1	9	8	4
8	13	12	9	14	17	10	8	16	15	11

Problem Set 2.4

1. (a) $3 \times 5 = 15$, set model (repeated addition)

   (d) $3 \times 6 = 18$, number-line model

2. (a) A set of dominoes, containing 11 stacks of 5 dominoes is modeled by a rectangular array.

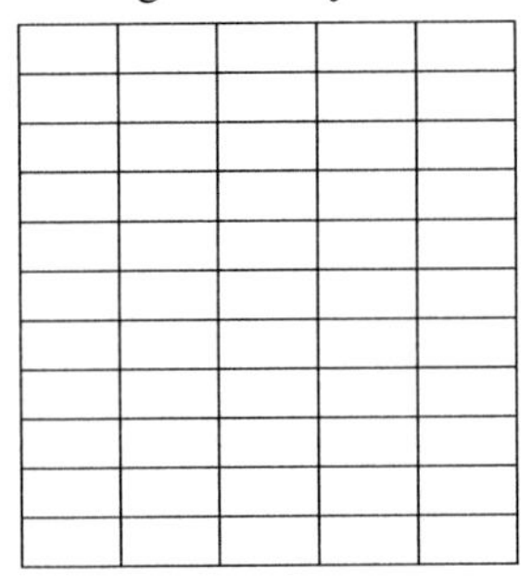

   $11 \times 5 = 55$
   The set contains 55 dominoes.

   (c) A hike of 10 miles each day for 5 days can be modeled by a number line.

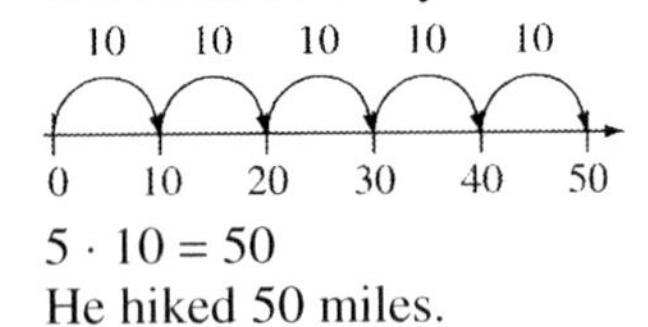

   $5 \cdot 10 = 50$
   He hiked 50 miles.

3. (a) Answers will vary.

   (b) (i) $4 \cdot 9 = 9 + 9 + 9 + 9 = 36$

   (ii) $7 \times 536 = 536 + 536 + 536 + 536 + 536 + 536 + 536 = 3752$

   (iii) $6 \times 47{,}819 = 47{,}819 + 47{,}819 + 47{,}819 + 47{,}819 + 47{,}819 + 47{,}819 = 286{,}914$

   (iv) Using the commutative property of multiplication,
   $56{,}108 \times 6 = 6 \times 56{,}108 = 56{,}108 + 56{,}108 + 56{,}108 + 56{,}108 + 56{,}108 + 56{,}108 = 336{,}648$

4. (a) Each of the a lines from set A intersects each of the b lines from set B. Since the lines for A are parallel and the lines for B are parallel, the intersection points will all be distinct.

   (b)

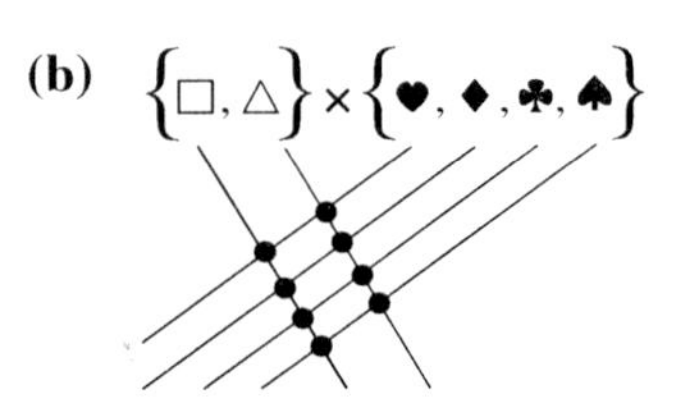

5. (a) Not closed. For example, $2 \times 2 = 4$, which is not in the set.

   (c) Not closed. For example, $2 \times 4 = 8$, which is not in the set.

   (e) Closed. The product of any two odd whole numbers is always another odd whole number.

   (g) Closed. $2^a \times 2^b = 2^{a+b}$ for any whole numbers a and b.

7. (a) Commutative property of multiplication

   (c) Multiplication-by-zero property

   (e) Associative property of multiplication

8. (a) Commutative property: $5 \times 3 = 3 \times 5$

9. (a)

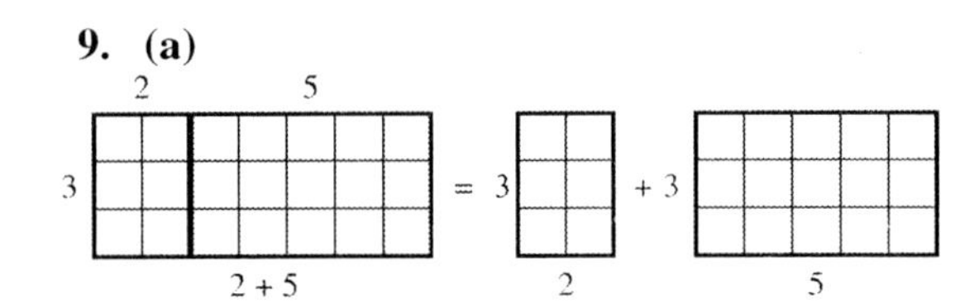

13. (a) Distributive property of multiplication over addition:
   $7 \cdot 19 + 3 \cdot 19 = (7 + 3) \cdot 19 = 10 \cdot 19 = 190$

14. (a) $18 \div 6 = 3$

15. (a) $4 \times 8 = 32$, $8 \times 4 = 32$, $32 \div 8 = 4$, $32 \div 4 = 8$

16. (a) Repeated subtraction

17. (a) $19 - 5 = 14$, which is greater than 5, so we must subtract 5 again. $14 - 5 = 9$, which is still greater than 5. Subtracting 5 again, $9 - 5 = 4$, which is less than 4, so we are done. Because we subtracted 5 three times, the quotient is 3. The remainder is 4. This is represented as
   $a = 19 \overset{-5}{\Rightarrow} 19 - 5 = 14 \overset{-5}{\Rightarrow} 14 - 5 = 9 \overset{-5}{\Rightarrow} 9 - 5 = 4$ (done)

(d) $14 - 7 = 7$. Since $7 \leq 7$, subtract 7 again to get $7 - 7 = 0$. The remainder is 0. We subtracted twice, so the quotient is 2.

19. (a) Since 78 – 13 = = = = = = gives 0, $78 \div 13 = 6$.

(c) Since 96 – 14 = = = = = = gives 12, $96 \div 14 = 6$ R 12.
(*Note*: Stop pressing = when the result is less than the divisor, 14.)

20. (a) $y = (5 \cdot 5) + 4 = 25 + 4 = 29$

21. (a) $3^{20} \cdot 3^{15} = 3^{20+15} = 3^{35}$

(e) $y^3 \cdot z^3 = (y \cdot z)^3$ or $(yz)^3$

22. (a) $8 = 2 \cdot 2 \cdot 2 = 2^3$

(c) $1024 = 2^{10}$

23. Use the guess and check method.

(a) $m = 4$; $3 \cdot 3 \cdot 3 \cdot 3 = 81$

(c) $p = 10$

24. Answers will vary. The following are sample answers.

(a) Have children hold hands in, say, groups of four. Have one group come to the front of the class, and verify that $1 \times 4 = 4$. Next have a second group come forward, and verify that $2 \times 4 = 8$. And so on.

(b) Have the children form rectangles of various sizes, and then count off row by row to obtain the total number of children in the array.

(c) Choose how many teams to be formed (the divisor) and see how large the teams will be, with an equal number on each team.

(d) For a given number of children on a team, add teams until the total number of children reaches a certain total. Then count how many teams were needed.

25. Answers will vary. These are sample problems:
Peter has a board 14 feet long, and each box he makes requires 3 feet of board. How many completed boxes can be made? Answer = 4.
Tina has a collection of 14 antique dolls. A display box can hold at most three dolls. How many boxes does Tina require to display her entire collection? Answer = 5.
Andrea has 14 one-by-one foot paving stones to place on her 3-foot-wide walkway. How many feet of walk can she pave, and how many stones will she have left over for a future project? Answer = 4 feet of walk paved, with 2 stones remaining.

29. Answers will vary, but you could first ask for her solution. She may say that since the total cost of one nut and one bolt together is \$1.00, and she is buying 18 of them, the total cost is \$18.00. This is a wonderful opportunity for the Mathematical Habit of the Mind because you could then ask her to justify her response:

$$18 \cdot 86 + 18 \cdot 14 = 18 \cdot (86 + 14)$$
$$= 18 \cdot 100 = 1800 \text{ cents or } \$18.00$$

You could then ask her a question that would lead to her saying it is the distributive property.

33. (a) The magic multiplication constant is $4096 = 2^{12}$.

(b) The exponents form a magic addition square with the magic addition constant 12. It is clear how the multiplication square has been formed. For example, the product of the upper row is $2^3 \times 2^8 \times 2^1 = 2^{3+8+1} = 2^{12}$. That is, the product is always 2^{12}.

2^3	2^8	2^1
2^2	2^4	2^6
2^7	2^0	2^5

3	8	1
2	4	6
7	0	5

(c)

$27 = 3^3$	$6561 = 3^8$	$3 = 3^1$
$9 = 3^2$	$81 = 3^4$	$729 = 3^6$
$2187 = 3^7$	$1 = 3^0$	$243 = 3^5$

34. (a) Since $2 \times 185 = 370, 500 - 370 = 130$, and $130 \div 2 = 65$, the question is "How many tickets must still be sold?"

38. **(a)** Note that $T_{n+1} = T_n + t_{n+1}$ for $n \geq 1$.

$$T_4 = 10 + 10 = 20 = \frac{4 \cdot 5 \cdot 6}{6}$$

$$T_5 = 20 + 15 = 35 = \frac{5 \cdot 6 \cdot 7}{6}$$

$$T_6 = 35 + 21 = 56 = \frac{6 \cdot 7 \cdot 8}{6}$$

$$T_7 = 56 + 28 = 84 = \frac{7 \cdot 8 \cdot 9}{6}$$

$$T_8 = 84 + 36 = 120 = \frac{8 \cdot 9 \cdot 10}{6}$$

$$T_9 = 120 + 45 = 165 = \frac{9 \cdot 10 \cdot 11}{6}$$

$$T_{10} = 165 + 55 = 220 = \frac{10 \cdot 11 \cdot 12}{6}$$

$$T_{11} = 220 + 66 = 286 = \frac{11 \cdot 12 \cdot 13}{6}$$

$$T_{12} = 286 + 78 = 364 = \frac{12 \cdot 13 \cdot 14}{6}$$

The pattern continues to hold.

(b) $T_{100} = \dfrac{100 \cdot 101 \cdot 102}{6} = 171{,}700$

Chapter 2 Review Exercises

1. **(a)** $S = \{4, 9, 16, 25\}$
$P = \{2, 3, 5, 7, 11, 13, 17, 19, 23\}$
$T = \{2, 4, 8, 16\}$

(b) $\overline{P} = \{4, 6, 8, 9, 10, 12, 14, 15, 16, 18, 20, 21, 22, 24, 25\}$

$S \cap T = \{4, 16\}$

$S \cup T = \{2, 4, 8, 9, 16, 25\}$

$S \cap \overline{T} = \{9, 25\}$

2.

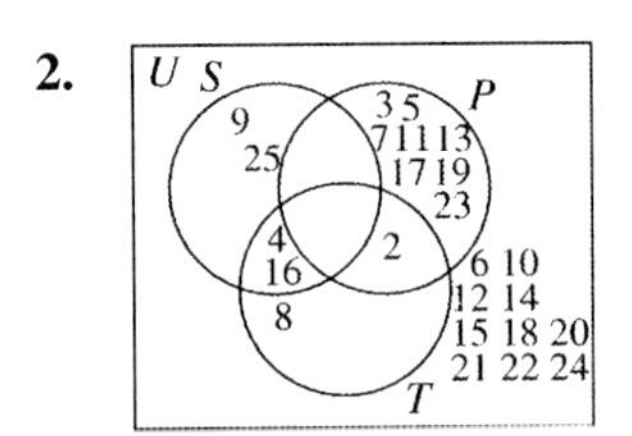

3. **(a)** $A \boxed{\subseteq} A \cup B$

(b) If $A \subseteq B$ and A is not equal to B, then $A \boxed{\subset} B$.

(c) $A \boxed{\cap} (B \cup C) = (A \cap B) \cup (A \cap C)$

(d) $A \boxed{\cup} \varnothing = A$

4. $n(S) = 3, n(T) = 6,$
$n(S \cup T) = n(\{\text{s, e, t, h, o, r, y}\}) = 7,$
$n(S \cap T) = n(\{\text{e, t}\}) = 2,$
$n(S \cap \overline{T}) = n(\{s\}) = 1,$
$n(T \cap \overline{S}) = n(\{\text{h, o, r, y}\}) = 4$

5.

1	4	9	16	25	36	49	64	81	100
↕	↕	↕	↕	↕	↕	↕	↕	↕	↕
a	b	c	d	e	f	g	h	i	j

6. There is a one-to-one correspondence between the set of cubes and a proper subset. For example,

1	8	27	64	125	…	K^3
↕	↕	↕	↕	↕		↕
1	2	3	4	5	…	K

7.

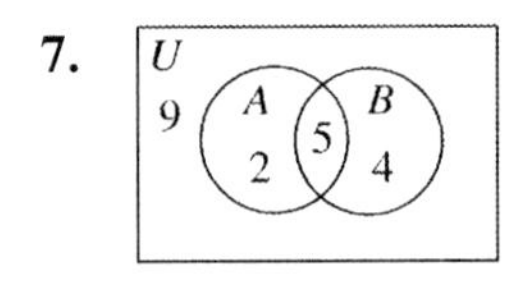

8. **(a)** Suppose $A = \{\text{a, b, c, d, e}\}$ and $B = \{■, ★\}$. Then $n(A) = 5$, $n(B) = 2$, $A \cap B = \varnothing$ and $n(A \cup B) = n(\{\text{a, b, c, d, e}, ■, ★\}) = 7$.

(b)

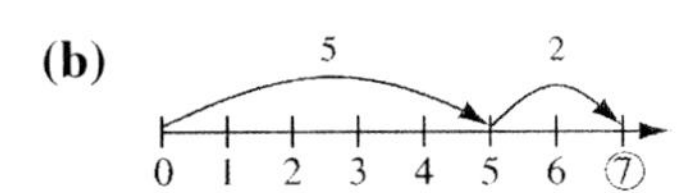

9. **(a)** Commutative property of addition:
$7 + 3 = 3 + 7$

(b) Additive-identity property of zero:
$7 + 0 = 7$

10. **(a)**

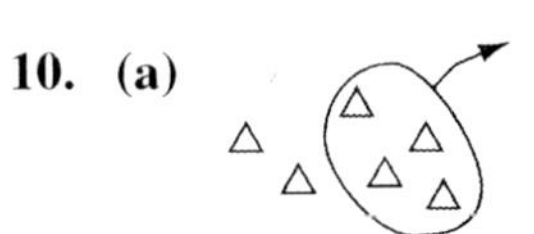

(b)

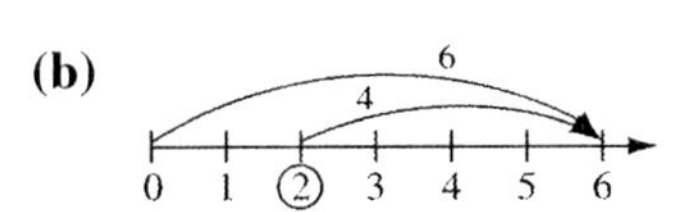

11. (a)

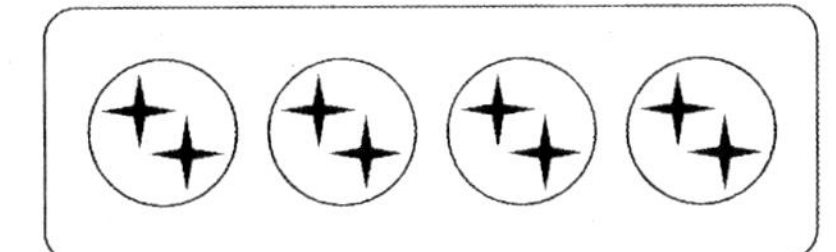

(b)

(c)

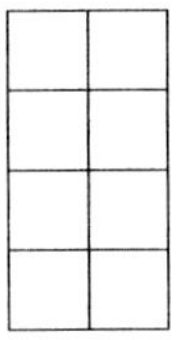

(d)

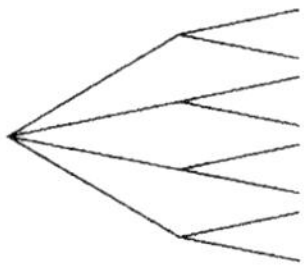

(e)

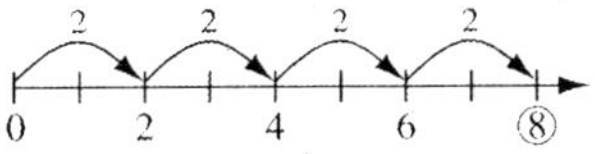

12. (a) $A \times B = \{(p, x), (p, y), (q, x), (q, y), (r, x), (r, y), (s, x), (s, y)\}$

(b) Since $n(A) = 4$, $n(B) = 2$, and $n(A \times B) = 8$, the Cartesian product models $4 \times 2 = 8$.

13. Answers will vary. Since $36 = 3 \times 3 \times 4$, one possibility is $6'' \times 6'' \times 8''$.

14. Since $92 \div 12 = 7$ R 8, there are eight rows (7 full rows, and a partial row of 8 soldiers in the back).

15. (a)

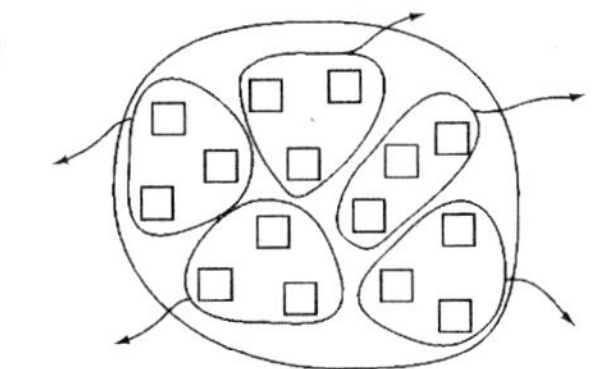

(b)

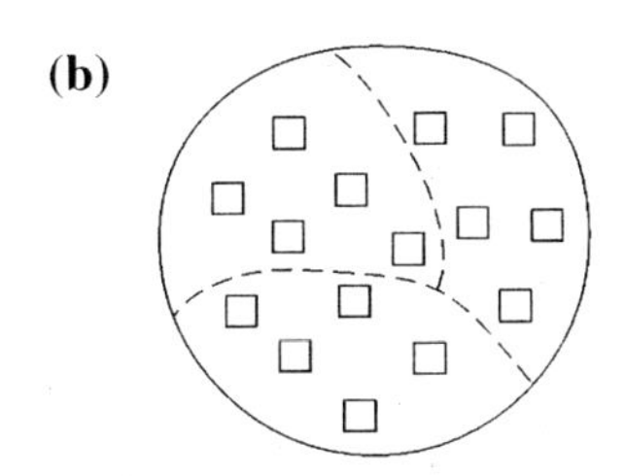

(c)

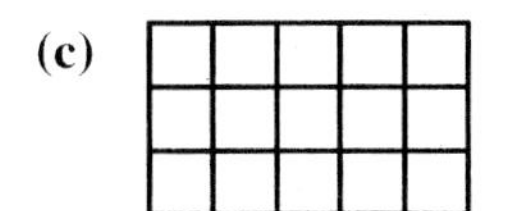

Chapter 2 Test

1. (a) $4 \times 2 = 8$

(b) $12 \div 3 = 4$

(c) $5 \cdot (9 + 2) = 5 \cdot 9 + 5 \cdot 2$

(d) $10 - 4 = 6$

2. 15th—ordinal
1040—nominal
\$253—cardinal

3. (a) Yes, because $2^a \cdot 2^b = 2^{a+b}$.

(b) No. For example, $2 \in S$ and $4 \in S$ but $4 + 2 \notin S$.

4. Answers can vary. For example, let $A = \{a, b, c, d, e\}$ and $B = \{a, b, c, d, e, f, g, h\}$. Then $A \subset B$ so $n(A) < n(B)$. That is, $5 < 8$.

5. (a) W is closed under &, since $a + b + ab$ is an element of W for any two whole numbers a and b.

(b) Using the properties of whole number arithmetic, $a \,\&\, b = a + b + ab = b + a + ba = b \,\&\, a$ for all whole numbers a and b, so & is commutative.

(c) Using the properties of whole number arithmetic,

$$\begin{aligned} a \,\&\, (b \,\&\, c) &= a + (b \,\&\, c) + a(b \,\&\, c) \\ &= a + b + c + bc + a(b + c + bc) \\ &= a + b + c + ab + bc + ac + abc \end{aligned}$$

Similarly,

$$\begin{aligned} (a \,\&\, b) \,\&\, c &= (a + b + ab) \,\&\, c \\ &= (a + b + ab) + c + (a + b + ab)c \\ &= a + b + c + ab + bc + ac + abc \end{aligned}$$

Therefore, for all a, b, and c, $a \,\&\, (b \,\&\, c) = (a \,\&\, b) \,\&\, c$ so & is associative.

(d) $0 \,\&\, a = 0 + a + 0(a) = a$ and $a \,\&\, 0 = a + 0 + (a)0 = a$ for all whole numbers a, showing that 0 is an identity for &.

6. **(a)** Number line

(b) Comparison

(c) Missing addend

7. **(a)** Associative property of addition

(b) Distributive property of multiplication over addition

(c) Additive-identity property of zero

(d) Associative property of multiplication

8. **(a)** $A \cap (B \cup C)$

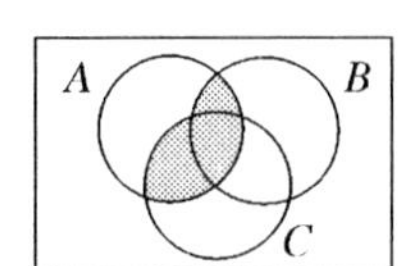

(b) $A \cup B$

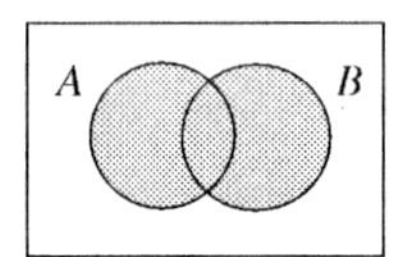

(c) $(A \cap \overline{B}) \cup (B \cap \overline{A})$

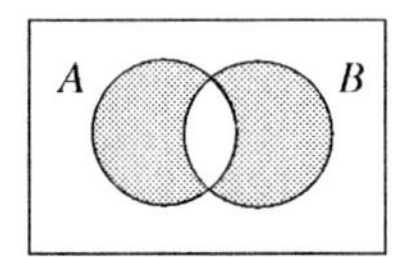

9. **(a)**

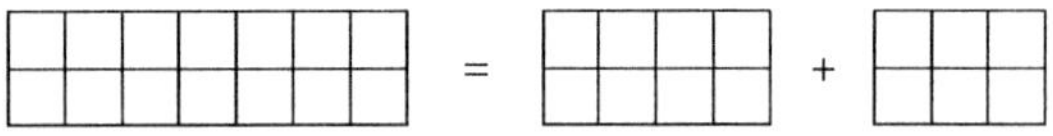

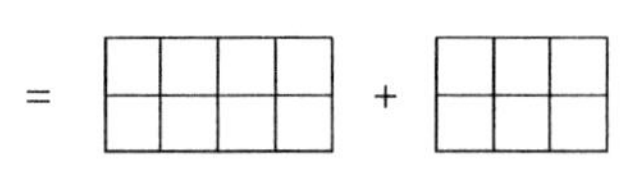

(b)

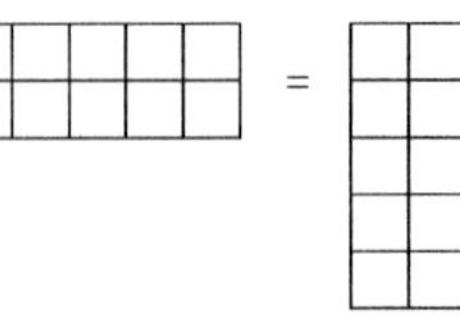

10. The possible values are the natural number factors of 21: 1, 3, 7, 21.

11. **(a)** $n(A \cup B) = n(\{w, h, o, l, e, n, u, m, b, r\}) = 10$

(b) $n(B \cap \overline{C}) = n(\{n, u, m, b\}) = 4$

(c) $n(A \cap C) = n(\{o, e\}) = 2$

(d) $n(A \times C) = n(A) \times n(C) = 5 \times 4 = 20$

12. Since $n(A \cup B) = 21 = 12 + 14 - 5$, there must be 5 elements that are in both sets. Thus, $n(A \cap B) = 5$, and $n(\overline{A \cap B}) = 26 - 5 = 21$.

13. Since $A \cap B = A$, we know that $A \subseteq B$. Therefore $A \cap \overline{B} = \varnothing$.

14. **(a)** Since 5 gallons = 20 quarts = 640 ounces, the number of bottles is $640 \div 10 = 64$.

(b) Grouping

Chapter 3 Numeration and Computation

Problem Set 3.1

1. **(a)** $2 \cdot 1000 + 1 \cdot 100 + 3 \cdot 10 + 7 = 2137$

(c) $1 \cdot 100{,}000 + 2 \cdot 10{,}000 + 3 \cdot 100 + 1 \cdot 10 = 120{,}310$

(e) $500 + 100 + 90 + 5 + 2 = 697$

(g) $1 \cdot 60^1 = 60$

(i) $2 \cdot 60^2 + 42 = 7200 + 42 = 7242$

(k) $2 \cdot 18 \cdot 20^2 + 6 \cdot 18 \cdot 20 + 18 \cdot 20 + 0$
$= 2 \cdot 7200 + 6 \cdot 360 + 360 + 0$
$= 16{,}920$

2. **(a)** $11 = 1 \cdot 10 + 1 \cdot 1 =$ ∩ |

3. **(a)** $9 = 10 - 1 =$ IX

4. **(a)** $251 = 4 \cdot 60 + 11 =$ ▼▼▼▼ ◀▼

5. **(a)** $12 = 12 \cdot 1$

••
═

9. **(a)** In Roman numerals 2002, 2003, and 2004 would be represented respectively by MMII, MMIII, and MMIV.

10. **(a)** 3795 **(c)** 6048

11. **(a)** ═ ╥ ⊥ ||| **(c)** ╥╥ ||||

12. **(a)** $374 + 281 = 655$, ⊤ ≣ |||||

(c) $6224 - 732 = 5492$, ≣ |||| ≘ ||

13. $452 = 4 \cdot 100 + 5 \cdot 10 + 2 \cdot 1$

□ □ □ □ ||||| :

17. First trade 10 of your units for a strip to get 3 mats, 25 strips, and 3 units. Then trade 20 strips for 2 mats to get 5 mats, 5 strips, and 3 units.

Problem Set 3.2

1.

Base Ten	Base 5
0	0
1	1
2	2
3	3
4	4
$5 = 1 \cdot 5 + 0$	10
$6 = 1 \cdot 5 + 1$	11
$7 = 1 \cdot 5 + 2$	12
$8 = 1 \cdot 5 + 3$	13
$9 = 1 \cdot 5 + 4$	14
$10 = 2 \cdot 5 + 0$	20
$11 = 2 \cdot 5 + 1$	21
$12 = 2 \cdot 5 + 2$	22
$13 = 2 \cdot 5 + 3$	23
$14 = 2 \cdot 5 + 4$	24
$15 = 3 \cdot 5 + 0$	30
$16 = 3 \cdot 5 + 1$	31
$17 = 3 \cdot 5 + 2$	32
$18 = 3 \cdot 5 + 3$	33
$19 = 3 \cdot 5 + 4$	34
$20 = 4 \cdot 5 + 0$	40
$21 = 4 \cdot 5 + 1$	41
$22 = 4 \cdot 5 + 2$	42
$23 = 4 \cdot 5 + 3$	43
$24 = 4 \cdot 5 + 4$	44
$25 = 1 \cdot 5^2 + 0 \cdot 5 + 0$	100

5. **(a)** $413_{\text{five}} = 4 \cdot 5^2 + 1 \cdot 5 + 3 = 108$

(c) $10_{\text{five}} = 1 \cdot 5 + 0 = 5$

(e) $1000_{\text{five}} = 1 \cdot 5^3 + 0 \cdot 5^2 + 0 \cdot 5 + 0 = 125$

6. **(a)** $2 \cdot 5 + 4 = 14$

(e) $1 \cdot 5^3 + 3 \cdot 5^2 + 3 \cdot 5 + 2$
$= 1 \cdot 125 + 3 \cdot 25 + 3 \cdot 5 + 2 = 217$

9. (a) $413_{six} = 4 \cdot 6^2 + 1 \cdot 6 + 3 = 153$

(c) $10_{six} = 1 \cdot 6 + 0 = 6$

(e) $1000_{six} = 1 \cdot 6^3 + 0 \cdot 6^2 + 0 \cdot 6 + 0 = 216$

10. (a) $413_{twelve} = 4 \cdot 12^2 + 1 \cdot 12 + 3 = 591$

(c) $10_{twelve} = 1 \cdot 12 + 0 = 12$

(e) $1000_{twelve} = 1 \cdot 12^3 + 0 \cdot 12^2 + 0 \cdot 12 + 0$
$= 1728$

11. (a) $362 = 2 \cdot 125 + 112$
$= 2 \cdot 125 + 4 \cdot 25 + 12$
$= 2 \cdot 125 + 4 \cdot 25 + 2 \cdot 5 + 2$
$= 2422_{five}$

(c) $5 = 1 \cdot 5 + 0 = 10_{five}$

12. (a) $342 = 1 \cdot 216 + 126$
$= 1 \cdot 216 + 3 \cdot 36 + 18$
$= 1 \cdot 216 + 3 \cdot 36 + 3 \cdot 6 + 0 = 1330_{six}$

(c) $6 = 1 \cdot 6 + 0 = 10_{six}$

13. (a) $2743 = 1 \cdot 1728 + 1015$
$= 1 \cdot 1728 + 7 \cdot 144 + 7$
$= 1 \cdot 1728 + 7 \cdot 144 + 0 \cdot 12 + 7$
$= 1707_{twelve}$

(c) $144 = 1 \cdot 144 + 0 \cdot 12 + 0 = 100_{twelve}$

14. (a)

One Thousand Twenty-Fours	1024	2^{10}
Five Hundred Twelves	512	2^9
Two Hundred Fifty-Sixes	256	2^8
One Hundred Twenty-Eights	128	2^7
Sixty-Fours	64	2^6
Thirty-Twos	32	2^5
Sixteens	16	2^4
Eights	8	2^3
Fours	4	2^2
Twos	2	2^1
Units	1	2^0

(c) (i) $24 = 16 + 8$
$= 1 \cdot 16 + 1 \cdot 8 + 0 \cdot 4 + 0 \cdot 2 + 0 \cdot 1$
$= 11{,}000_{two}$

(ii) $18 = 16 + 2$
$= 1 \cdot 16 + 0 \cdot 8 + 0 \cdot 4 + 1 \cdot 2 + 0 \cdot 1$
$= 10{,}010_{two}$

(iii) $2 = 1 \cdot 2 + 0 \cdot 1 = 10_{two}$

(iv) $8 = 1 \cdot 8 + 0 \cdot 4 + 0 \cdot 2 + 0 \cdot 1$
$= 1000_{two}$

19. (a) Note that this is similar to adding 99,999,999 + 1 in base ten.

$$\begin{array}{r} 11{,}111{,}111_{two} \\ +\quad 1_{two} \\ \hline 100{,}000{,}000_{two} \end{array}$$

20. (a) $7_{ten} = 111_{two}, 2^3 = 8,$ so row $n = 7$ has eight odd entries.

21. (a) Each 3-digit sequence of 0s and 1s must begin with either 0 or 1. The set of all those beginning with 0 can be obtained by appending a 0 to the left end of each of the different 2-digit sequences. In a similar way, we obtain all 3-digit sequences beginning with 1. Thus, appending a 0 and then a 1 to the left end of each 2-digit sequence, we obtain 000, 100, 010, 110, 001, 101, 011, and 111.

22. (a) They are 0, 1, 2, 3, 4, 5, 6, and 7 (but not in that order)—namely, the numbers that can be written with three or fewer digits in base 2.

(c) The whole numbers from 0 to $2^n - 1$. There are 2^n of these whole numbers, each with a different n-digit base 2 representation corresponding to one of the n-digit sequences of 0s and 1s.

Problem Set 3.3

1. (a) 36 + 75 = 111

36 + 75 = 111

2. (a)

$$\begin{array}{r} 23 \\ +44 \\ \hline 7 \\ 60 \\ \hline 67 \end{array}$$

3. (b) First exchange one of the tens in 275 to 10 ones.

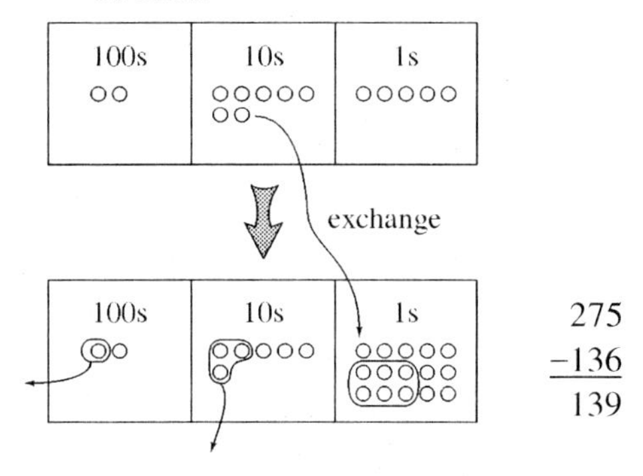

$$\begin{array}{r} 275 \\ -136 \\ \hline 139 \end{array}$$

4. (a)

$$\begin{array}{r} 78 \\ -35 \\ \hline 43 \end{array}$$

7. In these problems, we must exchange 60 seconds for one minute and 60 minutes for one hour or vice versa.

(a)

3 hours, 24 minutes, 54 seconds
+ 2 hours, 47 minutes, 38 seconds
5 hours, 71 minutes, 92 seconds
= 5 hours, 72 minutes, 32 seconds
= 6 hours, 12 minutes, 32 seconds

(c)

5 hours, 24 minutes, 54 seconds
− 2 hours, 47 minutes, 38 seconds
4 hours, 84 minutes, 54 seconds
− 2 hours, 47 minutes, 38 seconds
2 hours, 37 minutes, 16 seconds

9. (a)

Base ten	Base four
0	0
1	1
2	2
3	3
$4 = 1 \cdot 4 + 0$	10
$5 = 1 \cdot 4 + 1$	11
$6 = 1 \cdot 4 + 2$	12
$7 = 1 \cdot 4 + 3$	13
$8 = 2 \cdot 4 + 0$	20
$9 = 2 \cdot 4 + 1$	21
$10 = 2 \cdot 4 + 2$	22
$11 = 2 \cdot 4 + 3$	23
$12 = 3 \cdot 4 + 0$	30
$13 = 3 \cdot 4 + 1$	31
$14 = 3 \cdot 4 + 2$	32
$15 = 3 \cdot 4 + 3$	33

11. (a)

$$\begin{array}{r} {}^{1} \quad \\ 231_{\text{four}} \\ +121_{\text{four}} \\ \hline 1012_{\text{four}} \end{array} \qquad \text{Check: } \begin{array}{r} 45 \\ +25 \\ \hline 70 \end{array}$$

(c)

$$\begin{array}{r} {}^{1\,1\,1} \quad \\ 1223_{\text{four}} \\ +\ 231_{\text{four}} \\ \hline 2120_{\text{four}} \end{array} \qquad \text{Check: } \begin{array}{r} 107 \\ +45 \\ \hline 152 \end{array}$$

(e)

$$\begin{array}{r} {}^{1} \quad \\ 32_{\text{four}} \\ +13_{\text{four}} \\ \hline 111_{\text{four}} \end{array} \qquad \text{Check: } \begin{array}{r} 14 \\ +7 \\ \hline 21 \end{array}$$

(g)

$$\begin{array}{r} {}^{1\ 10} \quad \\ 21^{1}2_{\text{four}} \\ -\ 33_{\text{four}} \\ \hline 113_{\text{four}} \end{array} \qquad \text{Check: } \begin{array}{r} 38 \\ -15 \\ \hline 23 \end{array}$$

19. Even though it would be quite simple to line up the numbers and add, students are trying to do this problem mentally. While mental math is obviously encouraged, these kinds of errors do occur. In (a), Drew went to the millions place and simply added one in the millions, resulting in the number 10 being written out instead of carrying the one to the 10-millions digit. In (b), Alonzo went to the 10-millions spot and just added one, resulting in the number 3 in the 10-millions digit.

20. (a) Work from the right column to the left column as shown:

$$\begin{array}{r} 6\text{-}\text{-}3 \\ +\text{-}51\text{-} \\ \hline \text{-}2282 \end{array} \qquad \begin{array}{r} {}^{1}\quad \\ 6\text{-}\text{-}3 \\ +\text{-}519 \\ \hline \text{-}2282 \end{array} \qquad \begin{array}{r} {}^{1}\quad \\ 6\text{-}63 \\ +\text{-}519 \\ \hline \text{-}2282 \end{array}$$

$$\begin{array}{r} {}^{1\ \ 1}\quad \\ 6763 \\ +\text{-}519 \\ \hline \text{-}2282 \end{array} \qquad \begin{array}{r} {}^{1\ \ 1}\quad \\ 6763 \\ +5519 \\ \hline 12,282 \end{array}$$

(c) Work from right to left.

$$\begin{array}{r} ^{1} \\ 881 \\ +362 \\ \hline 1243 \end{array}$$

(e) Rewrite as an addition problem, then work from right to left.

$$\begin{array}{r} \text{-15-} \\ +\ 1843 \\ \hline \text{4--2} \end{array} \quad \text{gives} \quad \begin{array}{r} ^{111} \\ 2159 \\ +\ 1843 \\ \hline 4002 \end{array}$$

Solution: $\begin{array}{r} 4002 \\ -1843 \\ \hline 2159 \end{array}$

21. (a) Work from right to left.

$$\begin{array}{r} ^{111} \\ 2437 \\ 281 \\ +3476 \\ \hline 6194 \end{array}$$

(c) Work from right to left. Notice that the only way to get a 5 in the hundreds place of the sum is to "carry a 2," so the missing digits in the tens place of the addends must be 9s.

$$\begin{array}{r} 3891 \\ 2493 \\ +5125 \\ \hline 11{,}509 \end{array}$$

22. (a) Rewrite as an addition problem, then work from right to left.

$$\begin{array}{r} 594 \\ +\text{2-1} \\ \hline \text{- 3-} \end{array} \quad \text{gives} \quad \begin{array}{r} ^{1} \\ 594 \\ +241 \\ \hline 835 \end{array}$$

Solution: $\begin{array}{r} 835 \\ -241 \\ \hline 594 \end{array}$

(c) Rewrite as an addition problem, then work from right to left.

$$\begin{array}{r} 808 \\ +\text{-5-4} \\ \hline \text{7-4-} \end{array} \quad \text{gives} \quad \begin{array}{r} ^{11} \\ 808 \\ +6534 \\ \hline 7342 \end{array}$$

Solution: $\begin{array}{r} 7342 \\ -6534 \\ \hline 808 \end{array}$

23. (a) Five. Since 4s are used, the base is at least five. Since 1 + 4 gives a 0 in the units column, the base is five.

(c) Seven or greater. Since a 6 is used, the base is at least seven. No exchanges occur, so there is no other restriction.

(e) Seven. Exchanging is required in the units column. If the base is b, the calculation is $13_b - 4_b = 6$, which means $b + 3 - 4 = 6$, or $b - 1 = 6$, so $b = 7$.

(g) Twelve. Use the same reasoning as in (e): $b - 1 = 11$, so $b = 12$.

29. (a)

$$\begin{array}{r} 3'8'' \\ 4'2'' \\ 6'10'' \\ +5'11'' \\ \hline 18'31'' \\ =20'7'' \end{array}$$

Problem Set 3.4

1. (a)

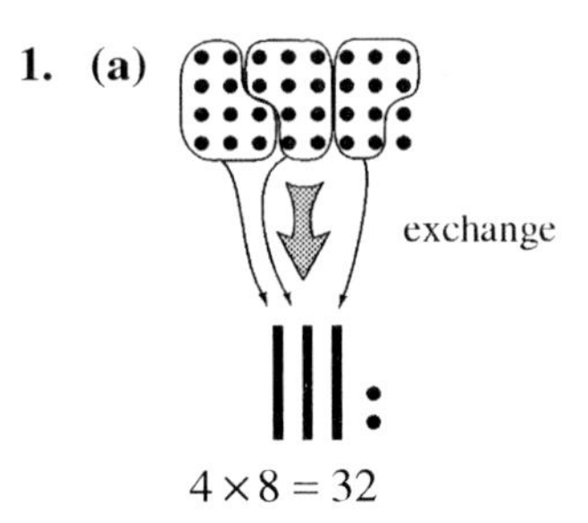

$4 \times 8 = 32$

3. (a) The red 2 represents the number of hundreds in $30 \times 70 + 100$. You can see this by thinking of the calculation 30×274. The 100 comes from $30 \times 4 = 120 = 100 + 20$.

(b) Twenty 10s are being exchanged for two 100s.

5. (a) Distributive property of multiplication over addition

(c) Associative property of addition

9. (a)

$$\begin{array}{r} 6\ \text{R}\,3 \\ 4\overline{)27} \\ \underline{24} \\ 3 \end{array}$$

$27 = 4 \cdot 6 + 3$

10. **(a)**

$$\begin{array}{r} \underline{21} \\ 1 \\ 20 \\ 351\overline{)7425} \\ \underline{7020} \\ 405 \\ \underline{351} \\ 54 \end{array}$$

The division checks because $7425 = 351 \cdot 21 + 54$.

11. **(a)** Follow these steps:

$8 \div 5 = 1 \text{ R } 3$

$37 \div 5 = 7 \text{ R } 2$

$23 \div 5 = 4 \text{ R } 3$

$$\begin{array}{r} 1\;\;7\;\;4\;\;\text{R } 3 \\ 5\overline{)8\,{}^{3}7\,{}^{2}3} \end{array}$$

Check: $873 = 5 \cdot 174 + 3$

14. **(a)** Reason as follows: $3 \times 3 = 9_{\text{ten}} = 14_{\text{five}}$, so write 4 and exchange 5 ones for 1 five. Then $3 \times 2 + 1 = 7_{\text{ten}} = 12_{\text{five}}$.

$$\begin{array}{r} {}^{1} \\ 23_{\text{five}} \\ \underline{\times 3_{\text{five}}} \\ 124_{\text{five}} \end{array} \qquad \text{Check:} \qquad \begin{array}{r} 13_{\text{ten}} \\ \underline{\times 3_{\text{ten}}} \\ 39_{\text{ten}} \end{array}$$

(d) See "checks" shown above.

15. **(a)** Base five: Check in base ten:

$$\begin{array}{r} 31\;\text{R } 2 \\ 4\overline{)231} \\ \underline{22} \\ 11 \\ \underline{4} \\ 2 \end{array} \qquad \begin{array}{r} 16\;\text{R } 2 \\ 4\overline{)66} \\ \underline{4} \\ 26 \\ \underline{24} \\ 2 \end{array}$$

18. **(a)** Yes. This is simply a rearrangement of the rows in the usual algorithm. Thus, we usually write

$$\begin{array}{r} 374 \\ \underline{\times\;23} \\ 1122 \\ \underline{748} \\ 8602 \end{array}$$

Here we multiply by 20 first and then 3 to obtain

$$\begin{array}{r} 374 \\ \underline{\times\;23} \\ 748 \\ \underline{1122} \\ 8602 \end{array}$$

(b)

$$\begin{array}{r} 285 \\ \underline{\times\;362} \\ 855 \\ 1710 \\ \underline{570} \\ 103{,}170 \end{array}$$

24. **(a)** They are the same because of the associative property of multiplication. $34 \cdot 54 = (17 \cdot 2) \cdot 54 = 17 \cdot (2 \cdot 54) = 17 \cdot 108$ since 2 evenly divides 34.

25. **(a)** The digits in the top number need to be written in decreasing order (to maximize the number of thousands, and so on). This eliminates all but five possibilities:

$$\begin{array}{r} 9753 \\ \underline{\times\;1} \\ 9753 \end{array} \quad \begin{array}{r} 9751 \\ \underline{\times\;3} \\ 29{,}253 \end{array} \quad \begin{array}{r} 9731 \\ \underline{\times\;5} \\ 48{,}655 \end{array}$$

$$\begin{array}{r} 9531 \\ \underline{\times\;7} \\ 66{,}717 \end{array} \quad \boxed{\begin{array}{r} 7531 \\ \underline{\times\;9} \\ 67{,}779 \end{array}}$$

The largest product is 7531×9.

33. Note, your calculator results may vary from those shown below due to the accuracy of your calculator.

(a) Without clearing the calculator and reentering the numbers after the equals signs we obtain the following results.
[ON/AC] 276,523 [÷] 511 [=] 541.1409
[−] 541 [=] 0.1409002 [×] 511
[=] 71.999997.
Therefore, $q = 541$ and $r = 72$. To check, note that $541 \cdot 511 + 72 = 276{,}523$.

Problem Set 3.5

1. Thought process may vary.

(c) $\overset{\frown}{27 + 42} + 23$

50, 90, 92

(e) $48 \cdot 5 = (50 - 2) \cdot 5 = 250 - 10 = 240$

2. Thought process may vary.

(c) $306 - 168 = 308 - 170$, 130, 138

(e) $479 + 97 = 476 + 100$, 576

3. Thought process may vary.

(a) 425 + 362
= (400 + 20 + 5) + (300 + 60 + 2)
= (400 + 300) + (20 + 60) + (5 + 2)
= 700 + 80 + 7 = 787
700, 780, 787

(e) $3 \cdot 342$
$= 3 \cdot (300 + 40 + 2)$
$= 3 \cdot 300 + 3 \cdot 40 + 3 \cdot 2$
$= 900 + 120 + 6 = 1026$
900, 1000, 1026

5. (a) Round down because there is a 1 in the thousands place. 630,000

6. (a) Round down because there is a 4 in the tens place. 900

(c) Round down because there is a 4 in the hundred thousands place. 27,000,000

8. (a) 17,000 + 7000 + 12,000 + 2000 + 14,000 = 52,000

(c) 28,000 + 1000 + 2000 + 5000 + 13,000 = 49,000

(e) 21,000 – 8000 = 13,000

9. (a) $3000 \cdot 30 = 90{,}000$

10. (a) $30{,}000 \div 40 = 750$

15. Sally recognized that the hundreds digit is a 4 and knew that means that the number would be rounded down. However, she should have been looking at the tens digit to make the rounding decision for the nearest hundred instead of looking at the hundreds digit. She rounded the number to the nearest thousand instead of rounding it to the nearest hundred. The correct answer is 8500 (using the 5 in the tens digit).

20. (a) (i) $\dfrac{500+400}{300} = 3$

(ii) $\dfrac{3 \cdot 300 + 500}{80} = 17.5$

(iii) $\dfrac{800 \cdot 200}{400} = 400$

21. (a) Since $3 \cdot 7 = 21$, the last digit should be 1. 27,451 is correct.

Chapter 3 Review Exercises

1. (a) 2000 + 300 + 50 + 3 = 2353

(b) $8 \cdot 7200 + 2 \cdot 360 + 0 \cdot 20 + 11 = 58{,}331$

(c) $1000 + (1000 - 100) + (100 - 10) + 5 + 3$
$= 1000 + 900 + 90 + 5 + 3 = 1998$

2.

$$\begin{array}{r} 1 \\ 144{,}000\overline{)234{,}572} \\ \underline{144\ 000} \\ 90\ 572 \end{array} \qquad \begin{array}{r} 12 \\ 7200\overline{)90{,}572} \\ \underline{72\ 00} \\ 18\ 572 \\ \underline{14\ 400} \\ 4\ 172 \end{array}$$

$$\begin{array}{r} 11 \\ 360\overline{)4172} \\ \underline{360} \\ 572 \\ \underline{360} \\ 212 \end{array} \qquad \begin{array}{r} 10 \\ 20\overline{)212} \\ \underline{200} \\ 12 \end{array}$$

$234{,}572 = 1 \cdot 144{,}000 + 12 \cdot 7200 + 11 \cdot 360 + 10 \cdot 20 + 12$
$= 1 \cdot 18 \cdot 20^3 + 12 \cdot 18 \cdot 20^2 + 14 \cdot 18 \cdot 20 + 11 \cdot 20 + 13 \cdot 1$

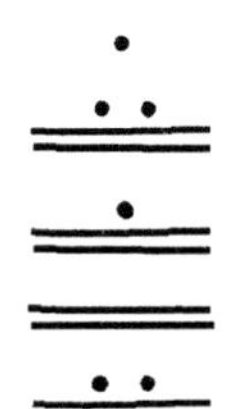

3. Exchange 30 units for 3 strips, then exchange all 30 strips for 3 mats. The result is 8 mats, 0 strips, and 2 units.

4. (a) $1 \cdot 2^5 + 0 \cdot 2^4 + 1 \cdot 2^3 + 1 \cdot 2^2 + 0 \cdot 2 + 1 = 45_{ten}$

(b) $3 \cdot 7^2 + 4 \cdot 7 + 6 = 181_{ten}$

(c) $2 \cdot 12^2 + 10 \cdot 12 + 9 = 417_{ten}$

5. (a) $287 = 2 \cdot 125 + 37$
$= 2 \cdot 125 + 1 \cdot 25 + 12$
$= 2 \cdot 125 + 1 \cdot 25 + 2 \cdot 5 + 2$
$= 2122_{five}$

(b)
$$\begin{aligned}287 &= 1\cdot 256+31\\ &= 1\cdot 256+0\cdot 128+0\cdot 64+0\cdot 32\\ &\quad +1\cdot 16+15\\ &= 1\cdot 256+0\cdot 128+0\cdot 64+0\cdot 32\\ &\quad +1\cdot 16+1\cdot 8+7\\ &= 1\cdot 256+0\cdot 128+0\cdot 64+0\cdot 32\\ &\quad +1\cdot 16+1\cdot 8+1\cdot 4+3\\ &= 1\cdot 256+0\cdot 128+0\cdot 64+0\cdot 32\\ &\quad +1\cdot 16+1\cdot 8+1\cdot 4+1\cdot 2+1\\ &= 100{,}011{,}111_{\text{two}}\end{aligned}$$

(c)
$$\begin{aligned}287 &= 5\cdot 49+42\\ &= 5\cdot 49+6\cdot 7+0\\ &= 560_{\text{seven}}\end{aligned}$$

6.

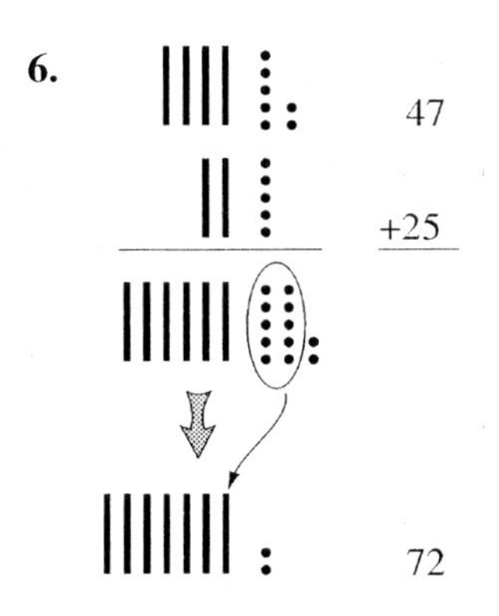

7. **(a)**
$$\begin{array}{r} 42\\ +54\\ \hline 6\\ 90\\ \hline 96\end{array}$$

(b)
$$\begin{array}{r} 47\\ +35\\ \hline 12\\ 70\\ \hline 82\end{array}$$

(c)
$$\begin{array}{r} 59\\ +63\\ \hline 12\\ 110\\ \hline 122\end{array}$$

8. **(a)**

100s	10s	1s

487 − 275 = 212

(b)

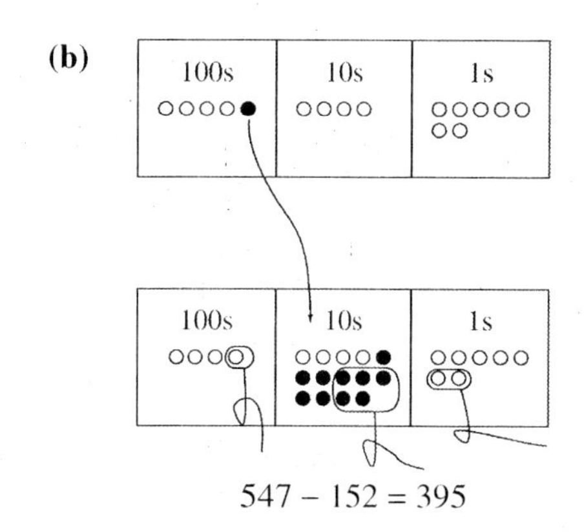

547 − 152 = 395

9. **(a)**
$$\begin{array}{r} {\scriptstyle 1\ 1}\ \ \\ 2433_{\text{five}}\\ +\ 141_{\text{five}}\\ \hline 3124_{\text{five}}\end{array}$$

(b)
$$\begin{array}{r} {\scriptstyle 3\ 13}\ \ \\ 2\not{4}\not{3}3_{\text{five}}\\ -\ 141_{\text{five}}\\ \hline 2242_{\text{five}}\end{array}$$

(c)
$$\begin{array}{r} 243_{\text{five}}\\ \times\ 42_{\text{five}}\\ \hline 1\,041\\ 21\,320\\ \hline 22{,}411_{\text{five}}\end{array}$$

10. **(a)**
$$\begin{array}{r} 357\\ \times\ 4\\ \hline 28\\ 200\\ 1200\\ \hline 1428\end{array}$$

(b)
$$\begin{array}{r} 642\\ \times\ 27\\ \hline 14\\ 280\\ 4\,200\\ 40\\ 800\\ 12{,}000\\ \hline 17{,}334\end{array}$$

11. **(a)**
$$\begin{array}{r} \underline{127}\\ 7\\ 20\\ 100\\ 7\overline{)895}\\ \underline{700}\\ 195\\ \underline{140}\\ 55\\ \underline{49}\\ 6\end{array}$$

(b)
$$\begin{array}{r} \underline{79}\\ 9\\ 70\\ 347\overline{)27483}\\ \underline{24290}\\ 3193\\ \underline{3123}\\ 70\end{array}$$

12. **(a)**
$$\begin{array}{r} 5\ 4\ 8\ 7\ \ \text{R1}\\ 5\overline{)2\ 7^{2}4^{4}3^{3}6}\end{array}$$

(b)

$$\begin{array}{r} 4\ 9\ 4\ 8\ \text{R0} \\ 8\overline{)3\ 9^7 5^3 8^6 4} \end{array}$$

13. a.

$$\begin{array}{r} \overset{2}{\underset{1}{}}\phantom{3_{five}} \\ 23_{five} \\ \times 42_{five} \\ \hline 101 \\ 202 \\ \hline 2121_{five} \end{array}$$

(b)

$$\begin{array}{r} 2\phantom{13_{five}} \\ 1\ 1\phantom{3_{five}} \\ 2413_{five} \\ \times 332_{five} \\ \hline 10331 \\ 13244 \\ 13244 \\ \hline 2023221_{five} \end{array}$$

14.

~~42~~	~~35~~
21	70
~~10~~	~~140~~
5	280
~~2~~	~~560~~
1	1120
	1470

15. (a) Round up because there is a 7 in the ten-thousands place. 300,000

(b) Round down because there is a 4 in the thousands place. 270,000

(c) Round up because there is a 5 in the hundreds place 275,000

16. 657 rounds to 700, 439 rounds to 400, 1657 rounds to 2000 and 23 rounds to 20. Thus,

(a) 657 + 439 is approximately 700 + 400 = 1100. The actual sum is 1096.

(b) 657 – 439 is approximately 700 – 400 = 300. The actual answer is 218.

(c) 657 · 439 is approximately 700 · 400 = 280,000. The actual answer is 288,423.

(d) 1657 ÷ 23 is approximately 2000 ÷ 20 = 100. The actual answer is approximately 72.04.

Chapter 3 Test

1. 74 + 48 = 122

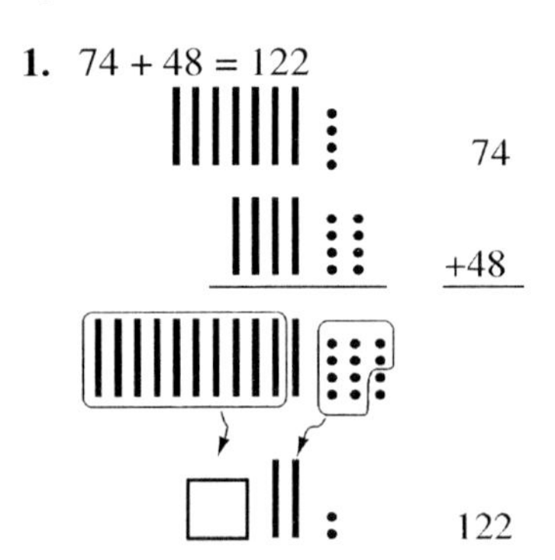

2. (a) Round down because a 3 is in the hundred-thousands place.
3,000,000

(b) Round up because a 7 is in the ten-thousands place.
3,400,000

(c) Round up because a 6 is in the thousands place. 3,380,000

(d) Round up because a 5 is in the hundreds place. 3,377,000

3. The digits in each number need to be written in decreasing order. There are ten possibilities.

531 × 97 = 51,507	731 × 95 = 69,445	**751 × 93 = 69,843**	753 × 91 = 68,523
931 × 75 = 69,825	951 × 73 = 69,423	953 × 71 = 67,663	971 × 53 = 51,463
973 × 51 = 49,623	975 × 31 = 30,225		

The largest product is 751×93.

4. **(a)** Let $G = 1+\sqrt{2}$. Store this value in your calculator's memory.

$f_1 = \frac{G}{\sqrt{8}} \approx 0.854$ to the nearest integer is 1.

$f_2 = \frac{G^2}{\sqrt{8}} \approx 2.061$ to the nearest integer is 2.

$f_3 = \frac{G^3}{\sqrt{8}} \approx 4.975$ to the nearest integer is 5.

$f_4 = \frac{G^4}{\sqrt{8}} \approx 12.010$ to the nearest integer is 12.

(b) f_5 to the nearest integer is 29 and f_6 to the nearest integer is 70.

(c) The rule is $f_n = 2f_{n-1} + f_{n-2}$ for $n \geq 3$.

5. **(a)** $2 \cdot 3^4 + 1 \cdot 3^3 + 0 \cdot 3^2 + 2 \cdot 3 + 2 = 197_{\text{ten}}$

(b) $3 \cdot 8^2 + 1 \cdot 8 + 7 = 207_{\text{ten}}$

(c) $4 \cdot 5^3 + 2 \cdot 5^2 + 1 \cdot 5 + 3 = 558_{\text{ten}}$

6. Work from right to left.

```
  1  1
  2 837
+ 7 224
 10,061
```

7. Rewrite as an addition problem and work from right to left.

```
 4-94            4694
+35-2   gives   +3542
 -23-            8236

           8236
Solution: -3542
           4694
```

8. 400 + 60 + 300 + 40 + 4000 = 4800

9. **(a)** $1^3 = 1^2 = 1$

$1^3 + 2^3 = (1+2)^2 = 9$

$1^3 + 2^3 + 3^3 = (1+2+3)^2 = 36$

$1^3 + 2^3 + 3^3 + 4^3 = (1+2+3+4)^2 = 100$

(b) $\sqrt{1} = 1, \sqrt{9} = 3, \sqrt{36} = 6, \sqrt{100} = 10$

(c) Since the answers to (b) are the triangular numbers $t_n = \frac{n(n+1)}{2}$, conjecture that both $1^3 + 2^3 + \cdots + n^3$ and $(1+2+3+\cdots+n)^2 = \left(\frac{n(n+1)}{2}\right)^2$.

10.

```
          5            9          12
7200)39,485    360)3485    20)245
     36 000        3240       240
      3 485         245         5
```

$39,485 = 5 \cdot 18 \cdot 20^2 + 9 \cdot 18 \cdot 20 + 12 \cdot 20 + 5$
$= 5 \cdot 7200 + 9 \cdot 360 + 12 \cdot 20 + 5$

11. Since $\frac{123-3}{5} = 24$, there are 25 terms and the [=] key must be pressed 24 times.

[ON/AC] 3 [M+] [+] 5 [=] [M+] [=] [M+] ... [=] [M+] [MR], 1575

12. The digits in each number must increase from left to right (except 0, which must be in the second position from the left). This leaves ten possibilities:

468	268	248	246
× 20	× 40	× 60	× 80
9360	10,720	14,880	19,680
608	408	406	
× 24	× 26	× 28	
14,592	10,608	11,368	
208	206	204	
× 46	× 48	× 68	
9568	9888	13,872	

The smallest product is 468 × 20.

13. Answers will vary. One possibility is [ON/AC] 2 [+] 5 [M+] [+] [M+] [MR] [+] [M+] [MR] with the repeating sequence [+] [M+] [MR]. The [M+] command repeats the preceding term, whereas the [+] and [MR] each give the next term in the sequence. A second possibility, more suitable for nonalgebraic calculators, is [ON/AC] 2 [M+] 5 [M+] [+] [MR] [M+] [+] [MR] [M+] with the repeating sequence [+] [MR] [M+]. For this algorithm the [MR] and [M+] commands give new terms of the sequence.

14. (a)

$$\begin{array}{r} {\scriptstyle 1\,1\,} \\ 242_{five} \\ +\ 43_{five} \\ \hline 340_{five} \end{array}$$

(b)

$$\begin{array}{r} {\scriptstyle 1\ 13\ 12} \\ \not{2}\,\not{4}\,\not{2}_{five} \\ -\ \ 4\,3_{five} \\ \hline 1\,4\,4_{five} \end{array}$$

(c)

$$\begin{array}{r} {\scriptstyle 3\,1\,} \\ {\scriptstyle 2\,1\,} \\ 242_{five} \\ \times\ \ 43_{five} \\ \hline 1\,331 \\ 21\,23 \\ \hline 23{,}111_{five} \end{array}$$

15. Since $1{,}171{,}875 = 3 \cdot 5^8$, there are 9 terms and the [=] key must be pressed 8 times. [ON/AC] 3 [M+] [×] 5 [=] [M+] [=] [M+] ... [=] [M+] [MR], 1,464,843

16. (a)

$$\begin{aligned} 281 &= 2 \cdot 125 + 31 \\ &= 2 \cdot 125 + 1 \cdot 25 + 6 \\ &= 2 \cdot 125 + 1 \cdot 25 + 1 \cdot 5 + 1 \\ &= 2111_{five} \end{aligned}$$

(b)

$$\begin{aligned} 281 &= 1 \cdot 256 + 25 \\ &= 1 \cdot 256 + 0 \cdot 128 + 0 \cdot 64 + 0 \cdot 32 \\ &\qquad + 1 \cdot 16 + 9 \\ &= 1 \cdot 256 + 0 \cdot 128 + 0 \cdot 64 + 0 \cdot 32 \\ &\qquad + 1 \cdot 16 + 1 \cdot 8 + 0 \cdot 4 + 0 \cdot 2 + 1 \\ &= 100011001_{two} \end{aligned}$$

(c)

$$\begin{aligned} 281 &= 1 \cdot 144 + 137 \\ &= 1 \cdot 144 + 11 \cdot 12 + 5 \\ &= 1E5_{twelve} \end{aligned}$$

Chapter 4 Number Theory

Problem Set 4.1

1. (a) 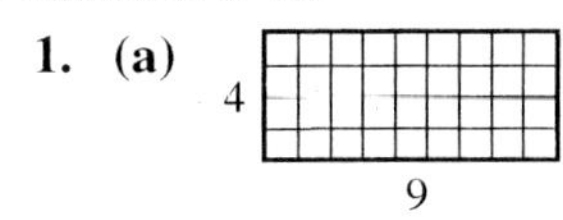

$36 = 4 \cdot 9$
4 divides 36.

3. (a) $1 \cdot 8 = 8, 2 \cdot 8 = 16, 3 \cdot 8 = 24, 4 \cdot 8 = 32,$
$5 \cdot 8 = 40, 6 \cdot 8 = 48, 7 \cdot 8 = 56,$
$8 \cdot 8 = 64, 9 \cdot 8 = 72, 10 \cdot 8 = 80$

6. Factor trees may vary, but the collection of prime numbers at the ends of the "branches" should agree in each case.

(a) 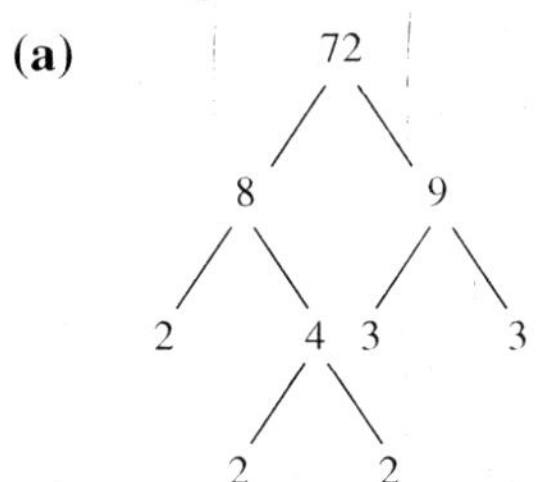

(c) 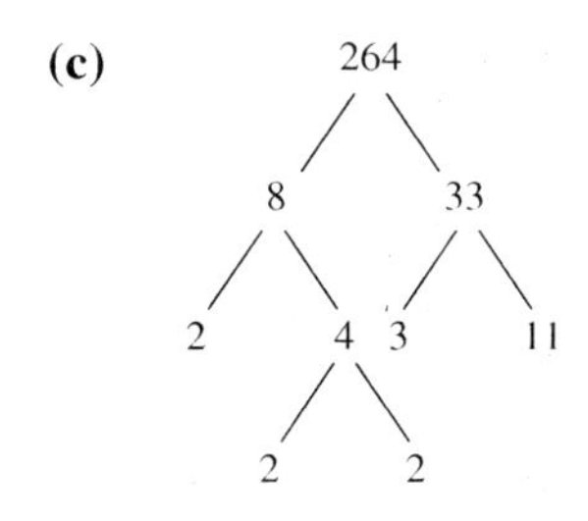

7. (a)
$$\begin{array}{r} 5 \\ 5\overline{)25} \\ 2\overline{)50} \\ 2\overline{)100} \\ 7\overline{)700} \end{array}$$
$700 = 2 \cdot 2 \cdot 5 \cdot 5 \cdot 7$

(c)
$$\begin{array}{r} 2 \\ 3\overline{)6} \\ 3\overline{)18} \\ 5\overline{)90} \\ 5\overline{)450} \end{array}$$
$450 = 5 \cdot 5 \cdot 3 \cdot 3 \cdot 2$

8. (a) 1, 2, 3, 4, 6, 8, 12, 16, 24, 48

9. (a) $136 = 2^3 \cdot 17^1$, $102 = 2^1 \cdot 3^1 \cdot 17^1$

(b) The divisors of 136 are
$2^0 \cdot 17^0 = 1$, $2^1 \cdot 17^0 = 2$, $2^2 \cdot 17^0 = 4$,
$2^3 \cdot 17^0 = 8$, $2^0 \cdot 17^1 = 17$, $2^1 \cdot 17^1 = 34$,
$2^2 \cdot 17^1 = 68$, $2^3 \cdot 17^1 = 136$.

10. (a)
$$\begin{array}{r} 3 \\ 2\overline{)6} \\ 2\overline{)12} \\ 2\overline{)24} \\ 2\overline{)48} \end{array}$$
$48 = 2 \cdot 2 \cdot 2 \cdot 2 \cdot 3 = 2^4 \cdot 3^1$

(c)
$$\begin{array}{r} 5 \\ 5\overline{)25} \\ 5\overline{)125} \\ 3\overline{)375} \\ 3\overline{)1125} \\ 2\overline{)2250} \end{array}$$
$2250 = 2 \cdot 3 \cdot 3 \cdot 5 \cdot 5 \cdot 5 = 2^1 \cdot 3^2 \cdot 5^3$

11. (a) Yes. $28 = 2^2 \cdot 7^1$. Since 2^3 and 7^2 appear in the factorization of a, all the factors of 28 appear in a and to at least as high a power.

(c) The exponent on 2 is $3 - 2 = 1$, since 2^3 appears in a and 2^2 appears in b. Likewise, the exponent on 3 is $1 - 1 = 0$, and the exponent on 7 is 2 (unchanged from a). Thus
$\frac{a}{b} = 2^1 \cdot 3^0 \cdot 7^2 = 2^1 \cdot 7^2 = 98$. This may be written as
$$\frac{a}{b} = \frac{2^3 \cdot 3^1 \cdot 7^2}{2^2 \cdot 3^1} = 2^{3-2} \cdot 3^{1-1} \cdot 7^2 = 2^1 \cdot 7^2.$$

15. (a) No. If $n = ab$, it need only be true that *one* of a or b is less than or equal to $\sqrt{n}$. For example, $10 = 2 \cdot 5$ with 2 and 5 both primes. Yet $5 > \sqrt{10} \approx 3.16$.

17. (a) True. $n \cdot 0 = 0$ for every natural number n.

(c) True. $1 \cdot n = n$ for every natural number n.

(e) False. The definition of b divides a specifies that $b \neq 0$. This is because $0 \div 0 = q$ if, and only if, $0 \cdot q = 0$ for a *unique* integer q. However, this is true for *every* integer q.

23. (a) $1,\ 3,\ 3^2 = 9$

24. (a) $496 = 2^4 \cdot 31^1$

(b) $1 + 2 + 4 + 8 + 16 + 31 + 62 + 124 + 248 = 496$

25. (a) The proper divisors of 10 are 1, 2, and 5. Since $1 + 2 + 5 = 8 < 10$, 10 is deficient.

(b) The proper divisors of 18 are 1, 2, 3, 6, and 9. Since $1 + 2 + 3 + 6 + 9 = 21 > 18$, 18 is abundant.

28. Yes. If p divides bc then p appears in the prime factorization of bc, which is the product of the prime factorization of b and the prime factorization of c. Therefore, p appears in the prime factorization of b or c (or both).

32. (a) If the prime p divides n, then $n = pk$ for some natural number k. But then $n + 1 = pk + 1$, which shows that p does not divide $n + 1$ since it leaves a remainder of 1.

(b) By part (a), if the prime p divides n, then p cannot divide $n + 1$. Therefore, if q is any prime divisor of $n + 1$, it follows that p and q are distinct divisors of $n(n + 1)$.

(c) By part (b), since 2 and 3 divide 6, then neither can divide $6 + 1 = 7$. Thus, any prime divisor of 7 (which is 7, of course) must be different from both 2 and 3, meaning that $6 \times 7 = 42$ has three different prime divisors: 2, 3, and 7.

33. (a) Since the sum is odd, one of the two numbers is even and the other is odd. But the only even prime is 2, so the other number is 311. Since $\sqrt{311} \doteq 17.6$ and no prime 2, 3, 5, 7, 11, 13, 17 is a divisor of 311, we see that 311 is prime. So, we conclude that the only pair of primes that sum to 311 is the pair 2 and 311.

38. (a) $548 = 2^2 \cdot 137^1$

(c) $274 = 2^1 \cdot 137^1$

(e) 548 does not divide any other number given because only 274 has a factor of 137, but its power of 2 is smaller. 936 divides 45,864 because 45,864 has powers of 2, 3, and 13 that are at least as great: $2^3 \cdot 3^2 \cdot 13^1$ divides $2^3 \cdot 3^2 \cdot 7^2 \cdot 13^1$. It does not divide 548 or 274, because these numbers do not have 13 in their prime factorizations. 274 divides 548 because 548 has powers of 2 and 137 that are at least as great: $2^1 \cdot 137^1$ divides $2^2 \cdot 137^1$. It does not divide 936 or 45,864 because these numbers do not have 137 in their prime factorization. 45,864 does not divide any other number given because the others do not have 7 in their prime factorization. The only answers are 274 divides 548 (or $2^1 \cdot 137^1$ divides $2^2 \cdot 137^1$) and 936 divides 45,864 (or $2^3 \cdot 3^2 \cdot 13^1$ divides $2^3 \cdot 3^2 \cdot 7^2 \cdot 13^1$).

39. (a) $894,348 = 2^2 \cdot 3^3 \cdot 7^2 \cdot 13^2$

(c) $1,265,625 = 3^4 \cdot 5^6$

(e) The exponents in the prime power representation of a square are even. This is because when you square a number, you double each exponent in its prime power representation.

Problem Set 4.2

1. (a) 1554 is divisible by 2 and 3: Divisible by 2 because units digit is even. Divisible by 3 because $1 + 5 + 5 + 4 = 15$ and 3 divides 15. Not divisible by 5 because units digit is not 0 or 5.

(c) 805 is divisible by 5. Not divisible by 2 because units digit is odd. Not divisible by 3 because $8 + 0 + 5 = 13$ and 3 does not divide 13. Divisible by 5 because units digit is 5.

2. (a) 1554. This is the only number that is divisible by 2 and 3.

(c) None. None of the numbers are divisible by 3 and 5.

4. (a) 253,799 is divisible by 7 and 13. Use the combined test: 799 – 253 = 546.
7 divides 546 because 546 = 7 · 78.
11 does not divide 546 because
(5 + 6) – 4 = 7 and 11 does not divide 7.
13 divides 546 because 546 = 13 · 42.

(c) 1,960,511 is divisible by 7. Use the combined test: 960 – (1 + 511) = 448.
7 divides 448 because 448 = 7 · 64.
11 does not divide 448 because
(4 + 8) – 4 = 8 and 11 does not divide 8.
13 does not divide 448 because
448 = 13 · 34 + 6.

5. (a) None; none of the numbers are divisible by 7 and 11.

(c) None. None of the numbers are divisible by 11 and 13.

10. (a) For any palindrome with an even number of digits, the digits in the odd positions are the same as the digits in the even positions, but with the order reversed. Thus, the difference of the sums of the digits in the even and odd positions is zero, which is divisible by 11.

16. (a) Since 2 is a divisor of both 11! and 2, it is also a divisor of the sum 11! + 2 by the Divisibility of Sums Theorem. The same reasoning shows that 3 divides the sum 11! + 3, and so on.

17. (a) $686 \leftrightarrow 68 - 12 = 56 = 7 \cdot 8$, so 686 is divisible by 7.

18. (a) $10m = 10q - 20r = 7k - r - 20r$
$= 7k - 21r = 7(k - 3r)$, so $10m$ is divisible by 7. Since the prime factors of 10 are 2 and 5, m must have 7 as a prime factor. That is, m is divisible by 7.

19. When using the combined test for divisibility by 7, 11, and 13 the digits of the number are broken up into 3-digit groups. If a number has form *abc*, *abc*, then the difference in the sums of the 3-digit numbers in odd positions and even positions will be zero, which is divisible by each of 7, 11, and 13. An alternate explanation is to note that *abc*, *abc* = 1001 · *abc* = 7 · 11 · 13 · *abc*, where *abc* denotes the number $100a + 10b + c$.

23. No. He or she could have made other kinds of errors that by chance resulted in the record being out of balance by an amount that is a multiple of 9. For example, the teller could have made a 5¢ error and later a 4¢ error, resulting in a total error of 9¢.

Problem Set 4.3

1. (a) Let D_{24} and D_{27} represent the sets of divisors of 24 and 27.
$D_{24} = \{1, 2, 3, 4, 6, 8, 12, 24\}$
$D_{27} = \{1, 3, 9, 27\}$
$D_{24} \cap D_{27} = \{1, 3\}$
GCD(24, 27) = 3

2. (a) Let M_{24} and M_{27} represent the set of multiples of 24 and 27.
$M_{24} = \{24, 48, 72, 96, 120, 144, 168, 192, 216, 240, \cdots\}$
$M_{27} = \{27, 54, 81, 108, 135, 162, 189, 216, 243, \cdots\}$
$M_{24} \cap M_{27} = \{216, \cdots\}$
LCM(24, 27) = 216

3. (a) GCD(24, 27) · LCM(24, 27) = 3 · 216 = 648 = 24 · 27

4. (a) For the GCD we choose the smaller of the two exponents with which each prime appears in r and s:
$r = 2^2 \cdot 3^1 \cdot 5^3$
$s = 2^1 \cdot 3^3 \cdot 5^2$
$\text{GCD}(r, s) = 2^1 \cdot 3^1 \cdot 5^2 = 150$
For the LCM we choose the larger of the two exponents with which each prime appears in r and s:
$\text{LCM}(r, s) = 2^2 \cdot 3^3 \cdot 5^3 = 13{,}500$.

8. Note that the x rod measures the y rod if and only if, x divides y.

(a) The 1-, 3-, and 9-rods. (The divisors of 9 are 1, 3, and 9.)

(d) Any rods or trains with length 1, 2, 3, 6, 9, or 18. (The divisors of 18 are 1, 2, 3, 6, 9, and 18.)

9. (a) The 4-rods can measure trains of length 4, 8, 12, 16, ⋯. The 6-rods can measure trains of length 6, 12, 18, ⋯. The shortest train that can be measured by 4-rods and by 6-rods has length 12.

(b) LCM(4, 6) = 12

16. (a) Since $60 = 2^2 \times 3 \times 5$ and $105 = 3 \times 5 \times 7$, we see that GCD(60, 105) $= 3 \times 5 = 15$. Thus, the largest square tile possible is 15″ by 15″.

(b) Since $60 \div 15 = 4$ and $105 \div 15 = 7$, there will be 4 rows of 7 tiles each. Thus, $4 \times 7 = 28$ tiles are required.

(c) Since

$$\text{GCD}(48, 216) = \text{GCD}\left(2^4 \times 3, 2^3 \times 3^3\right) = 2^3 \times 3 = 24,$$

the hallway can be tiled with 24″ by 24″ square tiles. Since $48 \div 24 = 2$ and $216 \div 24 = 9$, $2 \times 9 = 18$ tiles are required.

(d) The largest square tile is GCD(m, n) on each side. The number of tiles required is

$$\frac{m}{\text{GCD}(m, n)} \times \frac{n}{\text{GCD}(m, n)} = \frac{\text{LCM}(m, n)}{\text{GCD}(m, n)},$$

using the theorem $mn = \text{GCD}(m, n) \times \text{LCM}(m, n)$.

18. (a) $a = 2^2 \cdot 3^1 \cdot 5^2 \cdot 7^0$
$b = 2^1 \cdot 3^3 \cdot 5^1 \cdot 7^0$
$c = 2^0 \cdot 3^2 \cdot 5^3 \cdot 7^1$
For the GCD we choose the smallest of the three exponents with which each prime appears in a, b, and c:
$\text{GCD}(a, b, c) = 2^0 \cdot 3^1 \cdot 5^1 \cdot 7^0 = 15$. For the LCM we choose the largest of the three exponents with which each prime appears in a, b, and c:
$\text{LCM}(a, b, c) = 2^2 \cdot 3^3 \cdot 5^3 \cdot 7^1 = 94{,}500$.

19. (a) $D_{18} = \{1, 2, 3, 6, 9, 18\}$
$D_{24} = \{1, 2, 3, 4, 6, 8, 12, 24\}$
$D_{12} = \{1, 2, 3, 4, 6, 12\}$
$D_{18} \cap D_{24} \cap D_{12} = \{1, 2, 3, 6\}$
GCD(18, 24, 12) = 6

$M_{18} = \{18, 36, 54, 72, 90, \cdots\}$
$M_{24} = \{24, 48, 72, 96, \cdots\}$
$M_{12} = \{12, 24, 36, 48, 60, 72, 84, \cdots\}$
$M_{18} \cap M_{24} \cap M_{12} = \{72, \cdots\}$
LCM(18, 24, 12) = 72

21. (a)

	48	25	35
2	24	25	35
2	12	25	35
2	6	25	35
2	3	25	35
3	1	25	35
5	1	5	7

$\text{LCM}(48, 25, 35) = 2^4 \cdot 3^1 \cdot 5^2 \cdot 7^1$

23. (a) $F_{12} \div F_6 = 144 \div 8 = 18$
$F_{18} \div F_9 = 2584 \div 34 = 76$
$F_{30} \div F_{15} = 832{,}040 \div 610 = 1364$

(c) No. $F_{19} = 4181 = 37 \cdot 113$. The correct statement is: If n is prime, then F_n does not have any factors that are Fibonacci numbers, other than itself and 1.

(h) The conjecture predicts that $\text{GCD}(F_{16}, F_{20}) = F_{\text{GCD}(16, 20)} = F_4 = 3$, so the conjecture would be false if $\text{GCD}(F_{16}, F_{20}) = 4$.

26. (a) 1224 seconds. Since $72 = 2^3 \cdot 3^2$ and $68 = 2^2 \cdot 17^1$, $\text{LCM}(72, 68) = 2^3 \cdot 3^2 \cdot 17^1 = 1224$, so the first time that Hi and Sarah return to the starting point simultaneously is after 1224 seconds. To confirm that Hi has lapped Sarah exactly once at this time, it is necessary to note that 1224 ÷ 72 = 17 and 1224 ÷ 68 = 18, so Sarah has completed 17 laps and Hi has completed 18.

29. **(a)** The following keystrokes will give the remainder 21 of the division of 117 by 48. The quotient is the integer part of the result of the decimal division, 2.

Key in	117	÷	48	M+	=	−	2	=	×	MR	=
Display	117	117	48	48	2.4375	2.4375	2	0.4375	0.4375	48	21
Memory	0	0	0	48	48	48	48	48	48	48	48

(b) Since 3 is the last nonzero remainder, 3 = GCD(117, 48) by the Euclidean Algorithm.

30. **(a)** $10500 \div 6600 = 1 \text{ R } 3900$
$6600 \div 3900 = 1 \text{ R } 2700$
$3900 \div 2700 = 1 \text{ R } 1200$
$2700 \div 1200 = 2 \text{ R } 300$
$1200 \div 300 = 4 \text{ R } 0$
Therefore, GCD(6600, 10500) = 300.

(b) $$\begin{aligned}\text{LCM}(6600, 10500) &= \frac{6600}{300} \cdot 10{,}500 \\ &= 22 \cdot 10{,}500 \\ &= 231{,}000\end{aligned}$$

Chapter 4 Review Exercises

1.

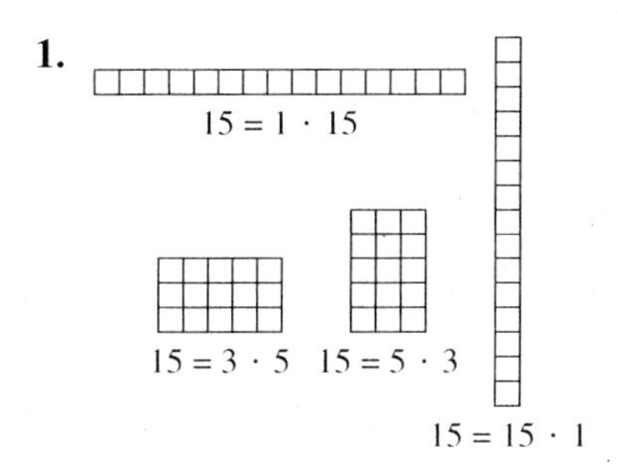

2. Answers will vary.

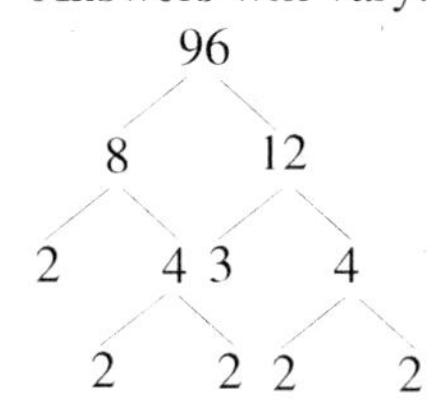

3. **(a)** $D_{60} = \{1, 2, 3, 4, 5, 6, 10, 12, 15, 20, 30, 60\}$

(b) $D_{72} = \{1, 2, 3, 4, 6, 8, 9, 12, 18, 24, 36, 72\}$

(c) $D_{60} \cap D_{72} = \{1, 2, 3, 4, 6, 12\}$ so GCD{60, 72} = 12

4. **(a)**

$$\begin{array}{r} 5 \\ 5\overline{)25} \\ 3\overline{)75} \\ 2\overline{)150} \\ 2\overline{)300} \\ 2\overline{)600} \\ 2\overline{)1200} \end{array}$$

$1200 = 2^4 \cdot 3^1 \cdot 5^2$

(b)

$$\begin{array}{r} 7 \\ 7\overline{)49} \\ 5\overline{)245} \\ 3\overline{)735} \\ 2\overline{)1470} \\ 2\overline{)2940} \end{array}$$

$2940 = 2^2 \cdot 3^1 \cdot 5^1 \cdot 7^2$

(c) $\text{GCD}(1200, 2940) = 2^2 \cdot 3^1 \cdot 5^1 \cdot 7^0 = 60$

$$\begin{aligned}\text{LCM}(1200, 2940) &= 2^4 \cdot 3^1 \cdot 5^2 \cdot 7^2 \\ &= 58{,}800\end{aligned}$$

5. The number 847 will be crossed out as a multiple of 7. (Note that $847 = 7^1 \cdot 11^2$.) Therefore, 847 is composite.

6. **(a)** Answers will vary. For example, $15 = 3 \cdot 5;\ 5 > \sqrt{15}$.

(b) Yes. $3 \le \sqrt{15}$.

7. Answers will vary. r and s must have a common factor other than 1. For example, 8 divides 16 and 4 divides 16, but 32 does not divide 16.

8. Let $n = 3 \cdot 5 \cdot 7 + 11 \cdot 13 \cdot 17$. Then none of 3, 5, 7, 11, 13, and 17 can divide n. Therefore, n is prime itself or it has prime factors other than 3, 5, 7, 11, 13 and 17. Indeed, $n = 2536 = 2^3 \cdot 317$, and 317 is prime (since $\sqrt{317} \approx 17.8$, 317 is odd, and none of the numbers 3, 5, 7, 11, 13, or 17 can divide 317. Thus, the number n can be used to generate the primes 2 and 317.

9. (a) 9310 is divisible by 2 and 5:
Divisible by 2 because the last digit is 0.
Not divisible by 3 because $9 + 3 + 1 + 0 = 13$, 3 does not divide 13.
Divisible by 5 because the last digit is 0 or 5. Not divisible by 11 because $(9 + 1) - (3 + 0) = 7$, and 11 does not divide 7.

(b) 2079 is divisible by 3 and 11:
Not divisible by 2 because the last digit is odd. Divisible by 3 because $2 + 0 + 7 + 9 = 18$, 3 divides 18. Not divisible by 5 because the last digit is not 0 or 5. Divisible by 11 because $(2 + 7) - (0 + 9) = 0$, and 11 divides 0.

(c) 5635 is divisible by 5: Not divisible by 2 because the last digit is odd. Not divisible by 3 because $5 + 6 + 3 + 5 = 19$, and 3 does not divide 19. Divisible by 5 because the last digit is 0 or 5. Not divisible by 11 because $(6 + 5) - (5 + 3) = 3$, and 11 does not divide 3.

(d) 5665 is divisible by 5 and 11: Not divisible by 2 because the last digit is odd. Not divisible by 3 because $5 + 6 + 6 + 5 = 22$, and 3 does not divide 22. Divisible by 5 because the last digit is 0 or 5. Divisible by 11 because $(5 + 6) - (6 + 5) = 0$, and 11 divides 0.

10. (a) 10,197 is divisible by 11.
Use the combined test: $197 - 10 = 187$.
7 does not divide 187 because $187 = 26 \cdot 7 + 5$.
11 divides 187 because $(1 + 7) - 8 = 0$, and 11 divides 0.
13 does not divide 187 because $187 = 14 \cdot 13 + 5$.

(b) 9373 is divisible by 7 and 13.
Use the combined test: $373 - 9 = 364$.
7 divides 364 because $364 = 52 \cdot 7$.
11 does not divide 364 because $(3 + 4) - 6 = 1$, and 11 does not divide 1.
13 divides 364 because $364 = 28 \cdot 13$.

(c) 36,751 is divisible by 11 and 13.
Use the combined test: $751 - 36 = 715$.
7 does not divide 715 because $715 = 102 \cdot 7 + 1$.
11 divides 715 because $(7 + 5) - 1 = 11$, and 11 divides 11.
13 divides 715 because $715 = 55 \cdot 13$.

11. (a) False, because $15 = 3 \cdot 5$, and 3 does not divide 9310.

(b) True, because $33 = 3 \cdot 11$, 3 divides 2079, and 11 divides 2079.

(c) False, because $55 = 5 \cdot 11$, and 11 does not divide 5635.

(d) True, because $55 = 5 \cdot 11$, 5 divides 5665, and 11 divides 5665.

12. (a) The exponents in the prime power representation are 4 and 2, and $(4 + 1)(2 + 1) = 15$.
There are 15 divisors.

(b) Make an orderly list:
$3^0 \cdot 7^0 = 1$, $3^0 \cdot 7^1 = 7$,
$3^0 \cdot 7^2 = 49$, $3^1 \cdot 7^0 = 3$,
$3^1 \cdot 7^1 = 21$, $3^1 \cdot 7^2 = 147$,
$3^2 \cdot 7^0 = 9$, $3^2 \cdot 7^1 = 63$,
$3^2 \cdot 7^2 = 441$, $3^3 \cdot 7^0 = 27$,
$3^3 \cdot 7^1 = 189$, $3^3 \cdot 7^2 = 1323$,
$3^4 \cdot 7^0 = 81$, $3^4 \cdot 7^1 = 567$,
$3^4 \cdot 7^2 = 3969$
In order, the divisors are 1, 3, 7, 9, 21, 27, 49, 63, 81, 147, 189, 441, 567, 1323, and 3969.

13. The sum of the odd digits is $2 + 6 + 3 + 1 + d = 12 + d$.
The sum of the even digits is $7 + 5 + 0 + 2 + 3 = 17$.
We require 11 to divide $[(12 + d) - 17]$ or 11 to divide $[17 - (12 + d)]$, which is true if $12 + d = 17$. Therefore, $d = 5$.

14. (a) $q = 2^3 \cdot 3^5 \cdot 7^2 \cdot 11^1 \cdot 13^0$

$m = 2^1 \cdot 3^0 \cdot 7^3 \cdot 11^3 \cdot 13^1$

$\text{LCM}(q, m) = 2^3 \cdot 3^5 \cdot 7^3 \cdot 11^3 \cdot 13^1$
$= 11{,}537{,}501{,}976$

(b) Divide q by its smallest divisor, 2:

$$\frac{q}{2} = \frac{2^3 \cdot 3^5 \cdot 7^2 \cdot 11^1}{2}$$
$$= 2^2 \cdot 3^5 \cdot 7^2 \cdot 11^1 = 523{,}908$$

15. (a) $D_{63} = \{1, 3, 7, 9, 21, 63\}$

$D_{91} = \{1, 7, 13, 91\}$

$D_{63} \cap D_{91} = \{1, 7\}$ so GCD(91, 63) = 7.

(b) $M_{63} = \{63, 126, 189, 252, 315, 378, 441, 504, 567, 630, 693, 756, 819, 882, 945, 1008, \ldots\}$

$M_{91} = \{91, 182, 273, 364, 455, 546, 637, 728, 819, 910, \ldots\}$

$M_{63} \cap M_{91} = \{819, 1638, \cdots\}$, so

LCM (63, 91) = 819.

16. $r = 2^1 \cdot 3^2 \cdot 5^1 \cdot 7^0 \cdot 11^3$

$s = 2^2 \cdot 3^0 \cdot 5^2 \cdot 7^0 \cdot 11^2$

$t = 2^3 \cdot 3^1 \cdot 5^0 \cdot 7^1 \cdot 11^3$

(a) $\text{GCD}(r, s, t) = 2^1 \cdot 3^0 \cdot 5^0 \cdot 7^0 \cdot 11^2 = 242$

(b) $\text{LCM}(r, s, t) = 2^3 \cdot 3^2 \cdot 5^2 \cdot 7^1 \cdot 11^3 = 16{,}770{,}600$

17. (a) $12{,}100\overline{)119{,}790}$ = 9 R 10,890

$10{,}890\overline{)12{,}100}$ = 1 R 1210

$1210\overline{)10{,}890}$ = 9 R 0

Thus, GCD(119,790, 12,100) = 1210

(b) $\text{LCM}(119{,}790, 12{,}100) = \frac{119{,}790 \cdot 12{,}100}{1210} = 1{,}197{,}900$

18. In 2192, because
LCM(17, 13) = 17 · 13= 221 and
1971 + 221 = 2192.

Chapter 4 Test

1. (a) 62,418 is divisible by 2 and 3.
Divisible by 2 because the last digit is even.
Divisible by 3 because
6 + 2 + 4 + 1 + 8 = 21, and 3 divides 21.
Not divisible by 9 because
6 + 2 + 4 + 1 + 8 = 21, and 9 does not divide 21.
Not divisible by 11 or 13 because
(6 + 4 + 8) – (2 + 1) = 15, and 11 or 13 do not divide 15.

(b) 222,789 is divisible by 3.
Not divisible by 2 because the last digit is odd.
Divisible by 3 because
2 + 2 + 2 + 7 + 8 + 9 = 30, and 3 divides 30.
Not divisible by 9 because
2 + 2 + 2 + 7 + 8 + 9 = 30, and 9 does not divide 30.
Not divisible by 11 or 13 because
(2 + 7 + 9) – (2 + 2 + 8) = 6, and 11 or 13 do not divide 6.

2. (a) No. The prime power representation of r contains two 7s, but the prime power representation of m contains only one 7, so r does not divide m.

(b) The exponents in the prime power representation of m are 3, 2, 1, and 4, so the number of divisors is
(3 + 1)(2 + 1)(1 + 1)(4 + 1) = 120.

(c) $m = 2^3 \cdot 5^2 \cdot 7^1 \cdot 11^4$

$n = 2^2 \cdot 5^0 \cdot 7^2 \cdot 11^3$

$\text{GCD}(m, n) = 2^2 \cdot 5^0 \cdot 7^1 \cdot 11^3 = 37{,}268$

(d) $\text{LCM}(m, n) = 2^3 \cdot 5^2 \cdot 7^2 \cdot 11^4 = 143{,}481{,}800$

3. (a) $N = (1 + 1)(1 + 1)(1 + 1)(1 + 1)(1 + 1) = 2^5 = 32$

(b) $N = (2 + 1)(2 + 1)(2 + 1)(2 + 1)(2 + 1) = 3^5 = 243$

4. (a) $2^1 \cdot 3^2 \cdot 5^1 = 90$

(b) $2^3 \cdot 3^3 \cdot 5^2 \cdot 7^1 = 37{,}800$

(c) 90 · 37,800 = 3,402,000

(d) Using a calculator, the product is 3,402,000.

5. (a)

$$\begin{array}{r} 7 \\ 5\overline{)35} \\ 5\overline{)175} \\ 2\overline{)350} \\ 2\overline{)700} \\ 2\overline{)1400} \end{array}$$

$1400 = 2^3 \cdot 5^2 \cdot 7^1$

(b)

$$
\begin{array}{r}
11 \\
11\overline{)121} \\
5\overline{)605} \\
3\overline{)1815} \\
3\overline{)5445}
\end{array}
$$

$5445 = 3^2 \cdot 5^1 \cdot 11^2$

(c)

$$
\begin{array}{r}
23 \\
11\overline{)253} \\
3\overline{)759} \\
3\overline{)2277} \\
2\overline{)4554}
\end{array}
$$

$4554 = 2^1 \cdot 3^2 \cdot 11^1 \cdot 23$

6. (a) False. For example 3 divides 12 and 6 divides 12, but 18 does not divide 12.

(b) True. If $s = ra$ and $t = sb$, then $t = (ra)b = r(ab)$, so r divides t.

(c) True. If $b = au$ and $c = av$, then $b + c = au + av = a(u + v)$, so a divides $(b + c)$.

(d) False. For example, 2 does not divide 7 and 2 does not divide 9, but 2 divides 16.

7. Since $\sqrt{281} \approx 16.7$, we must check for divisibility by 2, 3, 5, 7, 11, and 13. Standard divisibility tests immediately rule out divisibility by 2, 3, and 5. Also, $281 = 40 \cdot 7 + 1$, $281 = 11 \cdot 25 + 6$, and $281 = 13 \cdot 21 + 8$. Therefore, 281 is a prime.

8. Since $553 = 3 \cdot 154 + 91$, $154 = 1 \cdot 91 + 63$, $91 = 1 \cdot 63 + 28$, $63 = 2 \cdot 28 + 7$ and $28 = 4 \cdot 7$, it follows that GCD(154,553) = 7 and

$$\text{LCM}(154, 553) = \frac{154 \cdot 553}{7} = 12{,}166.$$

9. (a)

$$
\begin{array}{r}
73 \quad \text{R } 9494 \\
13{,}534\overline{)997{,}476}
\end{array}
$$

$$
\begin{array}{r}
1 \quad \text{R } 4040 \\
9494\overline{)13{,}534}
\end{array}
$$

$$
\begin{array}{r}
2 \quad \text{R } 1414 \\
4040\overline{)9494}
\end{array}
$$

$$
\begin{array}{r}
2 \quad \text{R } 1212 \\
1414\overline{)4040}
\end{array}
$$

$$
\begin{array}{r}
1 \quad \text{R } 202 \\
1212\overline{)1414}
\end{array}
$$

$$
\begin{array}{r}
6 \quad \text{R } 0 \\
202\overline{)1212}
\end{array}
$$

GCD(13,534, 997,476) = 202

(b)

$$\text{LCM}(13534, 997476) = \frac{13{,}534 \cdot 997{,}476}{202} = 66{,}830{,}892$$

10. (a) Answers will vary. One possibility is shown.

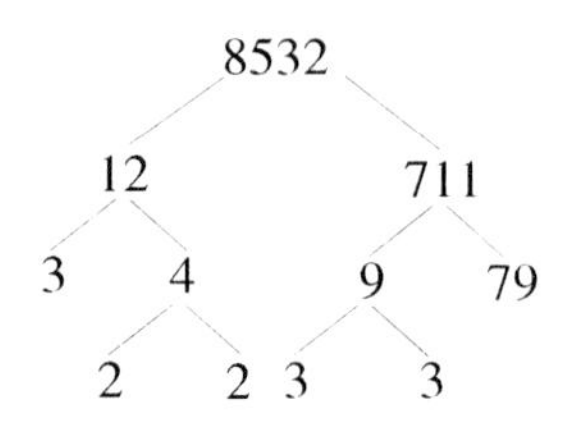

(b) $8532 = 2^2 \cdot 3^3 \cdot 79^1$

(c) Divide 8532 by its smallest divisor (2) to get 4266.

(d) $8532 \cdot 2 = 17{,}064$

Chapter 5 Integers

Problem Set 5.1

1. (a) The loop must have 5 more black counters than red counters. Two possibilities are

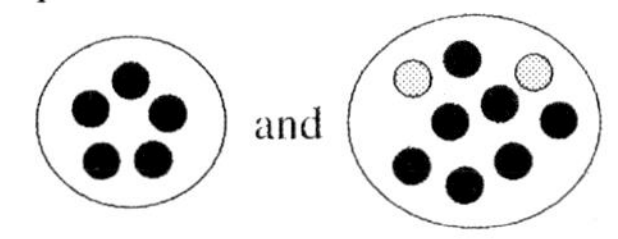

(c) The loop must contain the same number of red and black counters or no counters at all. Two possibilities are

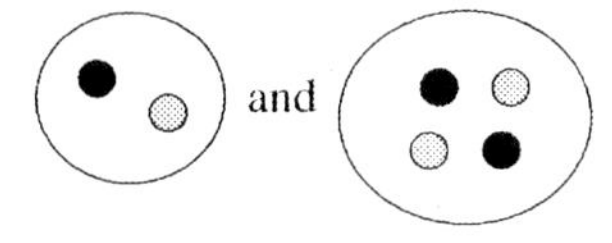

2. (a) The loop must contain 3 more black counters than red counters. To use the least number of counters, use 3 black counters only:

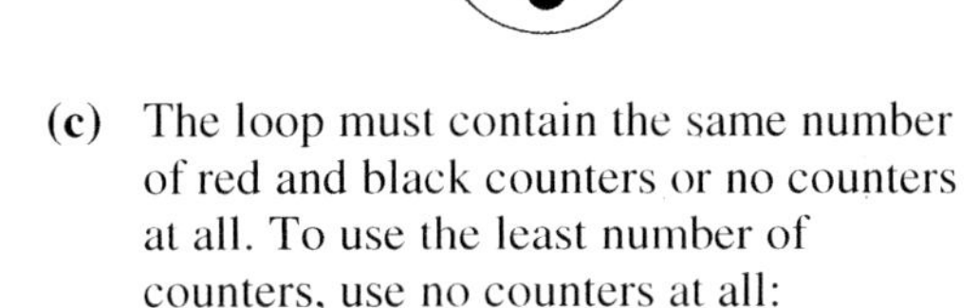

(c) The loop must contain the same number of red and black counters or no counters at all. To use the least number of counters, use no counters at all: no counters.

3. (a) At mail time, you are delivered a check for \$14.

6. (a) Since the bill is for \$15 more than the check you are poorer by \$15. The integer is –15.

7. (a) (i)

0

(iii)

–4 0

(v) $\frac{(4+8)}{2} = \frac{12}{2} = 6$

0 $\frac{4+8}{2}$

8. (a) The arrow points 4 units to the right, so it represents 4.

(c) The arrows point 6 units to the right, so each arrow represents 6.

9. (a) The arrow must point 7 units to the right. One possibility is shown.

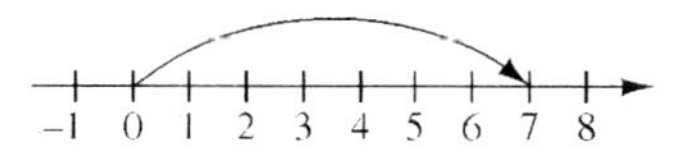

(c) The arrow must point 9 units to the left. One possibility is shown.

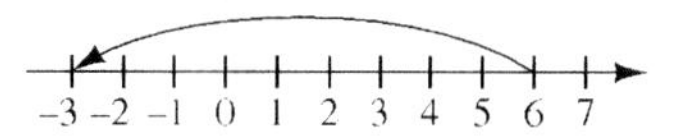

10. (a) (i) Since 34 is 34 units from 0 on a number line, $|34| = 34$.

(iii) Since –76 is 76 units from 0 on a number line, $|-76| = 76$.

11. (a) The equation $|x| = 13$ means that x is 13 units from 0 on a number line. Therefore, $x = 13$ or $x = -13$.

14. Assure the students that they are almost correct since the absolute value of any integer other than 0 is positive but $|0| = 0$. Remind them that the absolute value of a number is its distance on the number line from zero. Since n and $-n$ are both at distance n from 0, $|n| = |-n| = n$ if n is positive. However, if n is negative, then $-n$ is positive and so $|n| = -n$. It may even help to use an example. Clearly –8 is negative and is 8 units from 0, therefore $|-8| = 8 = -(-8)$.

17. (a) To represent –3, the loop needs to have 3 more red counters than black counters. Since there are 3 black counters, we could add red counters until there are 6 red counters—that is, add 4 red counters.

18. (a) 12 red: –12
11 red + 1 black: –10
10 red + 2 black: –8
9 red + 3 black: –6
8 red + 4 black: –4
7 red + 5 black: –2
6 red + 6 black: 0
5 red + 7 black: 2
4 red + 8 black: 4
3 red + 9 black: 6
2 red + 10 black: 8
1 red + 11 black: 10
12 black: 12

20. (a) From problem 19, the even integers from –12 to 12 can be represented using all 12 counters, and the odd integers from –11 to 11 can be represented using 11 counters. Furthermore, any number represented by 12 or fewer counters is an integer between –12 and 12 inclusive. So the set of integers that can be represented using 12 or fewer counters is $\{n \mid n \text{ is an integer and } -12 \le n \le 12\}$.

23. (a) Black: $1 + 3 + 5 + 7 + 9 + 11 + 13 + 15 + 17 + 19 = 100$
Red: $2 + 4 + 6 + 8 + 10 + 12 + 14 + 16 + 18 + 20 = 110$
To see this consider arrays with 1 row, 2 rows, 3 rows, ..., and look for a pattern.

Problem Set 5.2

1. (a)

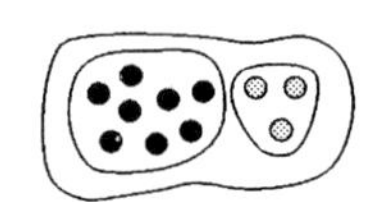

$8 + (-3) = 5$

(c)

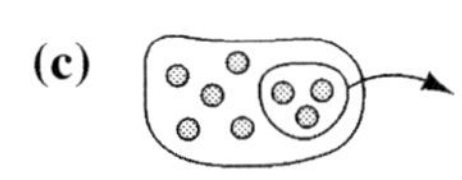

$-8 - (-3) = -5$

(e)

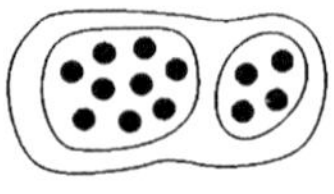

$9 + 4 = 13$

(g)

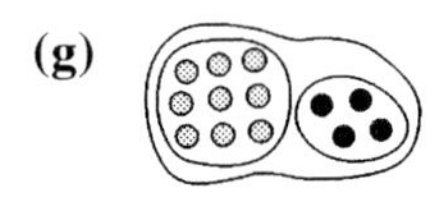

$(-9) + 4 = -5$

2. (a) At mail time you receive a bill for \$27 and a bill for \$13.
$(-27) + (-13) = -40$

(c) The mail carrier brings you a check for \$27 and a check for \$13.
$27 + 13 = 40$

(e) At mail time you receive a bill for \$41 and a check for \$13.
$(-41) + 13 = -28$

(g) At mail time you receive a bill for \$13 and a check for \$41.
$(-13) + 41 = 28$

3. (a)

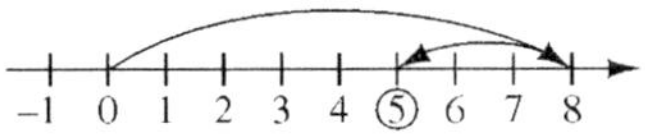

$8 + (-3) = 5$

(c)

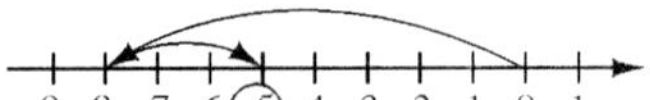

$(-8) + 3 = -5$

(e)

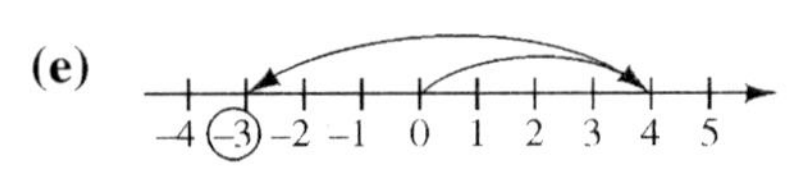

$4 + (-7) = -3$

(g)

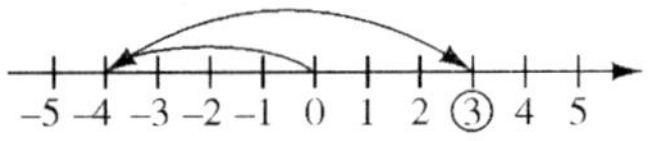

$(-4) + 7 = 3$

4. (a) $13 - 7 = 13 + (-7)$

(c) $(-13) - 7 = (-13) + (-7)$

(e) $3 - 8 = 3 + (-8)$

(g) $(-8) - 13 = (-8) + (-13)$

5. (a) $27 - (-13) = 27 + 13 = 40$

(c) $(-13) - 14 = (-13) + (-14) = -(13 + 14) = -27$

(e) $(-81) - 54 = (-81) + (-54) = -(81 + 54) = -135$

(g) $(-81) + (-54) = -(81 + 54) = -135$

10. (a) $314 - 208 = 106$
Sam's net worth was more, by \$106.

12. (a) $-117 < -24$

20. Only (a) and (c) are true.

22. No. If $a \ge b$, $a > b$ or $a = b$. Thus, $a \ge b$ does not imply that $a > b$ is true.

24. $|x| < 7$ means that, on a number line, x is less than 7 units away from 0. The integers that are less than 7 units away from 0 are –6, –5, –4, –3, –2, –1, 0, 1, 2, 3, 4, 5, and 6.

26. (a) (i) $5 - 11 = 5 + (-11) = -(11 - 5) = -6$, so $|5 - 11| = |-6| = 6$.

(iii) $8-(-7)=8+7=15$, so $|8-(-7)|=|15|=15$.

(b) (i)

4 5 6 7 8 9 10 11 12

Distance is 6.

(iii)

−7 −6 −5 −4 −3 −2 −1 0 1 2 3 4 5 6 7 8

Distance is 15.

(c) For any two integers a and b, the distance between a and b on a number line is $|a-b|$ (or, equivalently, $|b-a|$).

27. (a) (i) $|7+2|=|9|=9$ and $|7|+|2|=7+2=9$.

(iii) $|7+(-6)|=|1|=1$ and $|7|+|-6|=7+6=13$.

28. (a) Each sum must be $\dfrac{-4+(-3)+(-2)+(-1)+0+1+2+3+4}{3}=0$.

Since there are four sums that equal 0 [(−1) + 0 + 1, (−2) + 0 + 2, (−3) + 0 + 3, (−4) + 0 + 4], and just two sums involving 2 that equal 0 [(−2) + 0 + 2, (−3) + 2 + 1], it follows that 0 must be the middle number and 2 must appear as the middle number on one of the sides.

−1	4	−3
−2	0	2
3	−4	1

29. (a) We need to find x, y, and z such that

$x+y=-4$, $y+z=1$, and $x+z=-3$. Since $x+y+z=\dfrac{(-4)+1+(-3)}{2}=\dfrac{-6}{2}=-3$ and $x+y+z=x+(y+z)=x+1$, we know that $x+1=-3$, or $x=-4$. Therefore, $y=0$ and $z=1$.

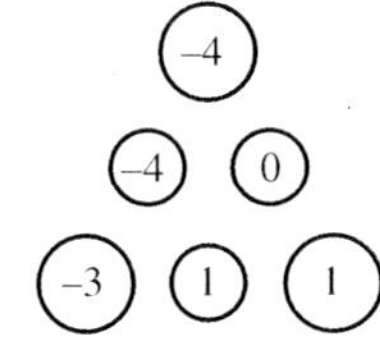

30. (a) (i) $7-(-3)=7+3=10$ and $(-3)-7=(-3)+(-7)=-(3+7)=-10$

31. No. $3-(2-1)=3-1=2$ and $(3-2)-1=1-1=0$, so $3-(2-1)\neq(3-2)-1$.

34. (a) First we write the first 10 Fibonacci numbers:

$F_1=1, F_2=1, F_3=2, F_4=3, F_5=5, F_6=8, F_7=13, F_8=21, F_9=34, F_{10}=55$

$k=1 \quad F_1=1 = F_2$

$k=2 \quad F_1+F_3=1+2=3 = F_4$

$k=3 \quad F_1+F_3+F_5=1+2+5=8 = F_6$

$k=4 \quad F_1+F_3+F_5+F_7=1+2+5+13=21 = F_8$

$k=5 \quad F_1+F_3+F_5+F_7+F_9=1+2+5+13+34=55 = F_{10}$

$k=n \quad F_1+F_3+F_5+F_7+F_9+\cdots+F_{2n-1} = F_{2n}$

The sums of the oddly indexed Fibonacci numbers are the values of the next evenly indexed Fibonacci number.

35. (a)

$$\begin{aligned} F_1 &= 1 &&= F_1^2 \\ F_1 - F_3 &= 1 - 2 = -1 &&= -F_2^2 \\ F_1 - F_3 + F_5 &= 1 - 2 + 5 = 4 &&= F_3^2 \\ F_1 - F_3 + F_5 - F_7 &= 1 - 2 + 5 - 13 = -9 &&= -F_4^2 \\ F_1 - F_3 + F_5 - F_7 + F_9 &= 1 - 2 + 5 - 13 + 34 = 25 &&= F_5^2 \end{aligned}$$

Additional rows are shown to clarify the pattern:

$F_1 - F_3 + F_5 - F_7 + F_9 - F_{11} = 1 - 2 + 5 - 13 + 34 - 89 = -64 = -F_6^2$

$F_1 - F_3 + F_5 - F_7 + F_9 - F_{11} + F_{13} = 1 - 2 + 5 - 13 + 34 - 89 + 233 = 169 = F_7^2$

General result: $F_1 - F_3 + F_5 - \cdots + (-1)^{n+1} F_{2n-1} = (-1)^{n+1}(F_n)^2$

The alternating sums of the first n oddly indexed Fibonacci numbers are the alternating signed squares of the Fibonacci numbers.

36. (a) $101 + 3 = 104$

39. (a)

t	h
0	$(-16)\cdot(0)^2 + 96(0) = 0$
1	$(-16)\cdot(1)^2 + 96(1) = 80$
2	$(-16)\cdot(2)^2 + 96(2) = 128$
3	$(-16)\cdot(3)^2 + 96(3) = 144$
4	$(-16)\cdot(4)^2 + 96(4) = 128$
5	$(-16)\cdot(5)^2 + 96(5) = 80$
6	$(-16)\cdot(6)^2 + 96(6) = 0$
7	$(-16)\cdot(7)^2 + 96(7) = -112$

41. (a) $1 - 2 + 3 - \cdots + 99 = 50$

42. (a) [ON/AC] 3742 [+] [(−)] 2167 [=] 1575

(c) [ON/AC] [(−)] 2167 [−] 3742 [=] −5909

(e) [ON/AC] [(−)] [(] 3571 [−] 5624 [)] [=] 2053

Problem Set 5.3

1. (a) $7 \cdot 11 = 77$

(c) $(-7) \cdot 11 = -(7 \cdot 11) = -77$

(e) $12 \cdot 9 = 108$

(g) $(-12) \cdot 9 = -(12 \cdot 9) = -108$

(i) $(-12) \cdot 0 = 0$

2. (a) $36 \div 9 = 4$

(c) $36 \div (-9) = -(36 \div 9) = -4$

(e) $(-143) \div 11 = -(143 \div 11) = -13$

(g) $(-144) \div (-9) = 144 \div 9 = 16$

(i) $72 \div (21 - 19) = 72 \div 2 = 36$

3. Multiplication: $(-25{,}753) \cdot (-11) = 283{,}283$
Division: $283{,}283 \div (-11) = -25{,}753$;
$283{,}283 \div (-25{,}753) = -11$

5. (a) Richer by \$78; $6 \cdot 13 = 78$

6. (a) $6 \cdot 3 = 18$

8. (a)

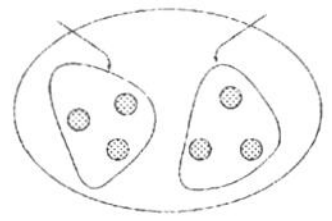

(b)

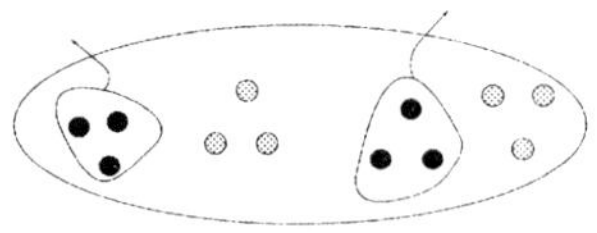

9. (a) $-15 \div 4 = -4$ R 1

11. (a) $-21 \times 18 \approx -20 \times 20 = -400$

13. (a) (i) Multiplicative property of zero; Distributive property of multiplication over addition; Definition of additive inverse

17. (a) Can't tell

20. (a) False. $-1 < 0$ but $(-1)^2 > 0$

22. (b) Complete the upper left circles by noting that $5 - 8 = -3$ and $(-3) + 5 = 2$. The remaining two small circles may be completed with any two numbers whose sum is 2, so the answer is not unique. One possibility is shown.

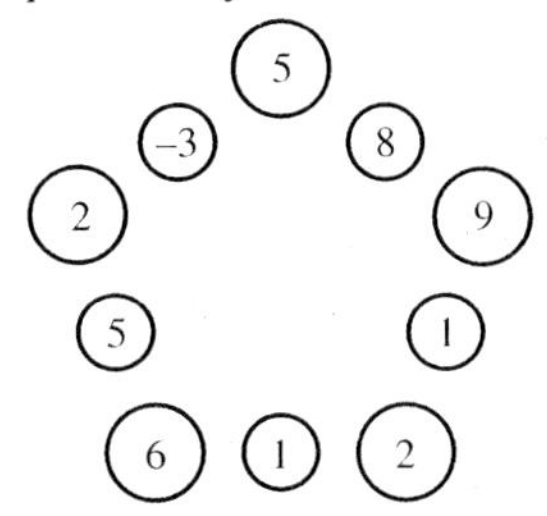

27. (a)

$+_{12}$	0	1	2	3	4	5	6	7	8	9	10	11
0	0	1	2	3	4	5	6	7	8	9	10	11
1	1	2	3	4	5	6	7	8	9	10	11	0
2	2	3	4	5	6	7	8	9	10	11	0	1
3	3	4	5	6	7	8	9	10	11	0	1	2
4	4	5	6	7	8	9	10	11	0	1	2	3
5	5	6	7	8	9	10	11	0	1	2	3	4
6	6	7	8	9	10	11	0	1	2	3	4	5
7	7	8	9	10	11	0	1	2	3	4	5	6
8	8	9	10	11	0	1	2	3	4	5	6	7
9	9	10	11	0	1	2	3	4	5	6	7	8
10	10	11	0	1	2	3	4	5	6	7	8	9
11	11	0	1	2	3	4	5	6	7	8	9	10

$\times_{12}$	0	1	2	3	4	5	6	7	8	9	10	11
0	0	0	0	0	0	0	0	0	0	0	0	0
1	0	1	2	3	4	5	6	7	8	9	10	11
2	0	2	4	6	8	10	0	2	4	6	8	10
3	0	3	6	9	0	3	6	9	0	3	6	9
4	0	4	8	0	4	8	0	4	8	0	4	8
5	0	5	10	3	8	1	6	11	4	9	2	7
6	0	6	0	6	0	6	0	6	0	6	0	6
7	0	7	2	9	4	11	6	1	8	3	10	5
8	0	8	4	0	8	4	0	8	4	0	8	4
9	0	9	6	3	0	9	6	3	0	9	6	3
10	0	10	8	6	4	2	0	10	8	6	4	2
11	0	11	10	9	8	7	6	5	4	3	2	1

29. (a) \$7

(b) $(-105) \div 15 = -7$. A loss of \$105, shared among 15 people results in each person losing \$7.

33. (a) Using the hint, we examine each of the starting parity patterns.

1. (e, e, e, e, e) is all even from the start.
2. $(d, e, e, e) \to (d, e, e, d) \to (d, e, d, e) \to (d, d, d, d) \to (e, e, e, e)$
3. $(d, d, e, e) \to (e, d, e, d) \to (d, d, d, d) \to (e, e, e, e)$
4. $(d, e, d, e) \to (d, d, d, d) \to (e, e, e, e)$
5. $(d, d, d, e) \to (e, e, d, d) \to (e, d, e, d) \to (d, d, d, d) \to (e, e, e, e)$
6. $(d, d, d, d) \to (e, e, e, e)$

(b) After four steps, 2 can be factored from each number of the 4-tuple. After four more steps played on the 4-tuple of quotients, all even numbers are obtained again. Since a 2 has already been factored out, the numbers in the 4-tuple after 8 steps are each divisible by $2 \times 2 = 4$. Similarly, after 12 steps, the resulting 4-tuple contains numbers that are divisible by 8, and after 16 steps, the numbers in the 4-tuple are each divisible by 16.
Starting with (81, 149, 274, 504), four steps later results in the 4-tuple of even numbers (40, 74, 136, 250), then four more steps gives the 4-tuple of even numbers (20, 36, 68, 124). Each number in this 4-tuple is divisible by 4. After 12 and 16 steps, the corresponding 4-tuples are (16, 16, 32, 64) and (0, 32, 0, 32), which are divisible by 8 and 16 respectively.

(c) After making 4, 8, 12, 16, 20, … steps, the corresponding 4-tuples contain only numbers divisible by 2, 4, 8, 16, 32, … At the same time, the numbers in the 4-tuples are nonnegative integers that are not increasing in size. The only integer divisible by all of the numbers 2, 4, 8, 16, 32, … is 0, so after sufficiently many steps, the 4-tuple (0, 0, 0, 0) must be reached. Note: it can be shown that there is no bound on the number of steps that may be required to reach (0, 0, 0, 0). For example, start with a 4-tuple of successive "tribonacci" numbers from the sequence 0, 1, 1, 2, 4, 7, 13, 24, 44, 81, 149, 274, 506, … in which each new term is the sum of the preceding three terms.

(d) The "Three Number Game" need not terminate with (0, 0, 0). Instead, it may reach a cycle like this:
$(0, 0, 1) \to (0, 1, 1) \to (1, 0, 1) \to (1, 1, 0)$ (and back to $(0, 1, 1)$)

It can be shown that an "*n* Number Game" terminates with all zeros if, and only if, *n* is a power of 2, such as 4, 8, 16, 32, … .

34. (b) [ON/AC] 57 [+] 165 [÷] [(−)] 11 [+] 17 [=] 59

Chapter 5 Review Exercises

1. (a) $7 - 8 = -1$

(b) 10 are black and 5 are red, so the number is $10 - 5 = 5$.

(c) Since there are 15 counters and 15 is an odd number, the counters can represent any odd integer from −15 to 15. The possible numbers are −15, −13, −11, ..., 11, 13, 15.

2. (a) Richer by \$12; 12

(b) Poorer by \$37; −37

3. (a) 12

(b) −24

4. (a) Any loop that shows 5 more red counters than black counters represents the integer −5. For example:
5 red only;
1 black + 6 red;
2 black + 7 red;
3 black + 8 red;
4 black + 9 red; etc.

(b) Any loop that shows 6 more black counters than red counters represents the integer 6. For example:
6 black only;
7 black + 1 red;
8 black + 2 red;
9 black + 3 red;
10 black + 4 red; etc.

5. (a) At mail time you receive a bill for \$114 and a check for \$29. $(-114) + 29 = -85$

(b) The mail carrier brings you a bill for \$19 and a check for \$66.
$(-19) + 66 = 47$

6. (a) The additive inverse of 44 is −44.

(b) The additive inverse of −61 is $-(-61) = 61$.

7. The loop on the left represents $5 - 3 = 2$ and the loop on the right represents $1 - 5 = -4$, so the diagram represents $2 + (-4) = -2$.

8. The entire loop represents $6 - 7 = -1$ and the counters that are removed represent $1 - 4 = -3$, so the diagram represents $-1 - (-3) = 2$.

9. (a) $45 + (-68) = -(68 - 45) = -23$.
You are poorer by \$23.

(b) $45 - (-68) = 45 + 68 = 113$.
You are richer by \$113.

10. (a) $6 + 3 = 9$

(b) $8 + (-4) = 4$

(c) $7 - 5 = 2$

(d) $3 - 9 = -6$

(e) $(-3) + (-4) = -7$

(f) $(-3) - (-7) = 4$

(g) $(-5) - 6 = -11$

11. (a) $5 + (-7) = -(7 - 5) = -2$

(b) $(-27) - (-5) = (-27) + 5 = -(27 - 5) = -22$

(c) $(-27) + (-5) = -(27 + 5) = -32$

(d) $5 - (-7) = 5 + 7 = 12$

(e) $8 - (-12) = 8 + 12 = 20$

(f) $8 - 12 = -(12 - 8) = -4$

12. (a) $(-15) - 12 = -(15 + 12) = -27$.
It is 27° below zero.

(b) $(-15) - 12 = -27$

13. (a) $(-12) + 37 = 37 - 12 = 25$.
Her balance is \$25.

(b) $(-12) + 37 = 25$

14. (a)

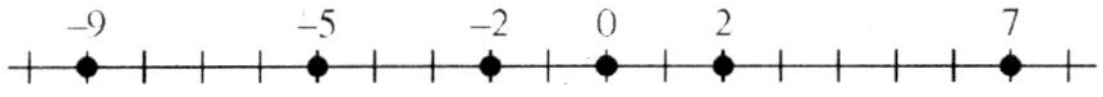

(b) $-9, -5, -2, 0, 2, 7$

(c) $-9 + 4 = -5,\ -5 + 3 = -2,\ -2 + 2 = 0,$
$0 + 2 = 2,\ 2 + 5 = 7$

15. (a) $3 \cdot 4 = 12$ **(b)** $3 \cdot (-4) = -12$

(c) $(-3) \cdot 4 = -12$ **(d)** $(-3) \cdot (-4) = 12$

16. (a) $3 \cdot 0 = 0$ by the multiplicative property of 0
$3 \cdot [5 + (-5)] = 0$ by the definition of additive inverse
$3 \cdot 5 + 3 \cdot (-5) = 0$ by the distributive property of multiplication over addition

(b) $3 \cdot (-5) = -(3 \cdot 5)$
by the definition of the additive inverse

(c) $0 \cdot (-5) = 0$
by the multiplicative property of 0
$[3 + (-3)] \cdot (-5) = 0$
by the definition of additive inverse
$3 \cdot (-5) + (-3) \cdot (-5) = 0$
by the distributive property of multiplication over addition
$-(3 \cdot 5) + (-3) \cdot (-5) = 0$
since $3 \cdot (-5) = -(3 \cdot 5)$ by part (b)

(d) $(-3) \cdot (-5) = -[-(3 \cdot 5)] = 3 \cdot 5$
by the definition of the additive inverse and the theorem on page 316, since $-[-(3 \cdot 5)] = 3 \cdot 5$

17. (a) $(-8) \cdot (-7) = 8 \cdot 7 = 56$

(b) $8 \cdot (-7) = -(8 \cdot 7) = -56$

(c) $(-8) \cdot 7 = -(8 \cdot 7) = -56$

(d) $84 \div (-12) = -(84 \div 12) = -7$

(e) $(-84) \div 7 = -(84 \div 7) = -12$

(f) $(-84) \div (-7) = 84 \div 7 = 12$

18. (a) At mail time you receive 7 checks, each for \$12.

(b) The mail carrier takes away 7 checks, each for \$13.

(c) The mail carrier takes away 7 bills, each for \$13.

19. If d divides n, there is an integer c such that $dc = n$. But then
$d \cdot (-c) = -n, (-d) \cdot (-c) = dc = n,$ and
$(-d) \cdot c = -dc = -n.$ Thus, d divides $-n$, $-d$ divides n, and $-d$ divides $-n$.

20. (a)

$$\begin{array}{l} 17 \\ 5\overline{)85} \\ 3\overline{)255} \end{array} \qquad \begin{array}{l} 13 \\ 3\overline{)39} \end{array}$$

Since $255 = 3 \cdot 5 \cdot 17$, $-39 = (-1) \cdot 3 \cdot 13$, and 3, 5, 13, and 17 are primes, GCD(255, –39) = 3.

(b)

$$\begin{array}{l} 13 \\ 11\overline{)143} \\ 7\overline{)1001} \end{array} \qquad \begin{array}{l} 241 \\ 11\overline{)2651} \end{array}$$

Since $-1001 = (-1) \cdot 7 \cdot 11 \cdot 13$, $2651 = 11 \cdot 241$, and 7, 11, 13, and 241 are primes, GCD(–1001, 2651) = 11.

21. By the division algorithm, n must be of one of these forms: $6q$, $6q + 1$, $6q + 2$, $6q + 3$, $6q + 4$, or $6q + 5$. If n is not divisible by 2, however, then n cannot be of any of the forms $6q$, $6q + 2$ or $6q + 4$. Likewise, if n is not divisible by 3, n cannot be of the form $6q$ or $6q + 3$. Thus, there must be an integer q such that either $n = 6q + 1$ or $n = 6q + 5$.

Case 1 $n = 6q + 1$

$$\begin{aligned} n^2 - 1 &= (6q + 1)^2 - 1 \\ &= (36q^2 + 12q + 1) - 1 \\ &= 36q^2 + 12q \\ &= 12q(3q + 1) \end{aligned}$$

If q is even, then 24 divides $12q$ and $n^2 - 1$ is divisible by 24. If q is odd, then $3q + 1$ is even, so 24 divides $12(3q + 1)$ and hence $n^2 - 1$ is divisible by 24.

(continued on next page)

(continued)

Case 2

$$\begin{aligned} n &= 6q+5 \\ n^2-1 &= (6q+5)^2-1 \\ &= (36q^2+60q+25)-1 \\ &= 36q^2+60q+24 \\ &= 12(3q^2+5q+2) \\ &= 12(3q+2)(q+1) \end{aligned}$$

If q is even, $3q + 2$ is even. If q is odd, $q + 1$ is even. In either case, it follows that $n^2 - 1$ is divisible by 24.

Chapter 5 Test

1. \$381; $129 + 341 - 13 - 47 - 29 = 381$

2. Richer by \$135; $(-5) \cdot (-27) = 135$

3. (a)

$$\begin{aligned} 1 &= 1 \\ 1-4 &= -3 \\ 1-4+9 &= 6 \\ 1-4+9-16 &= -10 \\ 1-4+9-16+25 &= 15 \end{aligned}$$

(b) $1-4+\cdots+n^2 = \dfrac{n(n+1)}{2} = t_n$ if n is odd.

$1-4+\cdots-n^2 = -\dfrac{n(n+1)}{2} = -t_n$ if n is even.

4. (a) First, note that we can obtain any of the numbers from 1 to 18.

1 = 1	10 = 10
2 = 2	10 + 1 = 11
5 + (–2) = 3	10 + 2 = 12
5 + (–1) = 4	10 + 5 + (–2) = 13
5 = 5	10 + 5 + (–1) = 14
5 + 1 = 6	10 + 5 = 15
5 + 2 = 7	10 + 5 + 1 = 16
10 + (–2) = 8	10 + 5 + 2 = 17
10 + (–1) = 9	10 + 5 + 2 + 1 = 18

The number 0 may be obtained as an "empty sum." The numbers –1 to –18 may be obtained by using the additive inverse of each addend shown above. Thus we may obtain any of the numbers –18, –17, –16, ..., –3, –2, –1, 0, 1, 2, 3, ..., 16, 17, 18.

(b) Yes, with the given conditions, there is only one way to represent each number.

5. (a) $(-7) + (-19) = -(7 + 19) = -26$

(b) $(-7) - (-19) = (-7) + 19 = 19 - 7 = 12$

(c) $7 - (-19) = 7 + 19 = 26$

(d) $7 + (-19) = -(19 - 7) = -12$

(e) $(-6859) \div 19 = -(6859 \div 19) = -361$

(f) $(-24) \cdot 17 = -(24 \cdot 17) = -408$

(g) $36 \cdot (-24) = -(36 \cdot 24) = -864$

(h) $(-1155) \div (-11) = 1155 \div 11 = 105$

(i) $0 \div (-27) = -(0 \div 27) = -0 = 0$

6. (a) $|a+2| = 5 \Rightarrow a+2 = 5$ or $a+2 = -5$

$a+2 = 5 \Rightarrow a = 3$

$a+2 = -5 \Rightarrow a = -7$

$a = 3$ or $a = -7$

(b)

$$\begin{aligned} 3m-4 &= 5m+6 \\ 3m-4+4 &= 5m+6+4 \\ 3m &= 5m+10 \\ 3m-5m &= 5m+10-5m \\ -2m &= 10 \\ \frac{-2m}{-2} &= \frac{10}{-2} \\ m &= -5 \end{aligned}$$

(c) $n \div 5 = -4 \text{ R } 1$

$n = 5(-4)+1 = -19$

(d)

$$\begin{aligned} t\left(t^2+8\right) &= 3t^3 \\ t^3+8t &= 3t^3 \\ 0 &= 2t^3-8t \\ 0 &= 2t\left(t^2-4\right) \\ 0 &= 2t(t-2)(t+2) \end{aligned}$$

$2t = 0$	$t-2=0$	$t+2=0$
$t = 0$	$t = 2$	$t = -2$

$t = 0$ or $t = 2$ or $t = -2$

7. At mail time the mail carrier delivers a check for \$7 and takes away a bill for \$4. You are \$11 richer.

8. (a) $(-5) + (-3) = -8$; $(-3) + (-8) = (-11)$; $(-8) + (-11) = -19$; $(-11) + (-19) = -30$. The sequence is –5, –3, –8, –11, –19, –30.

(b) $2 - 7 = -5$; $(-5) + 2 = -3$;
$2 + (-3) = -1$. $(-3) + (-1) = -4$.
The sequence is 7, –5, 2, –3, –1, –4.

(c) Use Guess and Check, make an orderly list and look for a pattern. Using the Fibonacci rule starting with a guess of 1 we obtain 6, 1, 7, 8, 15, 23.
Since 15 is too large, we next guess 0 to obtain 6, 0, 6 6, 12, 18.
Since 12 is still too large, we next guess –1 to obtain 6, –1, 5, 4, 9, 13.
Notice that 15, 12, and 9 get 3 smaller each time and guess that this pattern will continue. Since we want to arrive at –12 we guess that we should jump 7 more steps. Thus, we guess –8 and obtain 6, –8, –2, –10, –12, –22 as desired.

9. Poorer by $9; $(-27) \div 3 = -9$

10. (a)

–8 –7 –6 –5 –4 –3 –2 –1 0 1 2 (3)

$(-7) + 10 = 3$

(b)

0 2 4 6 8 10 12 14 (17)

$10 - (-7) = 17$

(c)

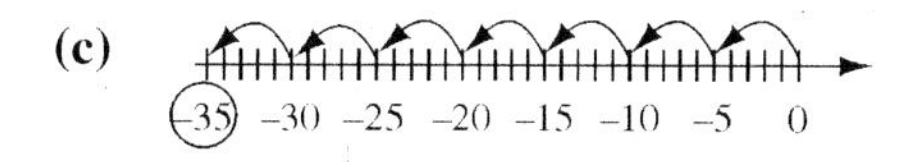

$7 \cdot (-5) = -35$

11.

$$\begin{array}{r} 5 \\ 3\overline{)15} \\ 2\overline{)30} \\ 2\overline{)60} \\ 2\overline{)120} \\ 2\overline{)240} \end{array} \qquad \begin{array}{r} 3 \\ 3\overline{)9} \\ 3\overline{)27} \\ 2\overline{)54} \end{array}$$

Since $-240 = (-1) \cdot 2^4 \cdot 3^1 \cdot 5^1$ and $54 = 2^1 \cdot 3^3$,
$\text{LCM}(-240, 54) = \text{LCM}(240, 54)$
$= 2^4 \cdot 3^3 \cdot 5 = 2160$.

Chapter 6 Fractions and Rational Numbers

Problem Set 6.1

1. (a) One of the six portions is shaded, so $\frac{1}{6}$

(c) None of the figure is shaded, so $\frac{0}{1}$

(e) Two of the six portions are shaded, so $\frac{2}{6}$

2. (a) Shade 1 of 8 equal subregions.
One way to do this is shown.

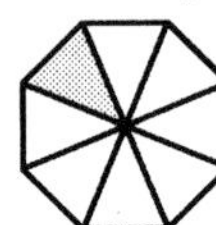

$\frac{1}{8}$

(c) Shade 3 of 4 equal subregions.
One way to do this is shown.

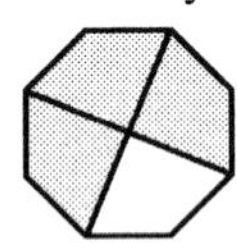

$\frac{3}{4}$

5. (a) Each tick mark represents $\frac{1}{4}$ of one unit.

$A: \frac{1}{4}, B: \frac{3}{4}, C: \frac{6}{4}$ or $\frac{3}{2}$

(c) Each tick mark represents $\frac{1}{5}$ of one unit.

$G: \frac{0}{5}$ or $0, H: \frac{2}{5}, I: \frac{8}{5}$

6. (a) Each tick mark represents $\frac{1}{3}$ of one unit.
From left to right, the arrows represent the fractions $-\frac{5}{3}, \frac{2}{3}, \frac{5}{3}$.

8. (a) Divide the original rectangle into two identical rectangles. The third section is equal in size to each of the original halves.

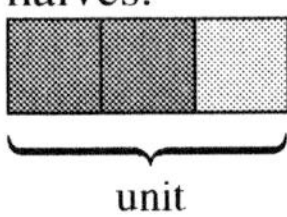

9. (a) There are three groups of two, so the unit consists of four groups of two.

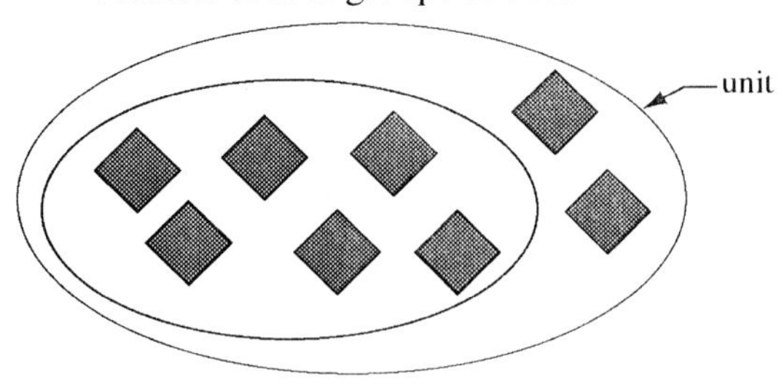

10. (a) One method is shown.

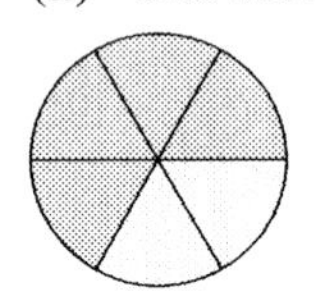

11. (a) $\frac{20}{60}$ or $\frac{1}{3}$ (since an hour is 60 minutes)

(c) $\frac{5}{7}$ (since a week is 7 days)

(e) $\frac{25}{100}$ or $\frac{1}{4}$ (since a quarter is 25¢ and a dollar is 100¢)

(g) $\frac{2}{3}$ (since a yard is 3 feet)

13. (a) $\frac{3}{6} = \frac{1}{2}$

15. (a)

$\frac{2}{4} = \frac{4}{8}$

16. (a) If $\frac{4}{5} = \frac{x}{30}$, then $4 \cdot 30 = 5 \cdot x$,
so $x = 120 \div 5 = 24$.
The missing integer is 24.

(c) If $\frac{-7}{25} = \frac{z}{500}$, then $(-7) \cdot 500 = 25 \cdot z$,
so $z = (-3500) \div 25 = -140$.
The missing integer is −140.

17. **(a)** A common denominator is LCM(42, 7) = 42. We can write the fractions as $\frac{18}{42}$ and $\frac{3 \cdot 6}{7 \cdot 6} = \frac{18}{42}$. They are equivalent.

(c) A common denominator is LCM(25, 500) = 500. We can write the fractions as $\frac{9 \cdot 20}{25 \cdot 20} = \frac{180}{500}$ and $\frac{140}{500}$. Since $\frac{180}{500} \neq \frac{140}{500}$, they are not equivalent.

18. **(a)** Since $78 \cdot 168 = 13{,}104$ and $24 \cdot 546 = 13{,}104$, we conclude that $78 \cdot 168 = 24 \cdot 546$ and the fractions are equivalent.

19. **(a)** Yes. $4 \cdot 3 = 12$ and $9 \cdot 3 = 27$.

(c) Yes. $4 + 3 = 7$ and $9 + 3 = 12$.

20. **(a)** $\frac{84}{144} = \frac{42}{72} = \frac{21}{36} = \frac{7}{12}$

(c) $\frac{-930}{1290} = \frac{-93}{129} = \frac{-31}{43}$

21. **(a)**

$$\begin{array}{r} 3 \\ 2\overline{)6} \\ 2\overline{)12} \\ 2\overline{)24} \\ 2\overline{)48} \\ 2\overline{)96} \end{array} \qquad \begin{array}{r} 3 \\ 3\overline{)9} \\ 2\overline{)18} \\ 2\overline{)36} \\ 2\overline{)72} \\ 2\overline{)144} \\ 2\overline{)288} \end{array}$$

$$\frac{96}{288} = \frac{2^5 \cdot 3^1}{2^5 \cdot 3^2} = \frac{1}{3^1} = \frac{1}{3}$$

22. Answers will vary. Possible answers (using the least common denominator) are shown.

(a) The least common denominator is LCM(11, 5) = 55. We write $\frac{3 \cdot 5}{11 \cdot 5} = \frac{15}{55}$ and $\frac{2 \cdot 11}{5 \cdot 11} = \frac{22}{55}$. The fractions are equivalent to $\frac{15}{55}$ and $\frac{22}{55}$.

(c) The least common denominator is LCM(3, 8, 6) = 24. We write $\frac{4 \cdot 8}{3 \cdot 8} = \frac{32}{24}$, $\frac{5 \cdot 3}{8 \cdot 3} = \frac{15}{24}$, and $\frac{1 \cdot 4}{6 \cdot 4} = \frac{4}{24}$. The fractions are equivalent to $\frac{32}{24}, \frac{15}{24}$, and $\frac{4}{24}$.

23. **(a)** The least common denominator is LCM(8, 6) = 24. We write $\frac{3 \cdot 3}{8 \cdot 3} = \frac{9}{24}$ and $\frac{5 \cdot 4}{6 \cdot 4} = \frac{20}{24}$. The fractions are equivalent to $\frac{9}{24}$ and $\frac{20}{24}$.

(c) The least common denominator is LCM(12,32) = 96. We write $\frac{17 \cdot 8}{12 \cdot 8} = \frac{136}{96}$ and $\frac{7 \cdot 3}{32 \cdot 3} = \frac{21}{96}$. The fractions are equivalent to $\frac{136}{96}$ and $\frac{21}{96}$.

25. **(a)** Since $2 \cdot 12 > 3 \cdot 7, \frac{2}{3} > \frac{7}{12}$. The rational numbers in order are $\frac{7}{12}, \frac{2}{3}$.

(c) Since $5 \cdot 36 > 6 \cdot 29, \frac{5}{6} > \frac{29}{36}$. The rational numbers in order are $\frac{29}{36}, \frac{5}{6}$.

27. **(a)** True. Given two fractions, two equivalent fractions with a common denominator may be found by finding a common multiple of the two original denominators. Once these fractions, say $\frac{a}{c}$ and $\frac{b}{c}$ are found, infinitely many more pairs of equivalent fractions can be found, namely, $\frac{a \cdot n}{c \cdot n}$ and $\frac{b \cdot n}{c \cdot n}$ for any integer n other than 0.

(c) False. Given any positive fraction, a smaller positive fraction may be found by multiplying the denominator by 2. So there cannot be a least positive fraction.

35. **(a)** 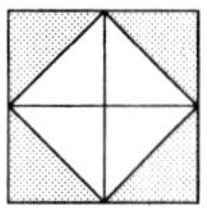 $\frac{4}{8}$

(d) 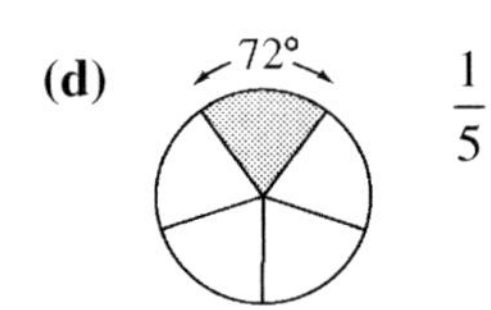

$\frac{1}{5}$

36. The four triangles can be reattached to form a figure with 5 squares, each congruent to the shaded inner square. Therefore, $\frac{1}{5}$ of the large square has been shaded.

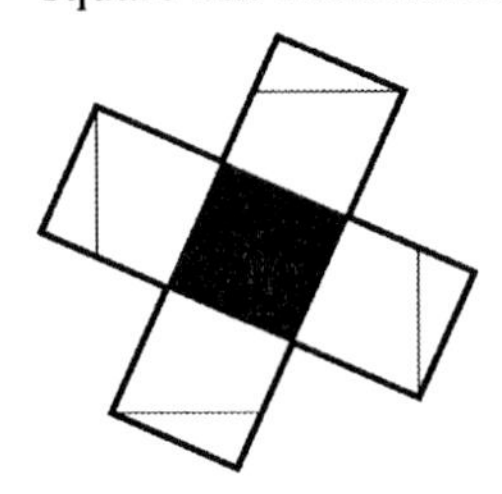

39. **(a)** Add tick marks to show that the rope was originally 60 feet long.

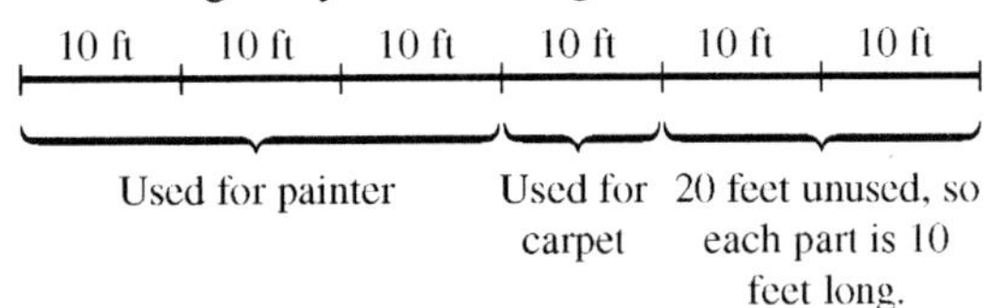

42. **(a)** trapezoid = $\frac{1}{2}$, triangle = $\frac{1}{6}$, rhombus = $\frac{1}{3}$

45. **(a)** Since $\frac{3}{4} = \frac{18}{24}$, the tank contains 18 gallons.

46. Answers may vary.

(a) Since $\frac{310}{498}$ is about $\frac{300}{500}$, a simpler fraction is $\frac{3}{5}$.

(c) Since $\frac{9}{35}$ is about $\frac{9}{36}$, a simpler fraction is $\frac{1}{4}$.

Problem Set 6.2

1. **(a)** $\frac{1}{3} + \frac{1}{2} = \frac{5}{6}$

2. **(a)** $\frac{2}{5} + \frac{6}{5} = \frac{8}{5}$

4. **(a)** $\frac{1}{8}$ $\frac{3}{8}$

0 $\frac{1}{8}$ $\frac{2}{8}$ $\frac{3}{8}$ $\frac{4}{8}$ $\frac{5}{8}$ $\frac{6}{8}$ $\frac{7}{8}$ $\frac{8}{8}$

5. **(a)** 0 $\frac{1}{4}$ $\frac{2}{4}$ $\frac{3}{4}$ $\frac{4}{4}$

$$\frac{3}{4} + \frac{-2}{4} = \frac{1}{4}$$

6. **(a)** $\frac{2}{7} + \frac{3}{7} = \frac{2+3}{7} = \frac{5}{7}$

(c) $\frac{3}{8} + \frac{11}{24} = \frac{9}{24} + \frac{11}{24} = \frac{9+11}{24} = \frac{20}{24} = \frac{5}{6}$

(e) $\frac{5}{12} + \frac{17}{20} = \frac{25}{60} + \frac{51}{60} = \frac{25+51}{60} = \frac{76}{60} = \frac{19}{15}$

(g) $\frac{-57}{100} + \frac{13}{10} = \frac{-57}{100} + \frac{130}{100} = \frac{(-57)+130}{100}$
$= \frac{73}{100}$

7. **(a)** Since $9 = 2 \cdot 4 + 1$,
$\frac{9}{4} = \frac{2 \cdot 4 + 1}{4} = 2 + \frac{1}{4} = 2\frac{1}{4}$.

(c) Since $111 = 4 \cdot 23 + 19$,
$\frac{111}{23} = \frac{4 \cdot 23 + 19}{23} = 4 + \frac{19}{23} = 4\frac{19}{23}$.

8. **(a)** $2\frac{3}{8} = \frac{2 \cdot 8 + 3}{8} = \frac{19}{8}$

(c) $111\frac{2}{5} = \frac{111 \cdot 5 + 2}{5} = \frac{557}{5}$

9. **(a)** $\frac{5}{6} - \frac{1}{4} = \frac{7}{12}$

10. **(a)** $\frac{3}{4} - \frac{1}{6} = \frac{7}{12}$

11. **(a)** $\frac{3}{4} - \frac{1}{3} = \frac{9}{12} - \frac{4}{12} = \frac{5}{12}$

12. (a) $\frac{5}{8}-\frac{2}{8}=\frac{5-2}{8}=\frac{3}{8}$

(c) $2\frac{2}{3}-1\frac{1}{3}=\frac{8}{3}-\frac{4}{3}=\frac{8-4}{3}=\frac{4}{3}=1\frac{1}{3}$

(e) $\frac{6}{8}-\frac{5}{12}=\frac{9}{12}-\frac{5}{12}=\frac{9-5}{12}=\frac{4}{12}=\frac{1}{3}$

(g) $\frac{137}{214}-\frac{-1}{3}=\frac{137\cdot 3-214\cdot(-1)}{214\cdot 3}=\frac{625}{642}$

14. (a) $\frac{2}{3}<\frac{3}{4}$ if, and only if, $\frac{3}{4}-\frac{2}{3}>0$.

$\frac{3}{4}-\frac{2}{3}=\frac{9}{12}-\frac{8}{12}=\frac{1}{12}>0$

22. (a) Top right: Using the top row,

$1-\frac{1}{2}-\frac{1}{12}=\frac{12}{12}-\frac{6}{12}-\frac{1}{12}=\frac{5}{12}.$

Bottom left: Using the left column,

$1-\frac{1}{2}-\frac{1}{4}=\frac{12}{12}-\frac{6}{12}-\frac{3}{12}=\frac{3}{12}=\frac{1}{4}.$

Middle middle: Using the diagonal,

$1-\frac{1}{4}-\frac{5}{12}=\frac{12}{12}-\frac{3}{12}-\frac{5}{12}=\frac{4}{12}=\frac{1}{3}.$

Bottom middle: Using the middle column,

$1-\frac{1}{12}-\frac{1}{3}=\frac{12}{12}-\frac{1}{12}-\frac{4}{12}=\frac{7}{12}.$

Bottom right: Using the diagonal,

$1-\frac{1}{2}-\frac{1}{3}=\frac{12}{12}-\frac{6}{12}-\frac{4}{12}=\frac{2}{12}=\frac{1}{6}.$

Middle right: Using the right column,

$1-\frac{5}{12}-\frac{1}{6}=\frac{12}{12}-\frac{5}{12}-\frac{2}{12}=\frac{5}{12}.$

Therefore, we obtain

$\frac{1}{2}$	$\frac{1}{12}$	$\frac{5}{12}$
$\frac{1}{4}$	$\frac{1}{3}$	$\frac{5}{12}$
$\frac{1}{4}$	$\frac{7}{12}$	$\frac{1}{6}$

To confirm that this is a Magic Fraction Square, we check the bottom two rows:

$\frac{1}{4}+\frac{1}{3}+\frac{5}{12}=\frac{3}{12}+\frac{4}{12}+\frac{5}{12}=\frac{12}{12}=1$ and

$\frac{1}{4}+\frac{7}{12}+\frac{1}{6}=\frac{3}{12}+\frac{7}{12}+\frac{2}{12}=\frac{12}{12}=1.$

24. (a) $\frac{1}{2}+\frac{1}{3}+\frac{1}{15}+\frac{1}{50}=\frac{75}{150}+\frac{50}{150}+\frac{10}{150}+\frac{3}{150}$
$=\frac{138}{150}=\frac{23}{25}$

25. (a) $\frac{1}{5}+\frac{1}{45}=\frac{9}{45}+\frac{1}{45}=\frac{10}{45}=\frac{5\cdot 2}{5\cdot 9}=\frac{2}{9}$

26. (a) It terminates.

$\frac{3}{4}$ $\frac{2}{3}$ $\frac{1}{2}$ $\frac{1}{4}$

$\frac{1}{12}$ $\frac{1}{6}$ $\frac{1}{4}$ $\frac{1}{2}$

$\frac{1}{12}$ $\frac{1}{12}$ $\frac{1}{4}$ $\frac{5}{12}$

0 $\frac{1}{6}$ $\frac{1}{6}$ $\frac{1}{3}$

$\frac{1}{6}$ 0 $\frac{1}{6}$ $\frac{1}{3}$

$\frac{1}{6}$ $\frac{1}{6}$ $\frac{1}{6}$ $\frac{1}{6}$

0 0 0 0

29. (a) The screw is sunk into the drywall by $\frac{1}{16}$", so the screw extends $1\frac{1}{4}+\frac{1}{16}=1\frac{4}{16}+\frac{1}{16}=1\frac{5}{16}$" through the drywall into the joist. Since the drywall is $\frac{5}{8}$" thick, the screw extends $1\frac{5}{16}-\frac{5}{8}=\frac{21}{16}-\frac{10}{16}=\frac{11}{16}$" into the joist.

Problem Set 6.3

1. (a) $4\times\frac{2}{5}=\frac{8}{5}$

2. (a)

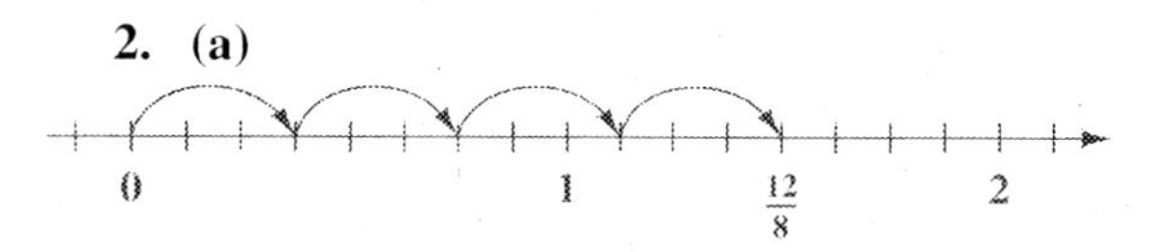

3. (a) $\frac{4}{6}\times 5=\frac{20}{6}=3\frac{2}{6}$

4. (a) $3\times\frac{5}{2}=\frac{15}{2}$

5. (a) $2\times\frac{3}{5}=\frac{6}{5}$

7. **(a)** The reciprocal of $\frac{3}{8}$ is $\frac{8}{3}$.

(c) The reciprocal of $2\frac{1}{4}=\frac{9}{4}$ is $\frac{4}{9}$.

(e) The reciprocal of $5=\frac{5}{1}$ is $\frac{1}{5}$.

9. **(a)** $\frac{5}{4}\div\frac{3}{4}=\frac{5}{3}$

(b) $\frac{7}{8}\div\frac{7}{11}=\frac{11}{8}$

10. **(a)** $\frac{2}{5}\div\frac{3}{4}=\frac{2}{5}\cdot\frac{4}{3}=\frac{8}{15}$

(c) $\frac{100}{33}\div\frac{10}{3}=\frac{100}{33}\cdot\frac{3}{10}=\frac{300}{330}=\frac{10}{11}$

(e) $3\div5\frac{1}{4}=\frac{3}{1}\div\frac{21}{4}=\frac{3}{1}\cdot\frac{4}{21}=\frac{12}{21}=\frac{4}{7}$

11. **(a)** $\frac{2}{3}\cdot\left(\frac{3}{4}+\frac{9}{12}\right)=\frac{2}{3}\cdot\left(\frac{9}{12}+\frac{9}{12}\right)=\frac{2}{3}\cdot\left(\frac{18}{12}\right)$

$=\frac{2\cdot18}{3\cdot12}=\frac{36}{36}=1$

12. **(a)** $2\frac{1}{2}\times3\frac{3}{4}=\frac{5}{2}\times\frac{15}{4}=\frac{75}{8}=9\frac{3}{8}$

The two towns are $9\frac{3}{8}$ miles apart.

13. **(a)** $\frac{2}{5}x-\frac{3}{4}=\frac{1}{2}$

Add $\frac{3}{4}$ to each side:

$\frac{2}{5}x-\frac{3}{4}+\frac{3}{4}=\frac{1}{2}+\frac{3}{4}$

$\frac{2}{5}x=\frac{2}{4}+\frac{3}{4}=\frac{5}{4}$

Multiply each side by $\frac{5}{2}$:

$\frac{5}{2}\cdot\frac{2}{5}x=\frac{5}{2}\cdot\frac{5}{4}$

$x=\frac{25}{8}=3\frac{1}{8}$

26. **(a)** $3\frac{1}{2}+1\frac{2}{5}=\frac{7}{2}+\frac{7}{5}=\frac{35}{10}+\frac{14}{10}=\frac{49}{10}$ and

$3\frac{1}{2}\times1\frac{2}{5}=\frac{7}{2}\times\frac{7}{5}=\frac{49}{10}$

31. **(a)** The process terminates when one row consists of all ones.

$\frac{2}{5}$ $\frac{6}{7}$ $\frac{3}{4}$ $\frac{1}{2}$

$\frac{15}{7}$ $\frac{8}{7}$ $\frac{3}{2}$ $\frac{5}{4}$

$\frac{15}{8}$ $\frac{21}{16}$ $\frac{6}{5}$ $\frac{12}{7}$

$\frac{10}{7}$ $\frac{35}{32}$ $\frac{10}{7}$ $\frac{35}{32}$

$\frac{64}{49}$ $\frac{64}{49}$ $\frac{64}{49}$ $\frac{64}{49}$

1 1 1 1

32. In each case, the number of votes must be rounded *up* to the next whole number.

3 members: $\frac{3}{4}\cdot3=\frac{9}{4}=2\frac{1}{4}$, so 3 votes are needed.

4 members: $\frac{3}{4}\cdot4=\frac{12}{4}=3$, so 3 votes are needed.

5 members: $\frac{3}{4}\cdot5=\frac{15}{4}=3\frac{3}{4}$, so 4 votes are needed.

6 members: $\frac{3}{4}\cdot6=\frac{18}{4}=4\frac{1}{2}$, so 5 votes are needed.

7 members: $\frac{3}{4}\cdot7=\frac{21}{4}=5\frac{1}{4}$, so 6 votes are needed.

8 members: $\frac{3}{4}\cdot8=\frac{24}{4}=6$, so 6 votes are needed.

35. Each bow requires $1\frac{1}{2}\div6=\frac{3}{2}\cdot\frac{1}{6}=\frac{3}{12}=\frac{1}{4}$ yard of ribbon. Since $5\frac{3}{4}\div\frac{1}{4}=\frac{23}{4}\cdot\frac{4}{1}=23$, 23 bows can be made.

37. Since \$28 was $\frac{2}{3}$ of the original price, the original price was $28\div\frac{2}{3}=\frac{28}{1}\cdot\frac{3}{2}=\frac{84}{2}=42$, or \$42.

45. Start with a unit area rectangle that is shaded to represent $\frac{3}{4}$. Next, draw a rectangle that encloses an area of $\frac{2}{3}$. Redistribute the shaded area to show that the number of regions of area $\frac{2}{3}$ covered by $\frac{3}{4}$ is $1\frac{1}{8}$.

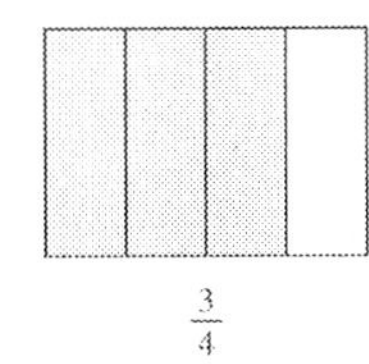

$\frac{3}{4}$

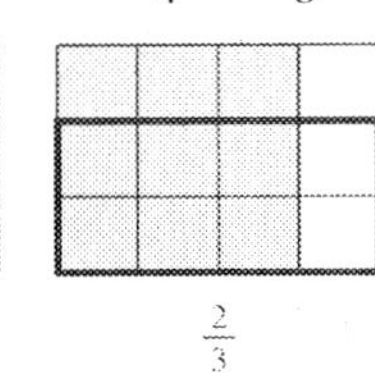

$\frac{2}{3}$

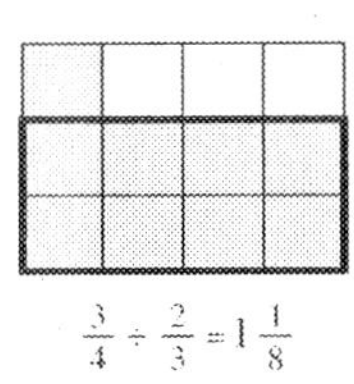

$\frac{3}{4} \div \frac{2}{3} = 1\frac{1}{8}$

Problem Set 6.4

1. Commutative and associative properties for addition:

$$\left(\left(3\frac{1}{5}+2\frac{2}{5}\right)8\frac{1}{5}\right) = (3+2+8)+\left(\frac{1}{5}+\frac{2}{5}+\frac{1}{5}\right)$$
$$= 13+\frac{4}{5} = 13\frac{4}{5}$$

3. **(a)** $\frac{-4}{5}$

(c) $\frac{8}{3}$

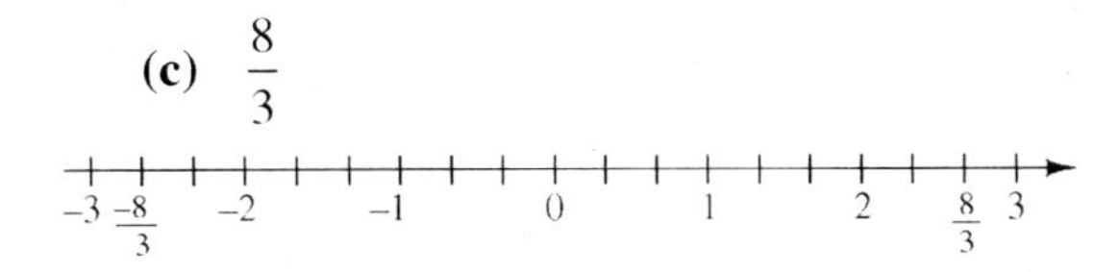

4. **(a)** $\frac{1}{6}+\frac{2}{-3} = \frac{1}{6}+\left(-\frac{2}{3}\right) = \frac{1}{6}-\frac{4}{6} = \frac{-3}{6} = \frac{-1}{2}$

(c) $\frac{9}{4}+\frac{-7}{8} = \frac{18}{8}+\frac{-7}{8} = \frac{11}{8}$

5. **(a)** $\frac{2}{5}-\frac{3}{4} = \frac{2\cdot4-5\cdot3}{5\cdot4} = \frac{-7}{20}$

(c) $\frac{3}{8}-\frac{1}{12} = \frac{9}{24}-\frac{2}{24} = \frac{7}{24}$

(e) $2\frac{1}{3}-5\frac{3}{4} = \frac{7}{3}-\frac{23}{4} = \frac{7\cdot4-3\cdot23}{3\cdot4}$
$= \frac{-41}{12} = -3\frac{5}{12}$

6. **(a)** $\frac{3}{5}\cdot\frac{7}{8}\cdot\left(\frac{5}{3}\right) = \frac{3\cdot7\cdot5}{5\cdot8\cdot3} = \frac{7}{8}$

(c) $\frac{-4}{3}\cdot\frac{6}{-16} = \frac{-24}{-48} = \frac{1}{2}$

(e) $\frac{14}{15}\cdot\frac{60}{7} = \frac{14\cdot60}{7\cdot15} = \frac{2\cdot4}{1} = \frac{8}{1}$ or 8

7. **(a)** $\frac{2}{3}$

(c) $\frac{-11}{-4}$ or $\frac{11}{4}$

(e) $-\frac{1}{2}$

8. **(a)** $\frac{2}{3}\cdot\frac{4}{7}+\frac{2}{3}\cdot\frac{3}{7} = \frac{2}{3}\cdot\left(\frac{4}{7}+\frac{3}{7}\right) = \frac{2}{3}\cdot1 = \frac{2}{3}$

(c) $\frac{4}{7}\cdot\frac{3}{2}-\frac{4}{7}\cdot\frac{6}{4} = \frac{4}{7}\cdot\frac{3}{2}-\frac{4}{7}\cdot\frac{3}{2} = \frac{12}{14}-\frac{12}{14} = 0$

9. **(a)** Addition of rational numbers—definition

(b) Multiplication of rational numbers—definition

(c) Distributive property of multiplication over addition (for whole numbers)

(d) Addition of rational numbers—definition

(e) Multiplication of rational numbers—definition

10. Multiply both sides by $\frac{7}{4}$ to get $\left(\frac{a}{b}\cdot\frac{4}{7}\right)\cdot\frac{7}{4}=\frac{2}{3}\cdot\frac{7}{4}$. By the associative property of multiplication and the multiplicative inverse property, $\left(\frac{a}{b}\cdot\frac{4}{7}\right)\cdot\frac{7}{4}=\frac{a}{b}\cdot\left(\frac{4}{7}\cdot\frac{7}{4}\right)=\frac{a}{b}\cdot 1=\frac{a}{b}$.

$\frac{a}{b}\cdot\frac{4}{7}=\frac{2}{3}$	Given
$\left(\frac{a}{b}\cdot\frac{4}{7}\right)\cdot\frac{7}{4}=\frac{2}{3}\cdot\frac{7}{4}$	Multiply both sides by $\frac{7}{4}$.
$\left(\frac{a}{b}\cdot\frac{4}{7}\right)\cdot\frac{7}{4}=\frac{7}{6}$	$\frac{2}{3}\cdot\frac{7}{4}=\frac{14}{12}=\frac{7}{6}$
$\frac{a}{b}\cdot\left(\frac{4}{7}\cdot\frac{7}{4}\right)=\frac{7}{6}$	Associative property of multiplication
$\frac{a}{b}\cdot 1=\frac{7}{6}$	Multiplicative inverse property
$\frac{a}{b}=\frac{7}{6}$	One is the multiplicative identity.

11. (a)

$4x+3=0$	Given
$4x=-3$	Definition of additive inverse
$\frac{1}{4}(4x)=\frac{1}{4}(-3)$	Multiply both sides by $\frac{1}{4}$.
$\left(\frac{1}{4}\cdot 4\right)x=\frac{1}{4}(-3)$	Associative property of multiplication
$1x=-\frac{3}{4}$	Multiplicative inverse property
$x=-\frac{3}{4}$	One is the multiplicative identity.

(c)

$\frac{2}{3}x+\frac{4}{5}=0$	Given
$\frac{2}{3}x=-\frac{4}{5}$	Definition of additive inverse
$\frac{3}{2}\cdot\left(\frac{2}{3}x\right)=\frac{3}{2}\cdot\left(\frac{-4}{5}\right)$	Multiply both sides by $\frac{3}{2}$.
$1x=\frac{3\cdot(-4)}{2\cdot 5}$	Associative property of multiplication, multiplicative inverse property, definition of multiplication of rational numbers
$x=\frac{-12}{10}$	One is the multiplication identity.
$x=\frac{6}{5}$	Equivalence of fractions

12. **(a)** Closure property for subtraction and the existence of a multiplicative inverse, as shown:

$$\frac{9}{5}C + 32 = F$$

$$\left(\frac{9}{5}C+32\right) - 32 = F-32$$

$$\frac{9}{5}C = F - 32$$

$$\frac{5}{9}\left(\frac{9}{5}C\right) = \frac{5}{9}(F-32)$$

$$C = \frac{5}{9}(F-32)$$

13. **(a)** $-\frac{1}{5}, \frac{2}{5}, \frac{4}{5}$

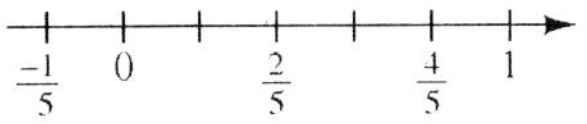

(c) $\frac{3}{8}, \frac{1}{2}, \frac{3}{4}$

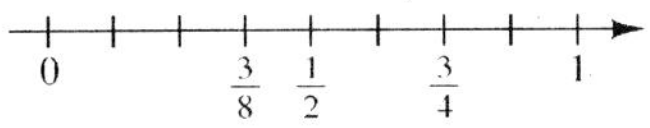

14. Note that in each case the fractions need to be written with positive denominators.

(a) $\frac{-4}{5} < \frac{-3}{4}$: $(-4)\cdot 4 = -16 < -15 = 5\cdot(-3)$

15. **(a)**

$$x + \frac{2}{3} > -\frac{1}{3}$$

$$x + \frac{2}{3} + \left(-\frac{2}{3}\right) > -\frac{1}{3} + \left(-\frac{2}{3}\right)$$

$$x > -1$$

(c)

$$\frac{3}{4}x < -\frac{1}{2}$$

$$\frac{4}{3}\cdot\left(\frac{3}{4}\right)\cdot x < \frac{4}{3}\cdot\left(\frac{-1}{2}\right)$$

$$x < \frac{-4}{6}$$

$$x < -\frac{2}{3}$$

16. Answers will vary.

(a) One answer is $\frac{1}{2}$, since $\frac{4}{9} < \frac{1}{2} < \frac{6}{11}$.

(c) One answer is $\frac{28}{47}$ since $\frac{7}{12} = \frac{14}{24} = \frac{28}{48}$ and $\frac{14}{23} = \frac{28}{46}$.

19. **(a)** $\frac{104}{391}$ is approximately $\frac{100}{400} = \frac{1}{4}$.

(c) $\frac{-193}{211}$ is approximately $\frac{-200}{200} = -1$.

20. **(a)** $3\frac{19}{40} + 5\frac{11}{19}$ is approximately $3\frac{1}{2} + 5\frac{1}{2} = 9$.

21. **(a)** Combine like terms:

$$\frac{1}{2} + \frac{1}{4} + \frac{3}{4} = \frac{1}{2} + 1 = \frac{3}{2}$$

(c) Use equivalence of fractions:

$$\frac{3}{4}\cdot\frac{12}{15} = \frac{3}{4}\cdot\frac{4}{5} = \frac{3}{5}$$

(e) Use the Distributive Property of Multiplication over addition:

$$2\frac{2}{3}\times 15 = 2\times 15 + \frac{2}{3}\times 15 = 30 + 10 = 40$$

(g) Group like terms and simplify:

$$6\frac{1}{8} - 8\frac{1}{4} = (6-8) + \left(\frac{1}{8} - \frac{1}{4}\right)$$

$$= -2 - \frac{1}{8} = -2\frac{1}{8}$$

22. **(a)** What is his new total acreage?

23. Answers will vary. Possible answers include:

(a) Two pizzas were ordered. $\frac{3}{4}$ of one pizza and $\frac{1}{2}$ of another were eaten. How many pizzas were eaten in all?

28.

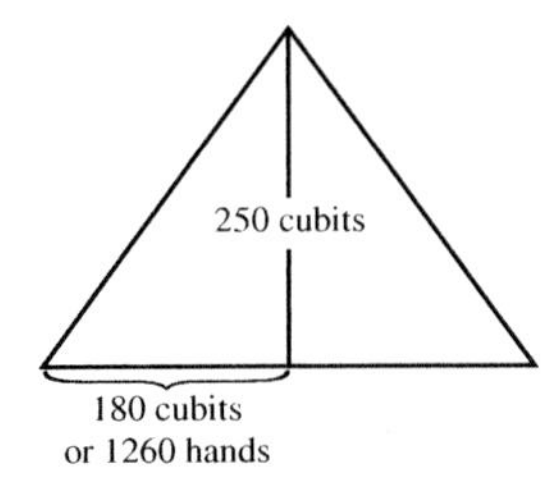

$$\text{Steepness} = \frac{\text{run in hands}}{\text{rise in cubits}} = \frac{1260}{250} = \frac{126}{25} = 5\frac{1}{25}$$

29. **(a)** $6 \cdot \frac{1}{3} = 2$ square yards

30. **(a)** $(-4) \cdot 6 + 3 \cdot W = 0 \Rightarrow 3 \cdot W = 24 \Rightarrow W = 8$

(c) $(-3) \cdot 2 + x \cdot 10 = 0$

$10 \cdot x = 6$

$x = \frac{6}{10} = \frac{3}{5}$

(e) $(-7) \cdot \frac{2}{3} + x \cdot \frac{3}{4} = 0$

$x \cdot \frac{3}{4} = \frac{14}{3} \Rightarrow x = \frac{14}{3} \cdot \frac{4}{3} = \frac{56}{9}$

35. **(a)** 4, since $\frac{1}{2} + \frac{2}{8} + \frac{1}{4} = \frac{2}{4} + \frac{1}{4} + \frac{1}{4} = \frac{4}{4}$

38. Cut the 8-inch side in half and the 10-inch side in thirds to get six $3\frac{1}{3}$-by-4 inch rectangles.

40. **(a)** The replacement rule gives $\frac{7+2\cdot 5}{7+5} = \frac{17}{12}$. This is a good approximation for $\sqrt{2}$ since $\left(\frac{17}{12}\right)^2 = \frac{289}{144} = 2 + \frac{1}{144}$.

(c) The next two rational numbers are $\frac{7}{4}$ and $\frac{19}{11}$ since $\frac{5}{3} \Rightarrow \frac{5+3\cdot 3}{5+3} = \frac{14}{8} = \frac{7}{4} \Rightarrow \frac{7+3\cdot 4}{7+4} = \frac{19}{11}$.

Chapter 6 Review Exercises

1. **(a)** Two of the four regions are shaded; $\frac{2}{4}$

(b) All six of the six regions are shaded; $\frac{6}{6}$

(c) None of the four regions is shaded; $\frac{0}{4}$

(d) Five of the three regions are shaded; $\frac{5}{3}$

2.

(a) (c) (b) (d)

$\frac{3}{4}$ 1 $\frac{12}{8}$ $2\frac{3}{8}$

0 1 2 3

3. **(a)** $\frac{27}{81} = \frac{9}{27} = \frac{3}{9} = \frac{1}{3}$

(b) $\frac{100}{825} = \frac{20}{165} = \frac{4}{33}$

(c) $\frac{378}{72} = \frac{189}{36} = \frac{63}{12} = \frac{21}{4}$

(d) $\frac{3^5 \cdot 7^2 \cdot 11^3}{3^2 \cdot 7^3 \cdot 11^2} = \frac{3^3 \cdot 1 \cdot 11^1}{1 \cdot 7^1 \cdot 1} = \frac{297}{7}$

4. $\frac{13}{30}, \frac{13}{27}, \frac{1}{2}, \frac{25}{49}, \frac{26}{49}$

5. **(a)** 36 because $36 \div 9 = 4$, $36 \div 12 = 3$

(b) 18 because $18 \div 18 = 1$, $18 \div 6 = 3$, $18 \div 3 = 6$

6.

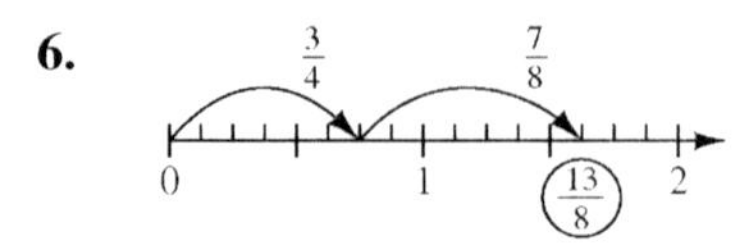

7.

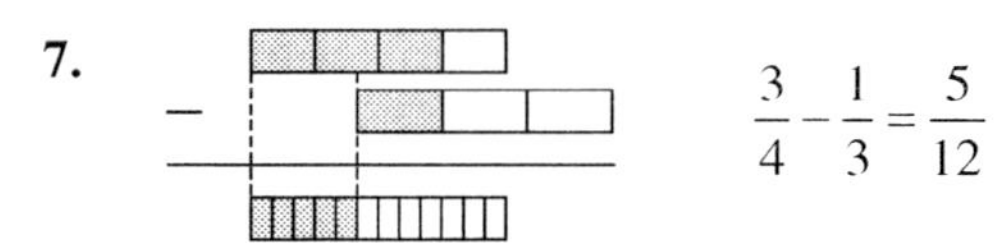

$\frac{3}{4} - \frac{1}{3} = \frac{5}{12}$

8. **(a)** $\frac{3}{8} + \frac{1}{4} = \frac{3}{8} + \frac{2}{8} = \frac{5}{8}$

(b) $\frac{2}{9} + \frac{-5}{12} = \frac{8}{36} + \frac{-15}{36} = \frac{-7}{36}$

(c) $\frac{4}{5} - \frac{2}{3} = \frac{4 \cdot 3 - 5 \cdot 2}{5 \cdot 3} = \frac{2}{15}$

(d) $5\frac{1}{4} - 1\frac{5}{6} = \frac{21}{4} - \frac{11}{6} = \frac{63}{12} - \frac{22}{12}$
$= \frac{41}{12}$ or $3\frac{5}{12}$

9. (a) $3 \times \frac{1}{3} = \frac{3}{3}$ or 1

(b) $\frac{2}{3} \times 4 = \frac{8}{3}$

(c) $\frac{5}{6} \times \frac{3}{2} = \frac{15}{12}$ or $\frac{5}{4}$

10. Using a diagram, we see that we have 9 quarters of a pizza to share.

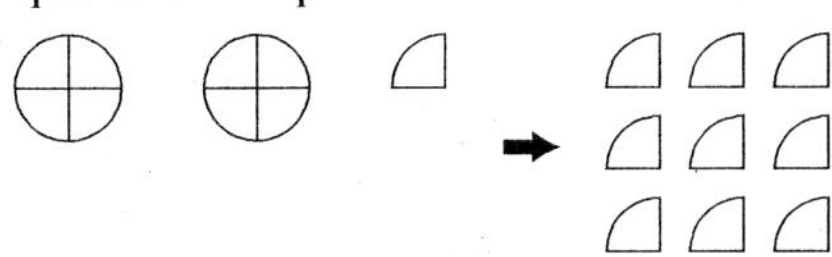

Since there are 9 quarters, each of the three should receive 3 of the quarters. That is, $2\frac{1}{4} \div 3 = \frac{3}{4}$. This problem corresponds to the sharing (partition model of division.)

11. Using a diagram, we see that we have 9 third-of-a-cup groups of beans.

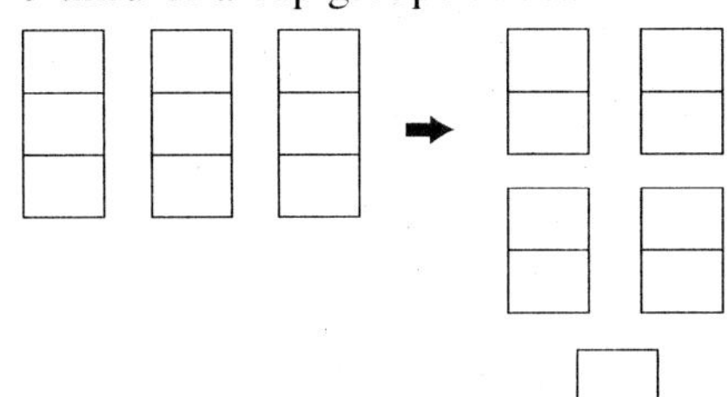

Next, group into pairs of thirds, giving 4 pairs and a half a pair. Therefore, $4\frac{1}{2}$ quarts of soup can be made. That is, $3 \div \frac{2}{3} = \frac{9}{2} = 4\frac{1}{2}$. This problem corresponds to the grouping (repeated subtraction) model of division.

12. $7\frac{1}{8} \div 1\frac{1}{4} = \frac{57}{8} \div \frac{5}{4} = \frac{57}{8} \cdot \frac{4}{5}$
$= \frac{57 \cdot 4}{8 \cdot 5} = \frac{57}{2 \cdot 5} = \frac{57}{10}$

The distance is $\frac{57}{10}$ miles or $5\frac{7}{10}$ miles.

13. (a) $\frac{-3}{4} + \frac{5}{8} = \frac{-6}{8} + \frac{5}{8} = \frac{-1}{8}$

(b) $\frac{4}{5} - \frac{-7}{10} = \frac{4}{5} + \frac{7}{10} = \frac{8}{10} + \frac{7}{10} = \frac{15}{10} = \frac{3}{2}$

(c) $\left(\frac{3}{8} \cdot \frac{-4}{27}\right) \div \frac{1}{9} = \left(\frac{3}{8} \cdot \frac{-4}{27}\right) \cdot \frac{9}{1}$
$= \frac{-4 \cdot 3 \cdot 9}{8 \cdot 27 \cdot 1} = \frac{-1}{2}$

(d) $\frac{2}{5} \cdot \left(\frac{3}{4} - \frac{5}{2}\right) = \frac{2}{5} \cdot \left(\frac{3}{4} - \frac{10}{4}\right)$
$= \frac{2}{5} \cdot \frac{-7}{4} = \frac{-14}{20} = \frac{-7}{10}$

14. (a)
$$3x + 5 = 11$$
$$3x + 5 + (-5) = 11 + (-5)$$
$$3x = 6$$
$$\frac{1}{3} \cdot (3x) = \frac{1}{3} \cdot 6$$
$$x = 2$$

(b)
$$x + \frac{2}{3} = \frac{1}{2}$$
$$x + \frac{2}{3} + \left(-\frac{2}{3}\right) = \frac{1}{2} + \left(-\frac{2}{3}\right)$$
$$x = \frac{1}{2} - \frac{2}{3} = -\frac{1}{6}$$

(c)
$$\frac{3}{5}x+\frac{1}{2}=\frac{2}{3}$$
$$\frac{3}{5}x+\frac{1}{2}+\left(-\frac{1}{2}\right)=\frac{2}{3}-\frac{1}{2}$$
$$\frac{3}{5}x=\frac{1}{6}$$
$$\frac{5}{3}\cdot\frac{3}{5}x=\frac{5}{3}\cdot\frac{1}{6}\Rightarrow x=\frac{5}{18}$$

(d)
$$-\frac{4}{3}x+1=\frac{1}{4}$$
$$-\frac{4}{3}x+1+(-1)=\frac{1}{4}+(-1)$$
$$-\frac{4}{3}x=-\frac{3}{4}$$
$$\left(-\frac{3}{4}\right)\cdot\left(-\frac{4}{3}\right)x=\left(-\frac{3}{4}\right)\cdot\left(-\frac{3}{4}\right)\Rightarrow x=\frac{9}{16}$$

15. Write $\frac{5}{6}=\frac{55}{66}$ and $\frac{10}{11}=\frac{60}{66}$. Then $\frac{56}{66}=\frac{28}{33}$ and $\frac{57}{66}$ are between $\frac{5}{6}$ and $\frac{10}{11}$. Other answers may be given.

16. **(a)** $1\frac{1}{3}+2\frac{5}{12}+\frac{1}{4}=1+2+\left(\frac{4}{12}+\frac{5}{12}+\frac{3}{12}\right)$
$=1+2+1=4$

(b) $\frac{6}{7}\cdot\frac{28}{3}\cdot\frac{5}{8}=\frac{6\cdot 28}{3\cdot 8\cdot 7}\cdot 5=\frac{168}{168}\cdot 5=1\cdot 5=5$

(c) $\frac{36}{5}\div\frac{9}{25}=\frac{36}{5}\cdot\frac{25}{9}=\frac{36}{9}\cdot\frac{25}{5}=4\cdot 5=20$

Chapter 6 Test

1. $\frac{a+b}{b}=\frac{c+d}{d}$ if, and only if,
$(a+b)d=b(c+d)$; that is, if, and only if,
$ad+bd=bc+bd$. But this is so if, and only if,
$ad=bc$. And this is so if, and only if, $\frac{a}{b}=\frac{c}{d}$.

2. **(a)** $\frac{1}{3}+\frac{5}{8}-\frac{5}{6}=\frac{8}{24}+\frac{15}{24}-\frac{20}{24}=\frac{3}{24}=\frac{1}{8}$

(b) $\left(\frac{2}{3}-\frac{5}{4}\right)\div\frac{3}{4}=\left(\frac{8}{12}-\frac{15}{12}\right)\div\frac{3}{4}$
$=\frac{-7}{12}\div\frac{3}{4}=\frac{-7}{12}\cdot\frac{4}{3}$
$=\frac{-28}{36}=\frac{-7}{9}$

(c) $\frac{4}{7}\cdot\left(\frac{35}{4}+\frac{-42}{12}\right)=\frac{4}{7}\cdot\left(\frac{35}{4}+\frac{-14}{4}\right)$
$=\frac{4}{7}\cdot\frac{21}{4}=\frac{21}{7}=3$

(d) $\frac{123}{369}\div\frac{1}{3}=\frac{1}{3}\div\frac{1}{3}=1$

3. **(a)** 0, since $\frac{2}{3}+\frac{-4}{6}=0$.

(b) 2, since $\frac{5}{6}\cdot\frac{36}{15}=\frac{5}{15}\cdot\frac{36}{6}=\frac{1}{3}\cdot 6=2$.

(c) 1, since $\frac{9}{5}-\frac{1}{5}=\frac{8}{5}$ is the reciprocal of $\frac{5}{8}$.

(d) $\frac{1}{3}$, since $\frac{2}{\not 3}\cdot\frac{\not 3}{\not 4}\cdot\frac{\not 4}{\not 5}\cdot\frac{\not 5}{6}=\frac{2}{6}=\frac{1}{3}$.

4. **(a)** $\frac{1}{8}\cdot\frac{1}{2}=\frac{1}{16}$ square miles
$\frac{1}{16}\div\frac{1}{640}=\frac{1}{16}\cdot\frac{640}{1}=\frac{640}{16}=40$ acres

(b) The area is $80\cdot\frac{1}{640}=\frac{1}{8}$ square miles, so the width is $\frac{1}{8}\div\frac{1}{2}=\frac{1}{8}\cdot\frac{2}{1}=\frac{1}{4}$ mile.

5. **(a)** If $\frac{a}{b}$ and $\frac{c}{d}$ are two rational numbers with $\frac{a}{b}<\frac{c}{d}$, then there is a rational number $\frac{e}{f}$ such that $\frac{a}{b}<\frac{e}{f}<\frac{c}{d}$.

(b) Write $\frac{3}{5}=\frac{18}{30}$ and $\frac{2}{3}=\frac{20}{30}$ to see that $\frac{19}{30}$ is between $\frac{3}{5}$ and $\frac{2}{3}$.

6. **(a)**

(b)

(c)

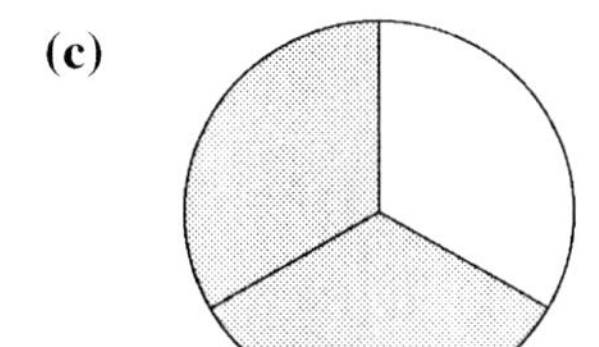

(d) One possibility is a set in which $\frac{2}{3}$ of the hearts have arrows through them.

7. (a) $2\frac{1}{48}+3\frac{1}{99}+6\frac{13}{25}$ is about $2+3+6\frac{1}{2}=11\frac{1}{2}$.

(b) $8\cdot\left(2\frac{1}{2}+3\frac{7}{15}\right)$ is about $8\cdot\left(2\frac{1}{2}+3\frac{1}{2}\right)=8\cdot 6=48$.

(c) $11\frac{9}{10}\div\frac{21}{40}$ is about $12\div\frac{1}{2}=12\cdot 2=24$. The closest answer is 23. (Note: the exact value of $11\frac{9}{10}\div\frac{21}{40}$ is $22\frac{2}{3}$, which really is closer to 23 than to 24.)

8. (a) If $\frac{a}{b}$ is a rational number with $a\neq 0$, then its multiplicative inverse is the rational number $\frac{b}{a}$.

(b) $\frac{2}{3}, -\frac{5}{4}, -\frac{1}{5}$

9. Answers will vary. Possibilities include $\frac{-6}{8}, \frac{-9}{12}, \frac{-12}{16}$.

10. (a)

$$\begin{aligned}2x+3&>0\\ 2x+3+(-3)&>0+(-3)\\ 2x&>-3\\ \frac{1}{2}\cdot 2x&>\frac{1}{2}\cdot(-3)\\ x&>\frac{-3}{2}\end{aligned}$$

(b)

$$\begin{aligned}\frac{3}{4}x+\frac{1}{2}&=\frac{1}{3}\\ \frac{3}{4}x+\frac{1}{2}+\left(-\frac{1}{2}\right)&=\frac{1}{3}+\left(-\frac{1}{2}\right)\\ \frac{3}{4}x&=-\frac{1}{6}\\ \frac{4}{3}\cdot\frac{3}{4}x&=\frac{4}{3}\cdot\left(\frac{-1}{6}\right)\\ x&=\frac{-4}{18}=\frac{-2}{9}\end{aligned}$$

(c)

$$\begin{aligned}\frac{5}{4}x&>-\frac{1}{3}\\ \frac{4}{5}\cdot\frac{5}{4}x&>\frac{4}{5}\cdot\frac{-1}{3}\\ x&>\frac{-4}{15}\end{aligned}$$

(d)

$$\begin{aligned}\frac{1}{2}&<4x+\frac{5}{6}\\ 4x+\frac{5}{6}&>\frac{1}{2}\\ 4x+\frac{5}{6}+\left(-\frac{5}{6}\right)&>\frac{1}{2}+\left(-\frac{5}{6}\right)\\ 4x&>\frac{-1}{3}\\ \frac{1}{4}\cdot 4x&>\frac{1}{4}\cdot\frac{-1}{3}\\ x&>\frac{-1}{12}\end{aligned}$$

11. (a) If $\frac{a}{b}$ and $\frac{c}{d}$ are rational numbers with $\frac{c}{d}\neq 0$, then $\frac{a}{b}\div\frac{c}{d}=\frac{e}{f}$ if, and only if, $\frac{a}{b}=\frac{c}{d}\cdot\frac{e}{f}$.

(b) If $\frac{a}{b} \div \frac{c}{d} = \frac{e}{f}$, then $\frac{a}{b} = \frac{c}{d} \cdot \frac{e}{f} = \frac{e}{f} \cdot \frac{c}{d}$.

So $\frac{a}{b} \cdot \frac{d}{c} = \frac{e}{f} \cdot \frac{c}{d} \cdot \frac{d}{c} = \frac{e}{f}$.

Thus, $\frac{a}{b} \div \frac{c}{d} = \frac{a}{b} \cdot \frac{d}{c}$.

12. Answers will vary.

(a) Find the area of a rectangular plot of land whose dimensions are $\frac{4}{5}$ mile by $\frac{2}{3}$ mile.

(b) A drink contains $\frac{3}{10}$ real fruit juice. If a jug of this drink contains $\frac{3}{8}$ of a quart of real juice, how much of the drink is in the jug?

13. **(a)** If $\frac{a}{b}$ is a rational number, then its additive inverse is the rational number $\frac{-a}{b}$.

(b) $\frac{-3}{4}, \frac{7}{4}, \frac{8}{2}$

14. $-3, -1\frac{1}{2}, 0, \frac{5}{8}, \frac{2}{3}, 3, \frac{16}{5}$

Chapter 7 Decimals, Real Numbers, and Proportional Reasoning

Problem Set 7.1

1. **(a)** $273.412 = 200 + 70 + 3 + \frac{4}{10} + \frac{1}{100} + \frac{2}{1000} = 2 \cdot 10^2 + 7 \cdot 10^1 + 3 \cdot 10^0 + 4 \cdot 10^{-1} + 1 \cdot 10^{-2} + 2 \cdot 10^{-3}$

2. **(a)** 0.21 **(b)** 0.235

(c) 0.278 **(d)** 0.302

4. **(a)** $\frac{1}{4} = 0.25$ (two strips and five small squares, or 25 small squares)

5. **(a)** $0.324 = \frac{324}{1000} = \frac{81}{250}$; $250 = 2 \cdot 5^3$

6. **(a)** $\frac{7}{20} = \frac{7 \cdot 5}{20 \cdot 5} = \frac{35}{100} = 0.35$

(c) $\frac{3}{75} = \frac{1}{25} = \frac{1 \cdot 2^2}{5^2 \cdot 2^2} = \frac{4}{100} = 0.04$

7. **(a)**

$$\begin{array}{r} 0.875 \\ 8\overline{)7.000} \\ \underline{6\,4} \\ 60 \\ \underline{56} \\ 40 \\ \underline{40} \\ 0 \end{array}$$

$\frac{7}{8} = 0.875$

8. **(a)**

$$\begin{array}{r} 0.833 \\ 6\overline{)5.000} \\ \underline{4\,8} \\ 20 \\ \underline{18} \\ 20 \\ \underline{18} \\ 2 \end{array}$$

$\frac{5}{6} = 0.8\overline{3}$

9. **(a)** Let $x = 0.321321\ldots$.
Then $1000x = 321.321321\ldots$
$1000x - x = 321.321321\ldots - 0.321321\ldots$
$999x = 321$
$x = \frac{321}{999} = \frac{107}{333}$

(e) Let $x = 0.142857142857\ldots$.
Then $1{,}000{,}000x = 142{,}857.142857142857\ldots$

$$1{,}000{,}000x - x = 142857.142857142857\ldots - 0.142857142857\ldots$$
$$999{,}999x = 142{,}857$$
$$x = \frac{142{,}857}{999{,}999} = \frac{1}{7}$$

10. (a)

$$x = 0.3\overline{54}$$
$$1000x = 354.\overline{54}$$
$$-(10x = 3.\overline{54})$$
$$990x = 351$$
$$x = \frac{351}{990} = \frac{39}{110}$$

(c)

$$x = 2.3\overline{402}$$
$$10{,}000x = 23{,}402.\overline{402}$$
$$-(10x = 23.\overline{402})$$
$$9990x = 23{,}379$$
$$x = \frac{23{,}379}{9990} = \frac{7793}{3330}$$

11. Answers will vary.

(a) $\frac{358}{999} = 0.\overline{358}$

12. (a) $0.007, 0.017, 0.01\overline{7}, 0.027$

14. (a) Assume $3-\sqrt{2}$ is rational, then $3-\sqrt{2} = q$ where q is rational. This implies that $\sqrt{2} = 3 - q$. But $3 - q$ is rational since the rational numbers are closed under subtraction. This can't be true since we know $\sqrt{2}$ is irrational. Thus, the assumption that $3-\sqrt{2}$ is rational must be false. So $3-\sqrt{2}$ is irrational.

20. Use a calculator or long division for parts (a), (b), and (c).

(a) $0.\overline{09}$ **(c)** $0.\overline{0009}$

21. (a) Let $x = 0.747474\ldots$.
Then $100x = 74.747474\ldots$

$$100x - x = 74.747474\ldots - 0.747474\ldots$$
$$99x = 74$$
$$x = \frac{74}{99}$$

22. Use the patterns discovered in problem 19(d) for parts (a) through (d).

(a) $\frac{5}{9} = 0.\overline{5}$

(d) $\frac{17}{33} = \frac{51}{99} = 0.\overline{51}$

23. Answers will vary.

(a) One example is $\sqrt{2} + \left(3-\sqrt{2}\right) = 3$. $\sqrt{2}$ and $\left(3-\sqrt{2}\right)$ are both irrational. (See the answer to problem 14.)

(b) One example is $\sqrt{3} + \sqrt{3} = 2\sqrt{3}$. $\sqrt{3}$ and $2\sqrt{3}$ are both irrational.

27. Answers will vary. For example, $\frac{\sqrt{2}}{2\sqrt{2}} = \frac{1}{2}$, and $\frac{1}{2}$ is rational.

Problem Set 7.2

1. (a)

$$\begin{array}{r} 32.174 \\ +\ 371.500 \\ \hline 403.674 \end{array}$$

(c)

$$\begin{array}{r} 0.057 \\ +\ 1.080 \\ \hline 1.137 \end{array}$$

2. (a)

$$\begin{array}{r} 37.1 \\ \times\ 4.7 \\ \hline 2597 \\ 1484 \\ \hline 174.37 \end{array}$$

(c)

$$\begin{array}{r} 26.1 \\ 5.3\overline{)138.33} \\ \underline{106} \\ 323 \\ \underline{318} \\ 53 \\ \underline{53} \\ 0 \end{array}$$

3. **(a)** Estimate: $4 + 31 = 35$
Answer: $4.112 + 31.3 = 35.412$

(c) Estimate: $4 \cdot 31 = 124$
Answer: $4.112 \cdot 31.3 = 128.7056$

6. **(a)** $34.796 \times 10^3 = 34,796$

(c) $34.796 \div 10^2 = 0.34796$

7. **(a)** $34,762 = 3.4762 \times 10^4$

(c) $0.009031 = 9.031 \times 10^{-3}$

8. **(a)** Since there is a 5 in the hundred-thousands place, round the 6 up to 7; 2.77×10^8

9. **(a)** ON/AC 1.27 EXP (−) 5 × 8.235 EXP (−) 6 = 1.05×10^{-10}

(c) ON/AC 1.27 EXP (−) 5 ÷ 98613428 = 1.29×10^{-13}

10. **(a)** Estimate:

$$\begin{aligned}(7 \times 10^5) \cdot (2 \times 10^4) &= (7 \cdot 2) \times (10^5 \cdot 10^4) \\ &= 14 \times 10^9 \\ &= 1.4 \times 10^{10}\end{aligned}$$

Answer:

$$(7.123 \times 10^5) \cdot (2.142 \times 10^4) \approx 1.53 \times 10^{10}$$

15. Since the given numbers are 0.123 times the digits 1, 2, 3, 4, 5, 6, 7, 8, and 9, respectively, we may obtain a magic square by multiplying each entry in the square in problem 14(a) of Problem Set 1.1 by 0.123. This gives

0.492	1.107	0.246
0.369	0.615	0.861
0.984	0.123	0.738

Aside from rotating and/or flipping the array, this is the only solution.

19. **(c)** Note that $2.415 - 0.041 = 2.374$, $7.723 - 2.415 = 5.308$, and $5.308 - 0.041 = 5.267$.

$$\begin{array}{ccccc} \underline{2.374} & & \underline{0.041} & & \underline{5.267} \\ & \underline{2.415} & & \underline{5.308} & \\ & & \underline{7.723} & & \end{array}$$

20. **(a)** Each number is 0.9 more than its predecessor. The sequence is 3.4, 4.3, 5.2, 6.1, 7.0, 7.9.

(c) If each number is d more than its predecessor, then $0.0114 + 3d = 0.3204$, so $d = 0.103$. The sequence is 0.0114, 0.1144, 0.2174, 0.3204, 0.4234, 0.5264.

21. **(a)** Each number is $\frac{2.321}{2.11} = \frac{11}{10}$ times its predecessor. The sequence is 2.11, 2.321, 2.5531, 2.80841, 3.089251.

24. **(a)** Note that $1.32 + 3.41 = 4.73$, $3.41 + 7.10 = 10.51$, and $1.32 + 7.10 = 8.42$.

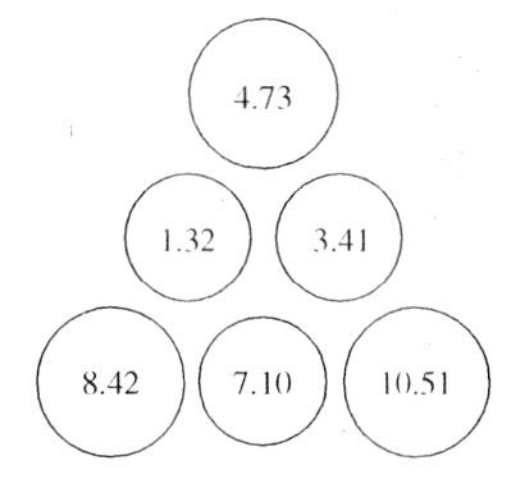

(c) If a, b, and c are the numbers in the small circles, then
$2(a + b + c) = (a + b) + (b + c) + (a + c)$
$= 2.341 + 7.133 + 4.012 = 13.486$,
so the sum of a, b, and c is 6.743. Thus the number opposite 7.133 is $6.743 - 7.133 = -0.39$ and the other missing numbers are $6.743 - 2.341 = 4.402$ and $6.743 - 4.012 = 2.731$.

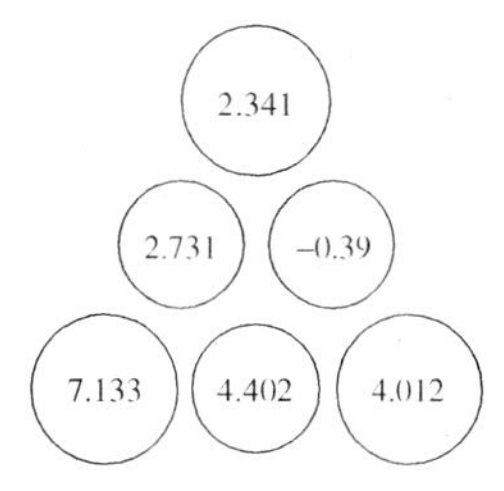

25. (a) Note that $0.92 + 0.41 = 1.33$, $0.41 + 1.23 = 1.64$, $1.23 + 0.72 = 1.95$, and $0.72 + 0.92 = 1.64$.

1.33	0.41	1.64
0.92		1.23
1.64	0.72	1.95

(c) Answers will vary. A solution may be calculated by placing any number at random in one of the small circles. For example, if 1 is placed in the middle of the bottom row, the remaining entries may be calculated by noting that $1.07 - 1 = 0.07$, $-1.41 - 0.07 = -1.48$, $2.53 - (-1.48) = 4.01$, and $4.01 + 1 = 5.01$. (All possible solutions have 5.01 in the lower right corner.)

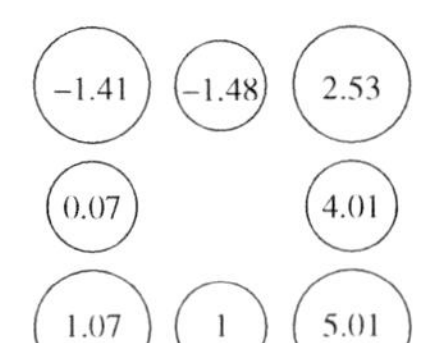

26. $3 \cdot 9.72 = 29.16$, so she spent \$29.16.

29. (a) It is helpful to draw a picture. Since $24.75 + 2 \cdot 2.25 = 29.25$ and $17.5 + 2 \cdot 2.25 = 22$, the drawing is as shown.

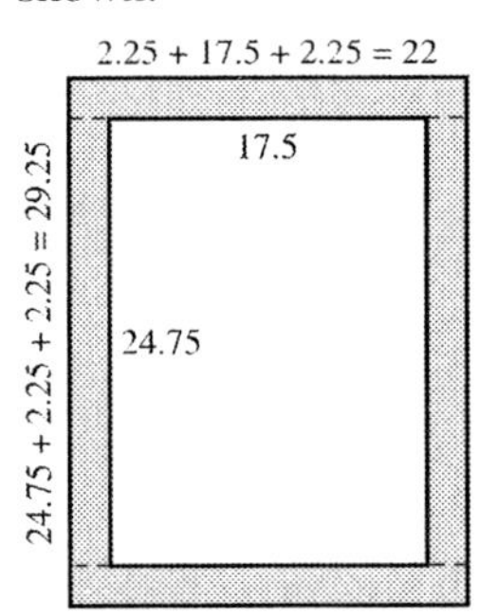

Thus, the area of the frame is $2(22 \cdot 2.25) + 2(2.25 \cdot 24.75) = 210.375$ square inches. Note that other equations are possible.

30. (a) $200,000,000,000 = 2 \times 10^{11}$, 1 significant digit

(b) $0.000000000753 = 7.53 \times 10^{-10}$, 3 significant digits

32. (a)

2.03	0.001	17.4	21.03
2.029	17.399	3.63	19
15.37	13.769	15.37	16.971
1.601	1.601	1.601	1.601
0	0	0	0

(c)

5	9	17	31
4	8	14	26
4	6	12	22
2	6	10	18
4	4	8	16
0	4	8	12
4	4	4	12
0	0	8	8
0	8	0	8
8	8	8	8
0	0	0	0

Problem Set 7.3

1. There are 10 girls, 14 boys, and $10 + 14 = 24$ students.

(a) boys to girls $= \dfrac{14}{10} = \dfrac{7}{5}$

(c) boys to students $= \dfrac{14}{24} = \dfrac{7}{12}$

(e) students to girls $= \dfrac{24}{10} = \dfrac{12}{5}$

2. (a) Yes. $2 \cdot 12 = 3 \cdot 8$

(c) No. $7 \cdot 31 \neq 28 \cdot 8$

(e) No. $14 \cdot 60 \neq 49 \cdot 18$

(g) Yes. $(1.5) \cdot (16.1) = (2.1) \cdot (11.5)$

3. (a)
$$\frac{6}{14} = \frac{r}{21}$$
$$6 \cdot 21 = 14 \cdot r$$
$$r = \frac{6 \cdot 21}{14} = 9$$

4. **(a)** $\frac{24}{16} = \frac{3}{2}$, so the ratio is 3 to 2.

(c) $\frac{248}{372} = \frac{2}{3}$, so the ratio is 2 to 3.

5. **(a)** $3\frac{1}{2} \cdot 5.50 = 19.25$, so she earned \$19.25.

(c) $\frac{3.5}{19.25} = \frac{5}{27.50}$ because
$3.5 \cdot 27.5 = 96.25$ and
$19.25 \cdot 5 = 96.25$.

9. Let x be the height of the flagpole.

$$\frac{x}{6\frac{1}{4}} = \frac{9\frac{2}{3}}{3\frac{1}{6}}$$

$$3\frac{1}{6}x = 6\frac{1}{4} \cdot 9\frac{2}{3}$$

$$x = \left(6\frac{1}{4} \cdot 9\frac{2}{3}\right) \div 3\frac{1}{6}$$

$$= \frac{25}{4} \cdot \frac{29}{3} \div \frac{19}{6} = \frac{25}{4} \cdot \frac{29}{3} \cdot \frac{6}{19}$$

$$= \frac{4350}{228} = \frac{725}{38} = 19\frac{3}{38}$$

To the nearest foot, the flagpole is 19 ft high.

13. **(a)** proportional: $y = kx$, where k is the price per gallon of gasoline

(c) not proportional: $y = x^2$

19. **(a)** $\frac{a}{b} = \frac{c}{d} \Rightarrow ad = bc \Rightarrow da = cb \Rightarrow \frac{d}{c} = \frac{b}{a}$

(c)

$$\frac{a}{b} = \frac{c}{d}$$

$$ad = bc$$

$$ac + ad = ac + bc$$

$$a(c + d) = (a + b)c$$

$$\frac{a}{a+b} = \frac{c}{c+d}$$

22. **(a)**

$$y = kx^2$$

$$27 = k(6^2) = 36k$$

$$k = \frac{27}{36} = \frac{3}{4}$$

When $x = 12$, $y = kx^2 = \frac{3}{4}(12^2) = 108$.

26. $\frac{100}{11.6} \div \frac{100}{11.8} = \frac{100}{11.6} \cdot \frac{11.8}{100} = \frac{1180}{1160} = \frac{59}{58}$

The ratio is 59 to 58.

28. Compare the cost per ounce (or cost per gallon) in each case.

(a) $\frac{0.90}{32} < \frac{1.20}{40}$ because $0.90 \cdot 40 < 32 \cdot 1.20$, so the better buy is 32 ounces for 90¢.

31. **(a)** The table is not quite a ratio table. For example $21\frac{1}{4} \div 6\frac{3}{4} = \frac{85}{27} \approx 3.15$, while $23\frac{1}{2} \div 7\frac{1}{2} = \frac{47}{15} \approx 3.13$. So the circumference of the hat is not quite proportional to the hat size.

33. Since the area of a circle is given by the formula $A = \pi r^2$, the price is proportional to the square of the radius of the pizza. Therefore, if P is the price of the large pizza, $\frac{P}{7^2} = \frac{9.56}{6^2}$. Hence, $P = \frac{9.56 \cdot 49}{36} \approx 13.01$.

The price should be \$13.01.

36. Let x be the cost of 7 shirts.

$$\frac{x}{7} = \frac{59.97}{3}$$

$$x \cdot 3 = 7 \cdot 59.97$$

$$x = \frac{7 \cdot 59.97}{3} = 139.93$$

The cost will be \$139.93.

40. **(a)**

Number of Teeth in Chainring	Number of Teeth in Cog: 34	28	23	19	16	13	11
24	0.71	0.86	1.04	1.26	1.50	1.85	2.18
35	1.03	1.25	1.52	1.84	2.19	2.69	3.18
51	1.50	1.82	2.22	2.68	3.19	3.92	4.64

Problem Set 7.4

1. **(a)** $100 \cdot \frac{3}{16} = 18.75$, so $\frac{3}{16} = 18.75\%$.

(c) $100 \cdot \frac{37}{40} = 92.5$, so $\frac{37}{40} = 92.5\%$.

(e) $100 \cdot \frac{3.24}{8.91} = 36.\overline{36}$, so $\frac{3.24}{8.91} = 36.\overline{36}\%$.

(g) $100 \cdot \frac{1.6}{7} \approx 22.86$, so $\frac{1.6}{7} \approx 22.86\%$.

2. **(a)** $100 \cdot 0.19 = 19$, so $0.19 = 19\%$.

(c) $100 \cdot 2.15 = 215$, so $2.15 = 215\%$

3. **(a)** $10\% = \frac{10}{100} = \frac{1}{10}$

(c) $62.5\% = \frac{62.5}{100} = \frac{625}{1000} = \frac{5}{8}$

4. **(a)** $70\% \times 280 = 0.7 \times 280 = 196$

(c) $38\% \times 751 = 0.38 \times 751 = 285.38$

(e) $0.02\% \times 27,481 = 0.0002 \times 27,481 = 5.4962$

5. **(a)** $50\% \times 840 = \frac{840}{2} = 420$

(c) $12.5\% \times 48 = \frac{48}{8} = 6$

(e) $200\% \times 56 = 2 \times 56 = 112$

6. **(a)**

$$105\% \times x = 50,400$$
$$1.05x = 50,400$$
$$x = \frac{50,400}{1.05} = 48,000$$

Michelle's former salary was $48,000.

9. **(a)**

15 mm

20 mm

(c)

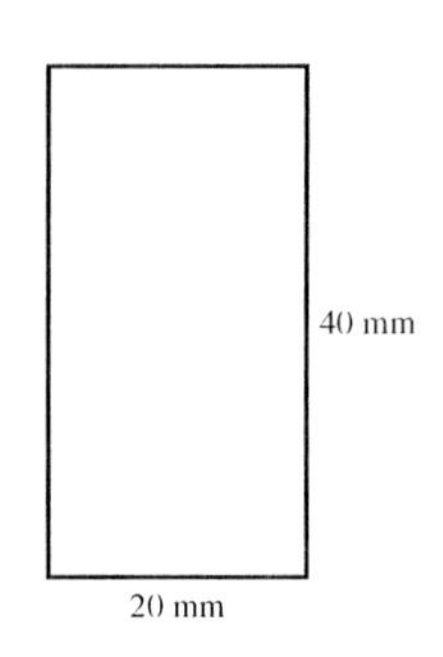

10. **(a)** $\frac{1}{8} = 0.125 = 12.5\%$

(c) $\frac{7}{18} = 0.3888\ldots \approx 39\%$

11. There are $10 \cdot 10 = 100$ total small squares.

(a) 100% of $100 = 1 \cdot 100 = 100$ small squares should be shaded.

(c) 25% of $100 = 0.25 \cdot 100 = 25$ small squares should be shaded.

12. There are $8 \cdot 5 = 40$ total small rectangles.

(a) 50% of $40 = 0.5 \cdot 40 = 20$ small rectangles should be shaded.

(c) 20% of $40 = 0.2 \cdot 40 = 8$ small rectangles should be shaded.

14. **(a)** $\frac{7}{28} = \frac{1}{4} = 25\%$

(c) $\frac{72}{144} = \frac{1}{2} = 50\%$

23. Assume that each married man is married to a woman in the same population and vice versa. If m is the number of men and w is the number of women, then $40\% \times m = 30\% \times w$; that is, $0.4m = 0.3w$ or $m = 0.75w$. The fraction of adults that are married is $\frac{0.4m + 0.3w}{m + w} = \frac{0.3w + 0.3w}{0.75w + w} = \frac{0.6w}{1.75w} = \frac{0.6}{1.75} \approx 0.3429$. About 34.29% of the adult population is married.

27. They made $0.6 \times 40 = 24$ field goals in the first half and $0.25 \times 44 = 11$ field goals in the second half. Since $\frac{24 + 11}{40 + 44} = \frac{35}{84} = \frac{5}{12} \approx 42$, their field goal shooting percentage for the game was about 42%.

30. **(a)** Let x be the wholesale cost. Then $2x = x + 100\%x$ is the regular price, and the sale price is $0.8(2x) = 1.6x$. Then the store's profit is $1.6x - x = 0.6x$ and the percent profit is $\frac{0.6x}{x} = 0.60 = 60\%$.

34. $11\% \times 158,000 = 0.11 \times \$158,000 = \$17,380$

36. **(a)** Let $P = 2500$, $r = 5.25$, $t = 1$, and $n = 7$. Then

$$P\left(1+\frac{r}{100t}\right)^{nt} = 2500\left(1+\frac{5.25}{100(1)}\right)^{7\cdot 1}$$

$= 2500(1.0525)^7 \approx 3576.80.$

The investment is worth \$3576.80.

37. **(a)** $x = 10$

5.0 MB is 10% of y MB, so

$5.0 = 0.1y \Rightarrow y = 50.$

At 2.50 MB/sec, it will take $45 \div 2.5 = 18$ seconds to copy the remaining 45 MB. Thus, $z = 18$.

38. Since $16,000 = P \cdot (1.06)^5$,

$$P = \frac{16,000}{(1.06)^5} \approx 11,956.13.$$

You would need to invest \$11,956.13.

40. **(a)** Calculate successive powers of 1.05: 1.05, 1.1025, 1.1576, 1.2155, 1.2763, 1.3401, 1.4071, 1.4775, 1.5513, 1.6289, 1.7103, 1.7959, 1.8856, 1.9799, 2.0789, ...

Since $1.05^{15} > 2$, it takes 15 years.

(c) Calculate successive powers of 1.14: 1.14, 1.2996, 1.4815, 1.6890, 1.9254, 2.1950, ... It takes 6 years.

41. **(a)** Since 2010 is 11 years after 1999, then $3,300,000 \cdot (1.04)^{11} = 5,080,198$ or about 5.1 million (5,100,000).

Chapter 7 Review Exercises

1. **(a)** $273.425 = 2\cdot 10^2 + 7\cdot 10^1 + 3\cdot 10^0 + 4\cdot 10^{-1} + 2\cdot 10^{-2} + 5\cdot 10^{-3}$

(b) $0.000354 = 3\cdot 10^{-4} + 5\cdot 10^{-5} + 4\cdot 10^{-6}$

2. **(a)** $\frac{7}{125} = \frac{56}{1000} = 0.056$

(b) $\frac{6}{75} = \frac{2}{25} = \frac{8}{100} = 0.08$

(c) $\frac{11}{80} = \frac{1375}{10,000} = 0.1375$

3. **(a)** $0.315 = \frac{315}{1000} = \frac{63}{200}$

(b) $1.206 = \frac{1206}{1000} = \frac{603}{500}$

(c) $0.2001 = \frac{2001}{10,000}$

4. Since $\frac{4}{12} = 0.\overline{3}$, $\frac{5}{13} \approx 0.38$, and $\frac{2}{66} \approx 0.03$, the numbers in order are $0.03, 0.33, 0.\overline{3}$, 0.3334, 0.38, or $\frac{2}{66}, 0.33, \frac{4}{12}, 0.3334, \frac{5}{13}$.

5. **(a)** Let $x = 10.\overline{363}$. Then

$1000x = 10,363.\overline{363}.$

$1000x - x = 10,363.\overline{363} - 10.\overline{363}$

$999x = 10,353$

$x = \frac{10,353}{999} = \frac{3451}{333}$

(b) Let $x = 2.1\overline{42}$. Then $10x = 21.\overline{42}$ and

$1000x = 2142.\overline{42}$

$1000x - 10x = 2142.\overline{42} - 21.\overline{42}$

$990x = 2121$

$x = \frac{2121}{990} = \frac{707}{330}$

6. Irrational. The decimal expansion of a does not have a repeating sequence of digits and it does not terminate.

7. **(a)** $0.\overline{2} = \frac{2}{9}$ **(b)** $0.\overline{36} = \frac{36}{99} = \frac{4}{11}$

8. **(a)**

$$\begin{array}{r} 21.7340 \\ 3.2145 \\ +71.2400 \\ \hline 96.1885 \end{array}$$

(b)

$$\begin{array}{r} 23.471 \\ -2.890 \\ \hline 20.581 \end{array}$$

(c)

$$\begin{array}{r} 35.4 \\ \times 2.37 \\ \hline 2\,478 \\ 10\,62 \\ 70\,8 \\ \hline 83.898 \end{array}$$

(d) $24.15 \div 3.45 = 2415 \div 345 = 7.0$

$$\begin{array}{r} 7 \\ 345\overline{)2415} \\ \underline{2415} \\ 0 \end{array}$$

9. (a) $31.47 + 3.471 + 0.0027 = 34.9437$

(b) $31.47 - 3.471 = 27.999$

(c) $31.47 \times 3.471 = 109.23237$

(d) $138.87 \div 23.145 = 6.0$

10. (a) Estimate: $50 + 10 = 60$
Answer: $47.25 + 13.134 = 60.384$

(b) Estimate: $50 - 10 = 40$
Answer: $52.914 - 13.101 = 39.813$

(c) Estimate: $50 \times 10 = 500$
Answer: $47.25 \times 13.134 = 620.5815$

(d) Estimate: $45 \div 15 = 3$
Answer: $47.25 \div 13.134 \approx 3.5975$

11. Suppose $3-\sqrt{2} = r$ where r is rational. Then $3 - r = \sqrt{2}$. But this implies that $\sqrt{2}$ is rational since the rationals are closed under subtraction. This is a contradiction since $\sqrt{2}$ is irrational. Therefore, by contradiction, $3-\sqrt{2}$ is irrational.

12. $3-\sqrt{2}$ and $\sqrt{2}$ are irrational, but $\left(3-\sqrt{2}\right)+\sqrt{2} = 3$, which is rational.

13. The decimal expansion of an irrational number is nonrepeating and is nonterminating.

14. (a) $8.25 \times 112.5 = 928.125\text{ ft}^2$

(b) $928.125 \div 110 = 8.4375$ qts which rounds to 9 qts.

15. Answers will vary. For example, $\frac{4123}{9999}$ is such a fraction.

16. (a) $\frac{5}{18} \approx 0.2777778$, so $\frac{5}{18} = 0.2\overline{7}$; the period starts in the second decimal place.
$\frac{41}{333} \approx 0.123231$, so $\frac{41}{333} = 0.\overline{123}$; the period starts right after the decimal point, in the first decimal place.
$\frac{11}{36} \approx 0.305556$, $\frac{11}{36} = 0.30\overline{5}$; the period starts in the third decimal place.
$\frac{7}{45} \approx 0.1555556$, so $\frac{7}{45} = 0.1\overline{5}$; the period starts in the second decimal place.
$\frac{13}{80} = 0.1625$; this decimal terminates but can be written as $0.1625\overline{0}$ or $0.1624\overline{9}$, so one might say that the period starts in the fifth decimal place.

(b) Look for a pattern based on the prime factorization of the denominator. (See table below.)The period begins in decimal place $r + 1$, where r is the highest power of 2 and/or 5 appearing in the prime factor representation of the denominator of the fraction in simplest form.

17. She had 11 successes and $20 - 11 = 9$ failures, so the ratio is 11 to 9.

Table for exercise 16b.

Fraction	$\frac{5}{18}$	$\frac{41}{333}$	$\frac{11}{36}$	$\frac{7}{45}$	$\frac{13}{80}$
Factorization	$2^1 \cdot 3^2$	$3^2 \cdot 37$	$2^2 \cdot 3^2$	$3^2 \cdot 5^1$	$2^4 \cdot 5$
Period starts at	2	1	3	2	5

18. **(a)** Yes. $775 \cdot 25 = 125 \cdot 155$

(b) No. $31 \cdot 32 \neq 64 \cdot 15$

(c) Yes. $9 \cdot 32 = 24 \cdot 12$

19. Let x be the cost of 5 pounds of candy.

$$\frac{x}{5} = \frac{3.15}{2}$$

$$x = \frac{5 \cdot 3.15}{2} = 7.875$$

It would cost \$7.88.

20. Let x be the number of gallons used to travel 300 miles.

$$\frac{x}{300} = \frac{7.5}{173}$$

$$x = \frac{300 \cdot 7.5}{173} \approx 13.006$$

He will need about 13 gallons.

21. $y = kx$

$7 = k \cdot 3$

$k = \frac{7}{3}$

If $x = 5$, $y = kx = \frac{7}{3} \cdot 5 = \frac{35}{3}$.

22. Let x be the height of the flagpole in feet, and note that the yardstick has length 3 feet and its shadow has length $\frac{5}{6}$ foot.

$$\frac{x}{3} = \frac{12}{\frac{5}{6}} = \frac{12 \cdot 6}{5}$$

$$x = 3 \cdot \frac{12 \cdot 6}{5} = 43.2$$

The flagpole is 43.2 feet tall.

23. **(a)** $\frac{5}{8} = 0.625 = 62.5\%$

(b) $2.115 = 211.5\%$

(c) $0.015 = 1.5\%$

24. **(a)** $28\% = \frac{28}{100} = 0.28$

(b) $1.05\% = \frac{1.05}{100} = 0.0105$

(c) $33\frac{1}{3}\% = \frac{33\frac{1}{3}}{100} = \frac{1}{3} = 0.\overline{3}$

25. $7.2\% \times \$49 = 0.072 \times \$49 \approx \$3.53$

26. $\frac{6.75}{84.37} \approx 0.08 = 8\%$

27. $\frac{11}{20} = 0.55 = 55\%$

28. **(a)** Alex received a \$3000 raise.

$$\frac{3000}{48000} = 0.0625 = 6.25\%$$

Alex received a 6.25% raise.

(b) Monique's stock is worth 90% of its original value. Let x = the original value of her stock.

$0.90x = 8100$

$x = 9000$

Monique's stock was worth \$9000.

29. $\$3000\left(1 + \frac{8}{100(4)}\right)^{2 \cdot 4} = \$3000(1.02)^8 \approx \$3514.98$

Chapter 7 Test

1. Your investment is multiplied by 1.0575 every year. By keying in 5000 + 1.0575 = = = =, you will find that it takes five presses of the = key to obtain approximately \$6612.60. It takes five years, so you invested the money 5 years ago.

2. Answers will vary. One example is $0.\overline{125} = \frac{125}{999}$.

3. Your investment is multiplied by 1.021 every six months. By keying in 2000 + 1.021 = = = ..., it takes 34 presses of the = key to obtain a number over 4000 (namely, 4054.20). Therefore, it will take 34 half-years, or 17 years.

4. **(a)** $\frac{84}{175} = \frac{12}{25} = \frac{48}{100} = 0.48$

(b) $\frac{24}{99} = 0.\overline{24}$

(c) $\frac{7}{11} = \frac{63}{99} = 0.\overline{63}$

5. $\$1000\left(1+\frac{9}{100(12)}\right)^{2\cdot 12} = \$1000(1.0075)^{24} \approx \$1196.41$

6. **(a)** Let $x = 0.\overline{45}$.
Then $100x = 45.\overline{45}$ and
$100x - x = 99x = 45$. So $x = \frac{45}{99} = \frac{5}{11}$.

(b) Let $x = 31.\overline{5}$. Then $10x = 315.\overline{5}$ and
$10x - x = 9x = 284$. So $x = \frac{284}{9}$.

(c) Let $x = 0.34\overline{9}$. Then $10x = 3.49\overline{9}$ and
$10x - x = 9x = 3.15$.
So $x = \frac{3.15}{9} = \frac{315}{900} = \frac{7}{20}$.

7. Since 21.432 has 3 digits to the right of the decimal point and 3.41 has 2 digits to the right of the decimal point, the number of digits to the right of the decimal point in the product should be 3 + 2 = 5.

8. **(a)** Since there are 17 wins and 32 – 17 = 15 losses, the ratio is 17 to 15.

(b) $\frac{17}{32} = 0.53125 = 53.125\%$

9. $(2.34 \times 10^{-6}) \cdot (3.12 \times 10^{5}) = 7.3008 \times 10^{-1} = 0.73008 \approx 0.730$

10. $\frac{1425}{9500} = 0.15 = 15\%$

Chapter 8 Algebraic Reasoning, Graphing, and Connections with Geometry

Problem Set 8.1

1. **(a)** Constant; the number of feet never changes.

(b) Variable; the number of hours of daylight increases and decreases during a year.

4. **(a)** $p + 5$

(c) $\frac{1}{2}(p+2)-2$

6. **(a)** $(3^2+5)\cdot 7 = 98$

(b) $(x^2+5)\cdot 7$

9. **(a)** Add $8x$ to both sides of the first equation to get $y = 8x + 2$. Subtract 4 from both sides of the second equation to get $y = 3x - 4$.

10. **(a)** Using distance = rate × time ($d = rt$), the average speed is distance divided by time. Therefore, her average speed was 15 mi ÷ 3 hr = 5 mph.

(b) Since $t = \frac{d}{r}$, it took 15 mi ÷ 15 mph = 1 hour to get home.

(f) The speed to Grandma's house was 5 mph and was 15 mph on the return trip. The average of 5 and 15 is 10, so this is not equal to the average speed of $7\frac{1}{2}$ mph for the round trip. This is because LRR spent much more time riding at 5 mph than riding at 15 mph.

11. **(a)** Not a function since element c is associated with more than one element of set B and element d is not associated with any elements of set B.

(b) A function, with range $\{p, q, r\} \subset B$.

12. **(a)**

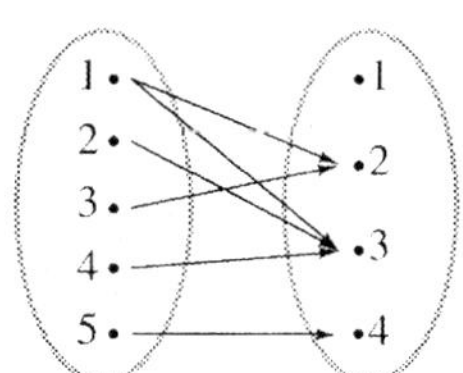

Not a function with domain A, since $\{1\} \subset A$ is associated with more than one element of set B.

13. **(a)** Not the graph of a function: for example, there are three y-values associated with the value $x = 3$.

(c) Not the graph of a function, since the two y-values 1 and 3 correspond to $x = 2$.

16. **(a)** $g(0) = 5 - 2(0) + 0^2 = 5$

$g(1) = 5 - 2(1) + 1^2 = 4$

$g(2) = 5 - 2(2) + 2^2 = 5$

$g(3) = 5 - 2(3) + 3^2 = 8$

$g(4) = 5 - 2(4) + 4^2 = 13$

(b) $\{4, 5, 8, 13\}$

18. **(a)** She left home about 0.75 hour past noon, or at 12:45 P.M.

(b) She realized she forgot her checkbook about 1 hour past noon, or at 1:00 P.M.

19. **(a)** G3; the temperature increases steadily until it levels off.

20. **(a)**

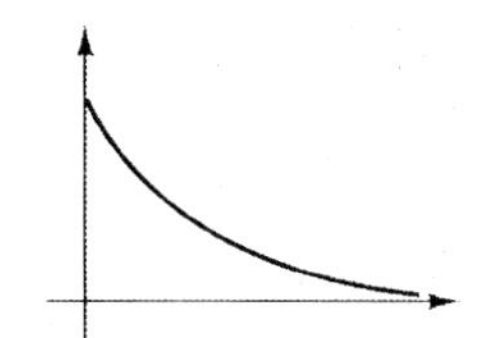

21. **(a)** $A = 2x \cdot x$ or $A = 2x^2$

(b) $P = 2x + 2(2x)$ or $P = 6x$

(c) Since $D^2 = x^2 + (2x)^2 = 5x^2$, $D = \sqrt{5}\,x$.

22. **(a)** Add 5: $y = x + 5$

24. **(a)**
$$\begin{aligned} F(0) &= f(g(0)) \\ &= f(0+3) \\ &= f(3) \\ &= 2\times 3 \\ &= 6 \end{aligned} \qquad \begin{aligned} F(1) &= f(g(1)) \\ &= f(1+3) \\ &= f(4) \\ &= 2\times 4 \\ &= 8 \end{aligned}$$
$$\begin{aligned} F(2) &= f(g(2)) \\ &= f(2+3) \\ &= f(5) \\ &= 2\times 5 \\ &= 10 \end{aligned} \qquad \begin{aligned} F(3) &= f(g(3)) \\ &= f(3+3) \\ &= f(6) \\ &= 2\times 6 \\ &= 12 \end{aligned}$$

31. **(a)** AnnElise correctly figured out that the town is growing by 38 people each year. However, she extended her data only one year, not two.

34. Let t and u denote the tens and units digit of n, respectively. Then $n = 10t + u$. The sum of the digits is $t + u$ and the product of the digits is tu, so we get the condition $10t + u = t + u + tu$. If t and u are subtracted from each side of the equation, we get $9t = tu$. When t is not the 0 digit, the equation can be divided by t to get $9 = u$. We conclude that *every* two digit number ending in 9 has the property. That is, the solution set is $\{19, 29, 39, 49, 59, 69, 79, 89, 99\}$.

36. We assume the commuter travels the same distance d in both directions. The time to work, using the $d = rt$ formula from Table 8.1, is therefore given by the expression $\frac{d}{u}$. Similarly, the time to return home is given by $\frac{d}{v}$. The round trip distance $2d$ is traveled in the total time of $\frac{d}{u}+\frac{d}{v}$, so the average speed of the roundtrip is given by $\frac{2d}{\frac{d}{u}+\frac{d}{v}}$. These steps can simplify this expression:
$$\frac{2d}{\frac{d}{u}+\frac{d}{v}} = \frac{2d}{\frac{dv+du}{uv}} = \frac{2duv}{dv+du} = \frac{2uvd}{(v+u)d} = \frac{2uv}{v+u}.$$
It wasn't necessary to know the value of the variable d to solve the problem. A very common error is to answer the problem by taking the average of the two speeds without thinking about the answer conceptually.

40. Since there are 9 square feet in a square yard, the cost of a remnant is $\frac{\$3.60}{9} = \$0.40$ per square foot. An L-by-W remnant will therefore cost $(\$0.40)LW$. The perimeter of the remnant is $2L + 2W$, so the cost to finish the edges is $(\$0.12)(2L + 2W)$. This gives a total cost of $(\$0.40)LW + (\$0.12)(2L + 2W)$.

42. **(a)** The approximate formula gives that area $\left(\frac{16}{9}\right)^2 100^2 = 31{,}604.9\ldots.$ The more exact value of π used by a calculator is $\pi 100^2 = 31{,}415.9\ldots.$ The error is about 189 square feet.

Problem Set 8.2

1. **(a), (c), (e), (g), (i)**

y, x, R(−5, 7), P(5, 7), T(0, 5), X(0, 0), V(0, −5.2)

4. **(a)**
$$\begin{aligned} \sqrt{(4-(-2))^2+(13-5)^2} &= \sqrt{6^2+8^2} \\ &= \sqrt{36+64} \\ &= \sqrt{100} \\ &= 10 \end{aligned}$$

(c)
$$\begin{aligned} \sqrt{(8-0)^2+(-8-7)^2} &= \sqrt{8^2+(-15)^2} \\ &= \sqrt{64+225} \\ &= \sqrt{289} \\ &= 17 \end{aligned}$$

5. **(a)** $m = \frac{8-4}{3-1} = \frac{4}{2} = 2$; upward to the right

(c) $m = \frac{-7-(-3)}{-4-(-2)} = \frac{-4}{-2} = 2$; upward to the right

(e) $m = \frac{-5-(-2)}{-2-1} = \frac{-3}{-3} = 1$; upward to the right

6. $\dfrac{7-3}{4-b} = 2$

$4 = 2(4-b)$

$2 = 4-b$

$b = 2$

8. Substitute a for x and 3 for y:

$2a + 3 \cdot 3 = 18$

$2a = 9$

$a = \dfrac{9}{2}$

10. (a), (c)

For part (a), $3x + 5y = 12$.
If $x = -1$, then $-3 + 5y = 12$, or $y = 3$, so $(-1, 3)$ is on the line.
If $y = 0$, then $3x = 12$, or $x = 4$, so $(4, 0)$ is on the line.
For part (c), $5y - 3x = 15$.
If $x = 0$, then $5y = 15$, or $y = 3$, so $(0, 3)$ is on the line.
If $y = 0$, then $-3x = 15$, or $x = -5$, so $(-5, 0)$ is on the line.

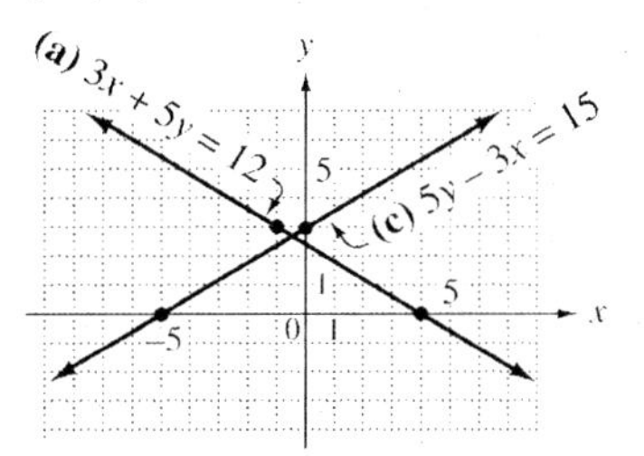

11. (a)

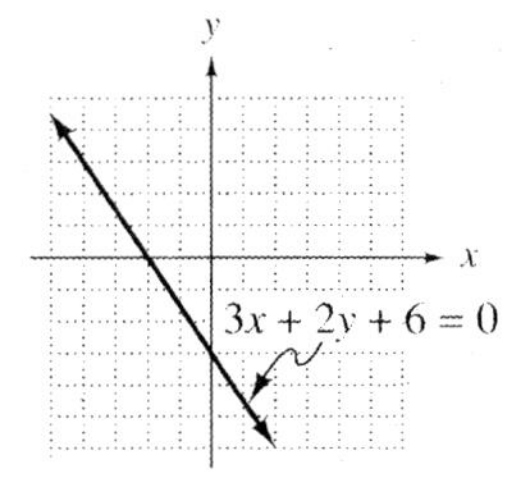

(b) Yes. The function is $y = -\dfrac{3}{2}x - 3$.

(d) Yes. The function is $y = \dfrac{5}{3}x - 5$.

13. (a) $y = x^2$

x	y
–3	9
–2	4
–1	1
0	0
1	1
2	4
3	9

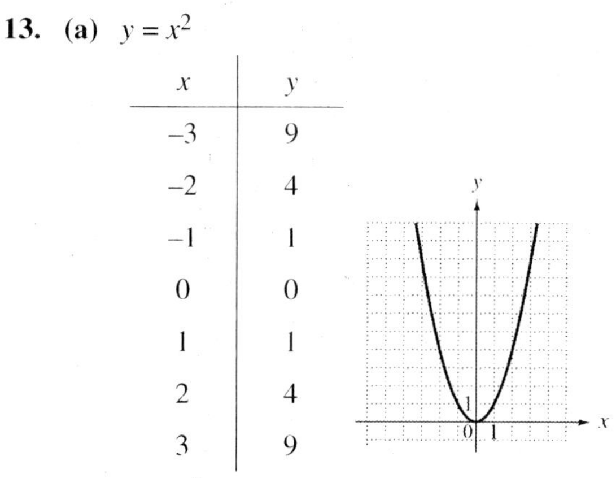

14. (a) $y = x^2 - 4x + 4$

x	y
–3	25
–2	16
–1	9
0	4
1	1
2	0
3	1

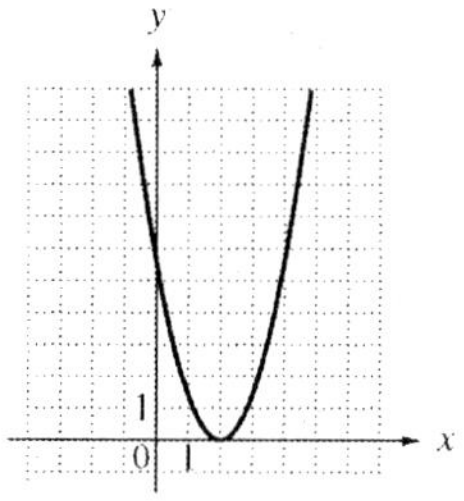

16. (a) $y = x^2 + 10x$

x	y
–8	–16
–7	–21
–6	–24
–5	–25
–4	–24
–3	–21
–2	–16
–1	–9
0	0
1	11
2	24

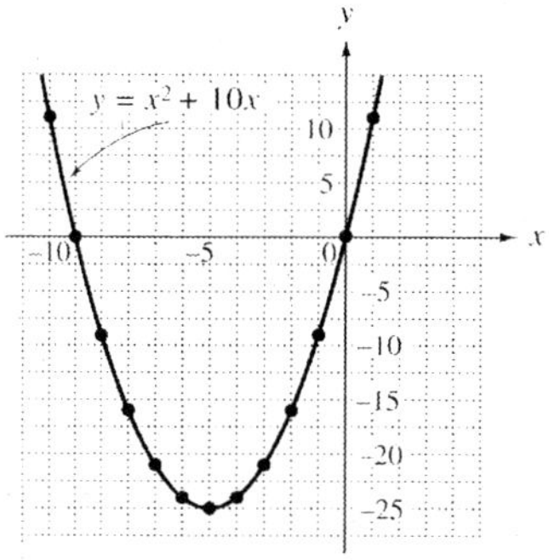

From the graph, the minimum value of $x^2 + 10x$ is –25. It occurs at $x = -5$.

19. (a) $x = 5$; the line is vertical and passes through the point $(5, 0)$.

(c) $y = \frac{1}{2}x + 1$; the slope is $m = \frac{2-1}{2-0} = \frac{1}{2}$, and the line crosses the y-axis at $y = 1$.

20. Note that B can be obtained by sliding A to the right 3 and up 5. Therefore, C can be obtained by sliding D to the right 3 and up 5. The coordinates of C are $(7 + 3, 0 + 5) = (10, 5)$. Hence $r = 10$ and $s = 5$.

26. (a) The run must be at least 12 times the 30 inch = 2.5 foot rise, so the minimum allowable run is $12 \times 2.5 = 30$ feet.

(b) The rise is $\frac{1800}{5280}$ miles in a run of 7 miles, so the slope is $\frac{1800}{5280} \div 7 = 0.05$, or about 5%.

28. $L = mw + b$
$10 = m \cdot 0 + b$, so $b = 10$.
$14 = m \cdot 2 + b = 2m + 10$, so $m = 2$.
The constants are $b = 10$ and $m = 2$.

29. $F = mC + b$
$32 = m \cdot 0 + b$, so $b = 32$.
$212 = m \cdot 100 + b = m \cdot 100 + 32$,
so $m = 1.8$ or $\frac{9}{5}$. The constants are $m = \frac{9}{5}$ and $b = 32$.

Problem Set 8.3

2. First, find the slope of $3x - 5y + 45 = 0$ by solving for y:

$$0 = 3x - 5y + 45$$
$$5y = 3x + 45$$
$$y = \frac{3}{5}x + 9$$

The slope is $\frac{3}{5}$.

(a) $7x + ky = 21$

$$ky = -7x + 21$$
$$y = \frac{-7}{k}x + \frac{21}{k}$$

The slope must be $\frac{3}{5}$.

$$\frac{-7}{k} = \frac{3}{5}$$
$$3k = -35$$
$$k = -\frac{35}{3}$$

(c) $y = kx + 5$

The slope must be $\frac{3}{5}$:

$$k = \frac{3}{5}$$

3. (a) $y = 2x - 3$

x	y
−3	−9
−2	−7
−1	−5
0	−3
1	−1
2	1
3	3

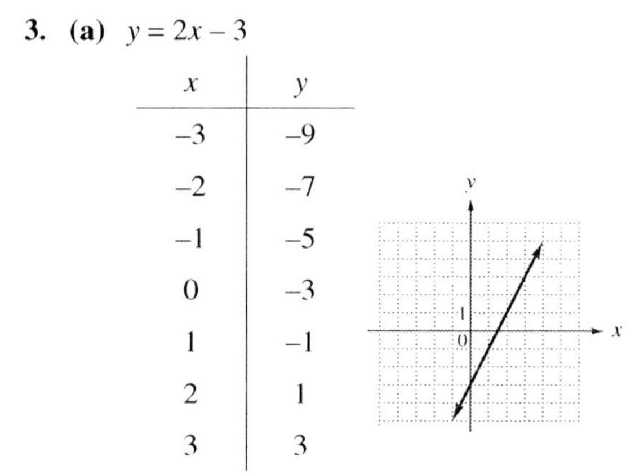

5. (a)

$$(RS)^2 = \left(\sqrt{(7-1)^2 + (10-2)^2}\right)^2$$
$$= \left(\sqrt{6^2 + 8^2}\right)^2 = \left(\sqrt{36+64}\right)^2$$
$$= \left(\sqrt{100}\right)^2 = 100$$

$$(RT)^2 = \left(\sqrt{(5-1)^2 + (-1-2)^2}\right)^2$$
$$= \left(\sqrt{4^2 + (-3)^2}\right)^2 = \left(\sqrt{16+9}\right)^2$$
$$= \left(\sqrt{25}\right)^2 = 25$$

$$(ST)^2 = \left(\sqrt{(5-7)^2 + (-1-10)^2}\right)^2$$
$$= \sqrt{(-2)^2 + (-11)^2} = \left(\sqrt{4+121}\right)^2$$
$$= \left(\sqrt{125}\right)^2 = 125$$

Since $(RS)^2 + (RT)^2 = 100 + 25 = 125 = (ST)^2$, by the Pythagorean theorem, $\triangle RST$ is a right triangle.

7. (a) $PB^2 = (x-1)^2 + (y-1)^2 = x^2 - 2x + 1 + y^2 - 2y + 1 = x^2 + y^2 - 2x - 2y + 2$

$PC^2 = (x-0)^2 + (y-1)^2 = x^2 + y^2 - 2y + 1 = x^2 + y^2 - 2y + 1$

$PD^2 = (x-0)^2 + (y-0)^2 = x^2 + y^2$

(b) $PA^2 + PC^2 = (x^2 + y^2 - 2x + 1) + (x^2 + y^2 - 2y + 1) = 2x^2 + 2y^2 - 2x - 2y + 2$

$PB^2 + PD^2 = (x^2 + y^2 - 2x - 2y + 2) + (x^2 + y^2) = 2x^2 + 2y^2 - 2x - 2y + 2$

These equations show that $PA^2 + PC^2 = PB^2 + PD^2$ for all x and y. That is, *all* points $P(x, y)$ in the plane satisfy the condition.

9. The slope of $\overline{RT}$ is $\frac{6-4}{7-1} = \frac{1}{3}$, so the slope of the altitude is –3. The altitude must pass through S (5, 0), so the equation of the altitude is $y - 0 = -3(x-5)$ or $y = -3x + 15$.

10. The slope of $\overline{AB}$ is $\frac{4-(-2)}{2-1} = 6$, so the slope of the perpendicular bisector is $-\frac{1}{6}$. The midpoint of $\overline{AB}$ is $\left(\frac{1+2}{2}, \frac{4+(-2)}{2}\right)$ or $\left(\frac{3}{2}, 1\right)$. The equation of the perpendicular bisector is $y - 1 = -\frac{1}{6}\left(x - \frac{3}{2}\right)$.

12. (a) By sketching the circle, it is clear that the equation of the tangent is $y = -3$.

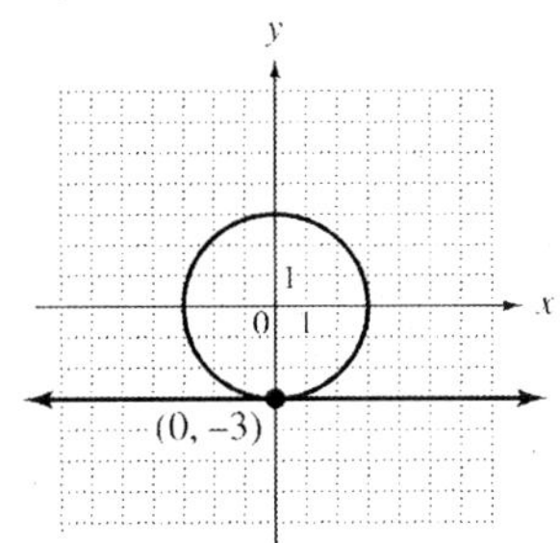

14. Responses will vary, but the conclusion is that l is parallel to n.

16. All are true except (d). (e) is true when the two circles are the same.

a.

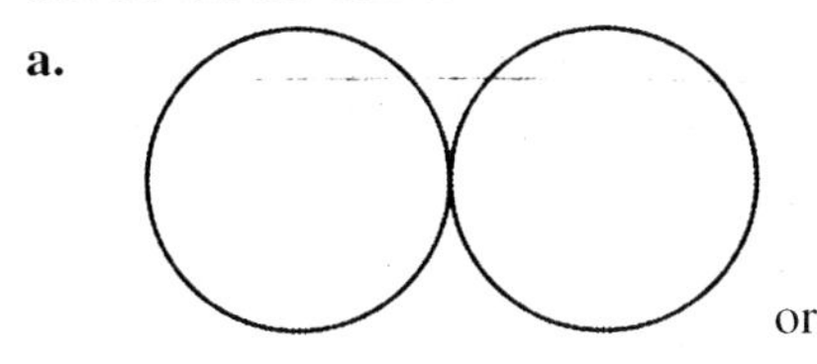

or

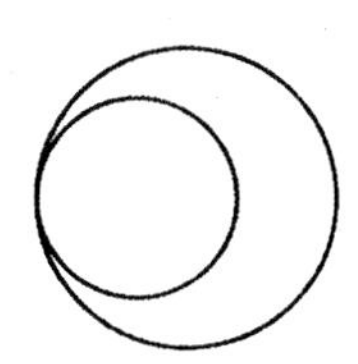

22. (a) By trial and error, these are 10 taxi segments. Notice that although there are infinitely many paths from A to B, the taxi segments correspond to the paths that accomplish the journey in 5 blocks—that is, no backtracking. Any such journey must include 3 eastward blocks and 2 northward blocks. Since these may be undertaken in any order, the number of paths may also be determined as the number of ways to arrange the symbols E, E, E, N, N. This is $C(5, 2) = 10$.

Chapter 8 Review Exercises

1. (a) $a + 5$ **(b)** $b < c$

(c) $c - b$ **(d)** $\frac{a+b+c}{3}$

2. (a) $a + 2 = 11$ **(b)** $b - 3 = \frac{1}{2}(c-1)$

(c) $a = \frac{b+c}{2}$ **(d)** $\frac{a+b+c}{3} = 10$

3. (a) Multiply the input by 3 and add 2; that is, $y = 3x + 2$

(b) Multiply the input by the value one larger than the input; that is, $y = x(x+1)$, or $y = x^2 + x$

4. (a) $f(3) = 2 \cdot 3 \cdot (3-3) = 0$
$f(0.5) = 2 \cdot 0.5 \cdot (0.5-3) = -2.5$
$f(-2) = 2 \cdot (-2) \cdot (-2-3) = 20$

(b) If $f(x) = 0$, then either $2x = 0$ or $x - 3 = 0$. Therefore, $x = 0$ or $x = 3$.

5. (a)

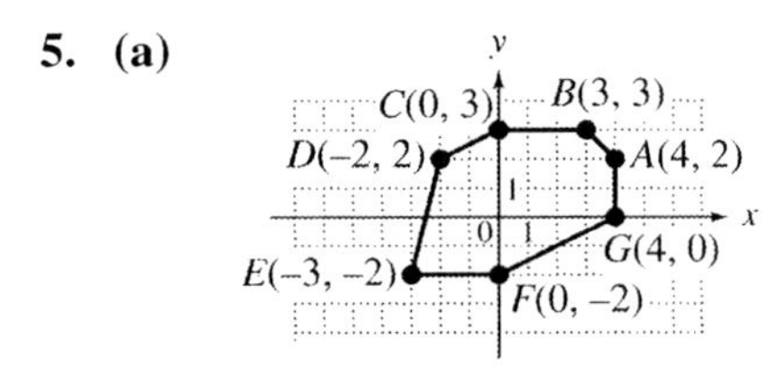

(b) B is in the first quadrant, C is on the positive y-axis, D is in the second quadrant, E is in the third quadrant, and F is on the negative y-axis.

(c) slope $\overline{BC} = \frac{0}{3} = 0$, slope $\overline{CD} = \frac{1}{2}$,

slope $\overline{DE} = \frac{4}{1} = 4$, slope $\overline{EF} = \frac{0}{3} = 0$,

slope $\overline{FG} = \frac{2}{4} = \frac{1}{2}$,

slope $\overline{GA}$ is undefined. Yes; $\overline{BC}$ is parallel to $\overline{EF}$, and $\overline{CD}$ is parallel to $\overline{FG}$.

6.

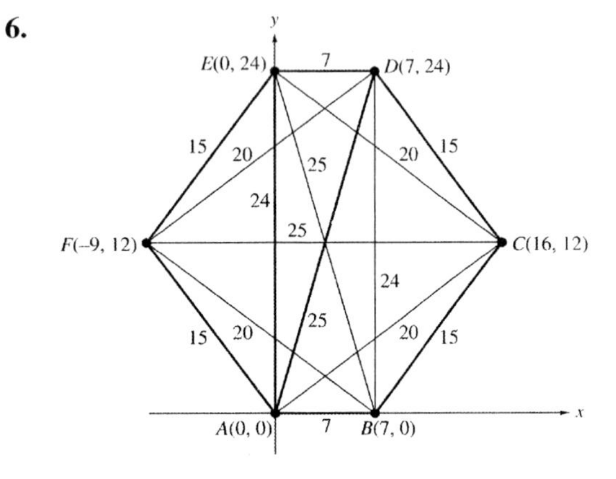

(a) $AB = 7$

$AC = \sqrt{16^2 + 12^2} = \sqrt{400} = 20$

$AD = \sqrt{7^2 + 24^2} = \sqrt{625} = 25$

$AE = 24$

$AF = \sqrt{(-9)^2 + 12^2} = \sqrt{225} = 15$

$CF = 16 - (-9) = 25$

(b) The lengths of all the sides and diagonals are positive integers.

7. (a) $y = 4 + 2(x - 3)$, by the point-slope form of the equation of a line. The equation can be rewritten in slope-intercept form as $y = 2x - 2$.

(b) $y = -1 + \frac{5-(-1)}{(-2)-6}(x-6)$, by the two-point form of the equation of a line. The equation can be rewritten in slope-intercept form as $y = -\frac{3}{4}x + \frac{7}{2}$.

(c) $y = 3x - 4$, using the slope-intercept form of the equation of a line.

8. First write $x + 7 = 2y$ in slope-intercept form. $x + 7 = 2y \Rightarrow y = \frac{1}{2}x + \frac{7}{2}$, so the slope of the perpendicular line is -2. The line passes through (7, 4), so the equation is $y - 4 = -2(x - 7)$ or $y = -2x + 18$.

9. The altitude through R is perpendicular to $\overline{ST}$. The slope of $\overline{ST}$ is $\frac{6-0}{7-5} = 3$, so the slope of the altitude is $-\frac{1}{3}$.

The altitude passes through $R(1, 4)$, so the equation of the altitude is $y - 4 = -\frac{1}{3}(x - 1)$.

10. $C_1 \cap C_1$ can have no points (they don't intersect), one point (they are tangent to each other), two points (they cross each other), or infinitely many points (the circles are concentric). The examples of specific sketches will vary.

Chapter 8 Test

1. (a)

x	$y = \left(\frac{1}{4}\right)^x$
0	1
1	0.25
2	0.0625
3	0.015625

(*continued on next page*)

(continued)

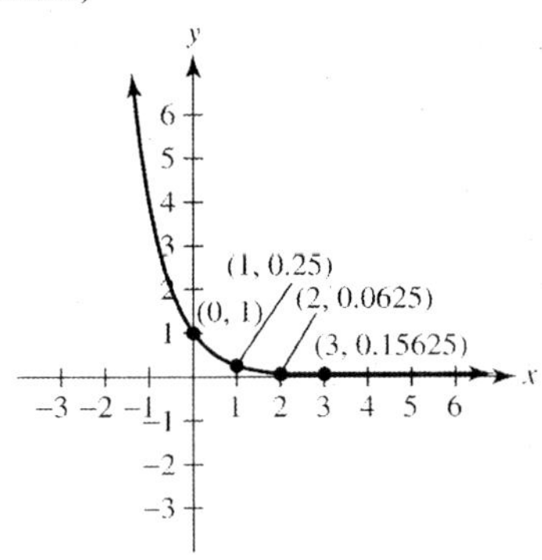

(b) Problem 25c of section 8.2 is a word problem whose math is identical with this one once you realize that $\$\frac{1}{4} = 25$ cents.

2. (a)

(b) The minimum function value is $y = -2$, the value of the function at $x = -1$.

3. (a) Linear, equivalent to $y = 3x - 4$, $m = 3$, $b = -4$

(b) Nonlinear, since x^2 appears in the equation. **(c)** Linear, equivalent to $y = x + 5$, $m = 1$, $b = 5$

4. (a) Let t denote the charged time, in total minutes for the month. Then $5 + 0.12t$ is the phone bill (in dollars) for that month.

(b) Let n = number of checks. Then the cost of checking is $8 + 0.15n$.

(c) Let a and c denote the number of adults and children, respectively. Then the admission fee is $4a + 2.50c$, in dollars.

5. (a) $y = 5$

(b) $y = -x - 2$, since slope is -1 and the line intersects the y-axis at -2.

(c) The line has slope $\frac{1}{3}$ and passes through $P(2, 2)$ so its equation in point-slope form is $y - 2 = \frac{1}{3}(x - 2)$.

6. (a) A hike up Mount Shasta and return to base camp involves increasing elevation and then decreasing elevation. Graph G3 matches this situation.

(b) A round-trip hike into the Grand Canyon involves decreasing elevation and then increasing elevation. Graph G1 matches this situation.

(c) A hike along the wilderness beach in Olympic National Park involves remaining at a fairly constant elevation. Graph G2 matches this situation.

7. From the drawing, it is clear to see that the equation of the tangent line is $x = 2$.

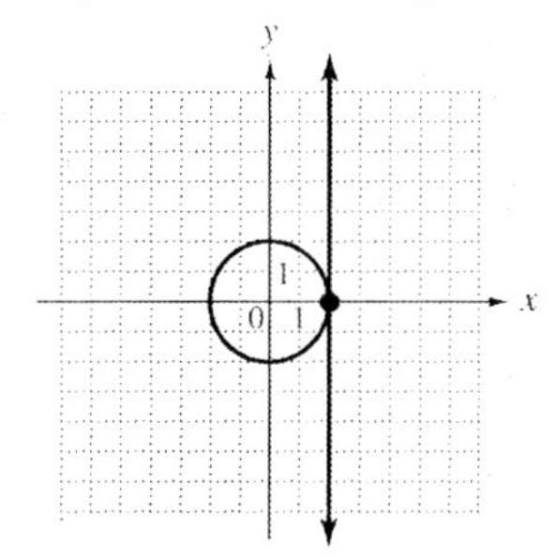

8. (a) $a = 0$, $b = 5$, since each side either has a horizontal run of 3 and a rise of 4 or a horizontal run of 4 and fall of 3.

(b) $\sqrt{4^2 + 3^2} = \sqrt{25} = 5$

9. $f(x) = x^2 - 4x + 5$

$f(1) = 1^2 - 4(1) + 5 = 2$
$f(2) = 2^2 - 4(2) + 5 = 1$
$f(3) = 3^2 - 4(3) + 5 = 2$
$f(4) = 4^2 - 4(4) + 5 = 5$
$f(5) = 5^2 - 4(5) + 5 = 10$

Therefore, the range is $\{1, 2, 5, 10\}$.

Chapter 9 Geometric Figures

Problem Set 9.1

1. **(a)** $\overrightarrow{AB}$ or $\overrightarrow{BA}$

2. **(a)** E U

3. **(a)** $\angle BAC$ will be a right angle if C is any of the circled points.

A B

4. $m(\angle AXB) = m(\angle AXE) - m(\angle BXE)$
$= 180° - 140° = 40°$
$m(\angle CXD) = m(\angle BXD) - m(\angle BXC)$
$= 90° - 45° = 45°$
$m(\angle DXE) = m(\angle BXE) - m(\angle BXD)$
$= 140° - 90° = 50°$

6. The three lines are concurrent. A possible drawing is shown.

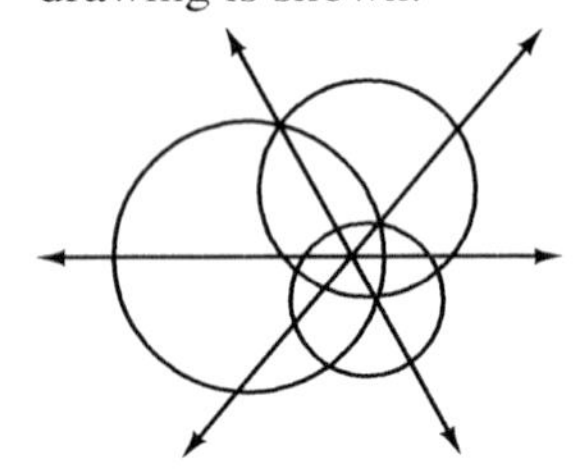

9. **(a)** For the figure shown in the textbook, $m(\angle A) \doteq 111°$, $m(\angle B) \doteq 88°$, $m(\angle C) \doteq 69°$, and $m(\angle D) \doteq 92°$. In general, opposite angles are supplementary: $m(\angle A) + m(\angle C) = 180°$ and $m(\angle B) + m(\angle D) = 180°$.

10. Ten times. A zero angle is formed at approximately 1:05, 2:11, 3:16, 4:22; 5:27, 6:33, 7:38, 8:44, 9:49, and 10:55.

11. **(a)** One complete turn, or 360°

(d) The hour hand moves 2 hours in 120 minutes. That is $\frac{2}{12}$ or $\frac{1}{6}$ of a complete rotation. $\frac{1}{6}(360°) = 60°$

12. **(a)** The minute hand is on the 12 and the hour hand is on the 4, so the hands form an angle of 120°.

(c) The minute hand is on the 6 and the hour hand is halfway between the 4 and the 5. The angle between two consecutive numbers is $\frac{1}{12}$ of a revolution, or 30°. So the angle is $(1.5)(30°) = 45°$.

14. Draw a horizontal ray $\overrightarrow{PQ}$ at P, in the opposite direction of $\overrightarrow{AB}$ and $\overrightarrow{CD}$ The opposite interior angles theorem then gives $m(\angle APQ) = 120°$, $m(\angle CPQ) = 150°$. Thus $m(\angle P) = 360° - 120° - 150° = 90°$.

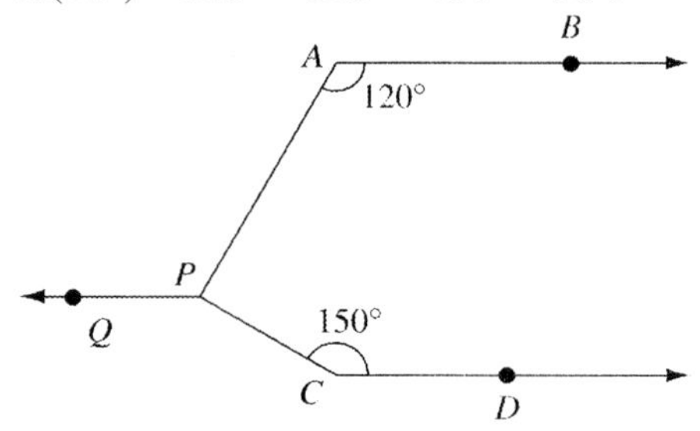

15. **(a)** The measures of the interior angles of a triangle add up to 180°. Therefore, $m(\angle 1) = 180° - 70° - 70° = 40°$.

(c) The interior angles of a triangle add up to 180°, and a right angle has measure 90°, so $m(\angle 3) + 90° + 41° = 180°$. Therefore, $m(\angle 3) = 49°$.

16. **(a)** $x + x + 30° = 180°$ so $x = 75°$. The interior angles measure 75°, 75°, and 30°.

17. **(a)** No, because an obtuse angle has measure greater than 90°, and adding two such measures would exceed 180°, which is the sum of all three interior angle measures for any triangle.

(b) No, because a right angle has measure of 90°, so adding together the measures of two right angles and a third angle would exceed 180°.

(c) The sum of the three angle measures must be 180°, so each angle must be exactly 60°.

28. **(a)** Zero intersection points, if the five lines are parallel to each other.

29. **(a)** 6 lines

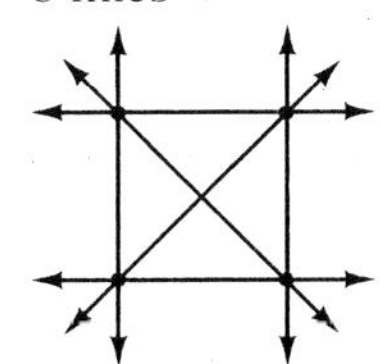

31. The pencil turns through each interior angle of the triangle. Notice that each turn is in the same direction (counterclockwise). Since the pencil faces the opposite direction when it returns to the starting side, it has turned a total of 180°. This demonstrates that the sum of measures of the interior angles of a triangle is 180°.

35. The earliest clocks were sundials used in the northern hemisphere. The shadow cast by the gnomon follows an arc in the direction we call clockwise.

38. **(a)** A full revolution takes 24 hours. In 1 hour, the earth turns $\frac{1}{24}$ of a revolution, that is, $\frac{1}{24}\cdot(360°) = 15°$.

39. Extend the line $\overleftrightarrow{PS_2}$ downward to form a right triangle as shown.
Then $m(\angle 2) = 180° - 90° - 37° = 53°$, and $m(\angle 1) = 180° - 90° - 53° = 37°$. In general, the angle of latitude ($\angle 1$) has the same measure as the angle of elevation of Polaris—37°, in this case.

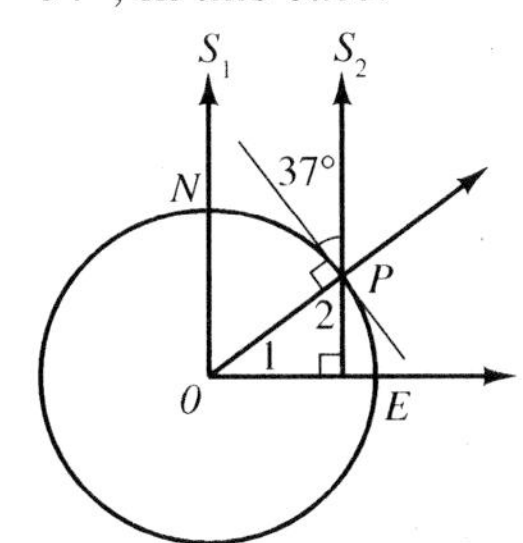

40. **(a)** $58°\ 36'\ 45''$

$$= 58° + \left(\frac{36}{60}\right)^{\circ} + \left(\frac{45}{3600}\right)^{\circ} = 58.6125°$$

(c) $71.32° = 71° + (0.32)(60)' = 71° + 19.2'$
$= 71° + 19' + (0.2)(60)''$
$= 71°\ 19'\ 12''$

42. **(a)** The lines $\overleftrightarrow{BB'}$ and $\overleftrightarrow{DD'}$ intersect at a right angle at P.

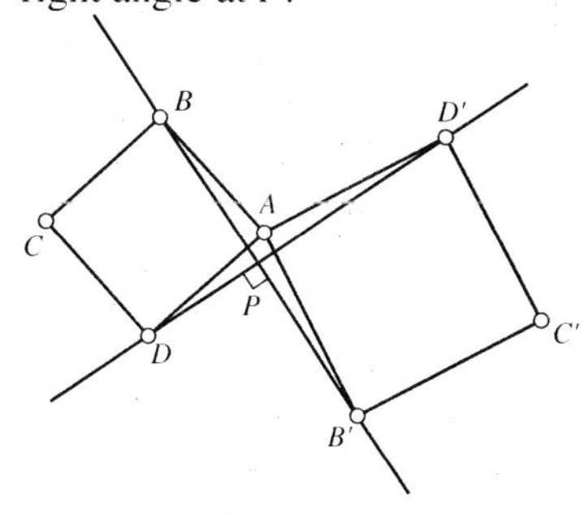

Problem Set 9.2

1.

	(a)	(b)	(c)	(d)	(e)	(f)	(g)	(h)	(i)	(j)	(k)	(l)
Simple Curve		✓			✓	✓	✓	✓	✓	✓	✓	
Closed Curve		✓	✓	✓			✓	✓		✓	✓	✓
Polygonal Curve	✓	✓			✓		✓	✓	✓		✓	✓
Polygon		✓					✓	✓			✓	

2. **(a)** A nonsimple closed four-sided polygonal curve is a figure that has 4 line segments which when traced without lifting your pencil, at least one point will be touched more than once and the trace will end at the same point at which it started.

(c) An equiangular quadrilateral has 4 sides and all angles have the same measure. Any rectangle is acceptable.

4. **(a)** Convex

(b) Concave, because there is a segment whose endpoints are in the figure that contains points not in the figure.

5. **(a)**

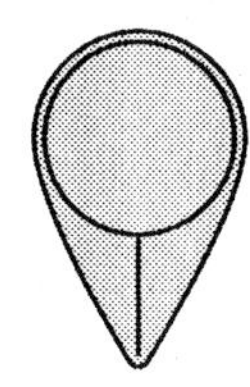

(c) 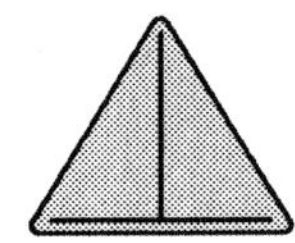

The shaded region is always convex.

6. **(a)** 6 regions: the interior of the circle, the exterior of the square, and the four corner regions.

(c) 8 regions: the interior of the hexagon, the six triangular regions, and the exterior region.

7. The polygon has 6 sides, so the sum of the interior angle measures is $(6-2)(180°) = 720°$. Therefore,

$$2x+5x+5x+5x+5x+2x = 720°$$
$$24x = 720°$$
$$x = 30°.$$

Since $2x = 60°$ and $5x = 150°$, the angles measure 60°, 150°, 150°, 150°, 150°, and 60°.

10. **(a)** The polygon has 5 sides, so the sum of the interior angle measures is $(5-2)(180°) = 540°$.

(c) The polygon has 6 sides, so the sum of the interior angle measures is $(6-2)(180°) = 720°$.

11. **(a)** $(n-2)(180°) = 180°$

$$n-2 = 1$$
$$n = 3$$

The lattice polygon can be any triangle.

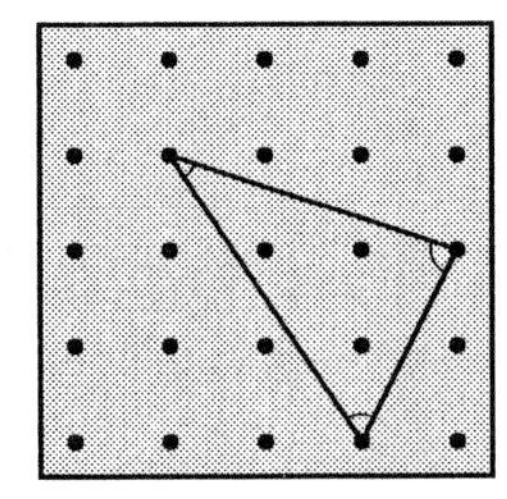

(c) $(n-2)(180°) = 1440°$

$$n-2 = 8 \Rightarrow n = 10$$

The lattice polygon can be any decagon.

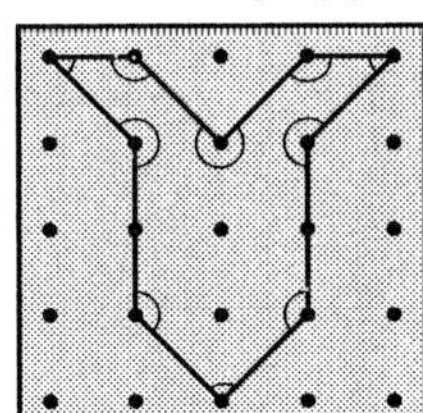

14. **(a)** $1 \times 360° = 360°$

(c) $360° - 360° - 180° + 180° = 0°$

16. **(a)** An equilateral triangle has all three sides of equal length, so the triangles are DAF and DBC.

18. In each case, the measure of the interior angle is given by $\dfrac{(n-2)(180°)}{n}$, and the measures of the exterior angle and central angle are each given by $\dfrac{360°}{n}$ (or 180° minus the exterior measure).

	n	Interior Angle	Exterior Angle	Central Angle
(a)	5	108°	72°	72°
(c)	7	$128\frac{4}{7}°$	$51\frac{3}{7}°$	$51\frac{3}{7}°$

19. **(a)** $\dfrac{360°}{n} = 15°$ so $n = 24$.

31. **(a)** The boat can drift to any position inside the circle centered at A, where the radius of the circle is the length of the anchor rope.

32. **(a)** The right angle is located at A or B. Therefore, C could be any point (other than A or B) on either of the two lines drawn through A and through B which are perpendicular to $\overline{AB}$.

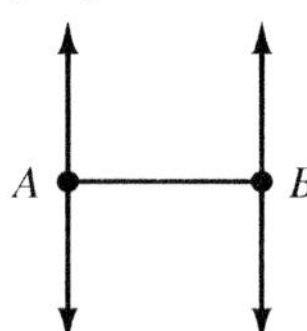

(b) The right angle is located at C. Therefore, C could be any point (other than A or B) on the circle with $\overline{AB}$ as diameter.

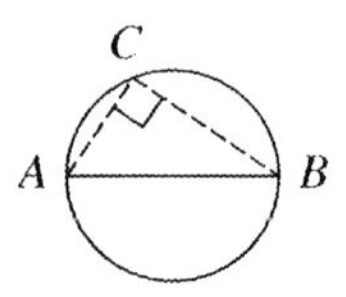

33. At a vertex, the interior angle and the conjugate angle add up to 360°. For an n-gon, the sum of all interior and all conjugate angles is $n \cdot 360°$. All the interior angles add up to $(n-2) \cdot 180°$, so all the conjugate angles add up to $360°n - (n-2)\ 180°$

$= 360°n - 180°n + 360°$
$= 180°n + 360° = (n + 2)\cdot 180°$.

35. Such a point S allows for n triangles to be formed, all with the vertex S. The sum of all the interior angles of these n triangles is $n \cdot 180°$ which is equal to the sum of all the interior angles of the n-gon plus 360° for the angles that surround the point S. Thus the sum of the interior angles of the n-gon is $n \cdot 180° - 360° = (n - 2) \cdot 180°$.

37. **(a)** The total turn made when tracing the star is $7 \times 360°$, so the turn made at each point of the star is $7 \times \frac{360°}{16} = 157.5°$.
Therefore, the angle at each point measures $180° - 157.5° = 22.5°$.

38. Each new circle creates a new region each time it intersects a previously drawn circle. Since the new circle intersects each of the old circles in at most two points, this creates the following pattern:

Number of	
Circles	**Regions**
1	2
2	$2 + 2 \cdot 1 = 4$
3	$4 + 2 \cdot 2 = 8$
4	$8 + 2 \cdot 3 = 14$
5	$14 + 2 \cdot 4 = 22$
6	$22 + 2 \cdot 5 = 32$
7	$32 + 2 \cdot 6 = 44$
8	$44 + 2 \cdot 7 = 58$
9	$58 + 2 \cdot 8 = 74$
10	$74 + 2 \cdot 9 = 92$

41. **(a)** The sum of the interior angles is 360°.
$360° = m(\angle P) + m(\angle Q) + m(\angle R) + m(\angle S)$
We also know $m(\angle P) = m(\angle R)$ and $m(\angle Q) = m(\angle S)$, so
$360° = m(\angle P) + m(\angle Q) + m(\angle P) + m(\angle Q) = 2 \cdot (m(\angle P) + m(\angle Q))$.
Thus, $180° = m(\angle P) + m(\angle Q)$.

(b) Draw $PQRS$ and extend $\overline{PQ}$ to form $\angle q$, as shown. $m(\angle Q) + m(\angle q) = 180°$ and $m(\angle P) + m(\angle Q) = 180°$ so $m(\angle q) = m(\angle P)$. $\angle q$ and $\angle P$ are corresponding angles, so segments $\overline{PS}$ and $\overline{QR}$ are parallel. $m(\angle q) = m(\angle P)$ and $m(\angle P) = m(\angle R)$ so $m(\angle q) = m(\angle R)$. $\angle q$ and $\angle R$ are alternate interior angles, so their congruence gives $\overline{PQ}$ parallel to $\overline{SR}$. Hence the figure is a parallelogram.

42. **(a)** Only even numbers of intersection points are possible.

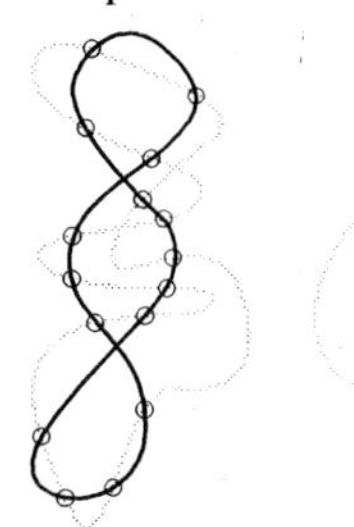

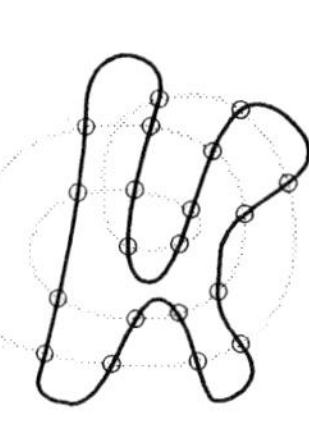

(b) Suppose the black curve has been drawn. It will be seen that it is easy to shade some of the regions in such a way that exactly one of any two regions sharing a common boundary is shaded and the other region is left unshaded. Now imagine adding the red closed curve. Each time the red curve crosses the black curve, it alternates between shaded and unshaded regions. Since the red curve is closed, the red curve closes upon itself in a region of the same type as its starting point. Thus, the number of color alternations must be even. That is, the number in intersection points of the red and black curves is necessarily even.

43. The number of regions interior to both simple curves is always equal to the number of regions exterior to both curves. The number of regions inside the red and outside the black curve is always equal to the number of regions inside the black and outside the red curve. There is no connection between the two pairs of equal numbers.

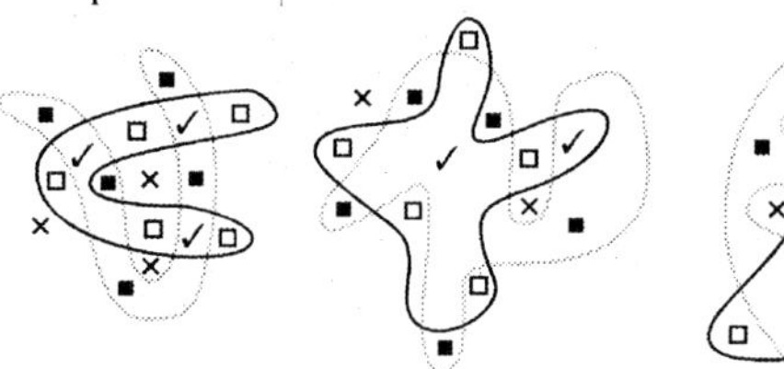

46. **(a)** *SQRE* is a square.

(b) No. As in the example shown here, *SQRE* will still be a square.

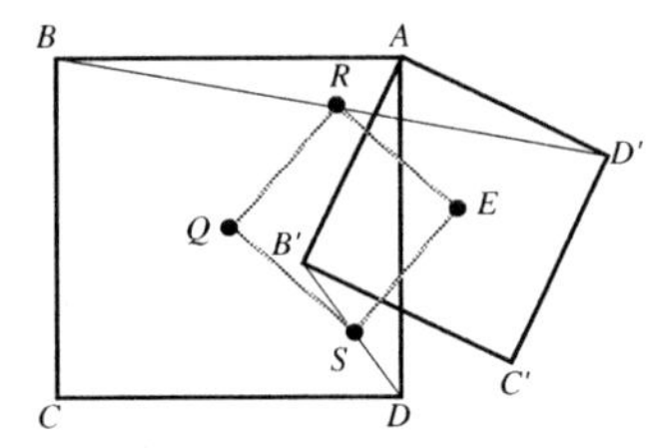

Problem Set 9.3

1. **(a)** Polyhedron. It is a simple closed surface formed from planar polygonal regions.

(c) Polyhedron. Note that a polyhedron need not be convex.

(e) Not a polyhedron. It is not a simple closed surface because it has two interior regions of space.

2. **(a)** Pentagonal prism (and possibly a right pentagonal prism)

(c) Oblique circular cone

(e) Right rectangular prism (assuming all angles are right angles)

3. **(a)** 4. There are 4 faces, and each is in a different plane.

(c) A, B, C, D

5. **(a)** Note that the tetrahedron includes the lower back left corner.

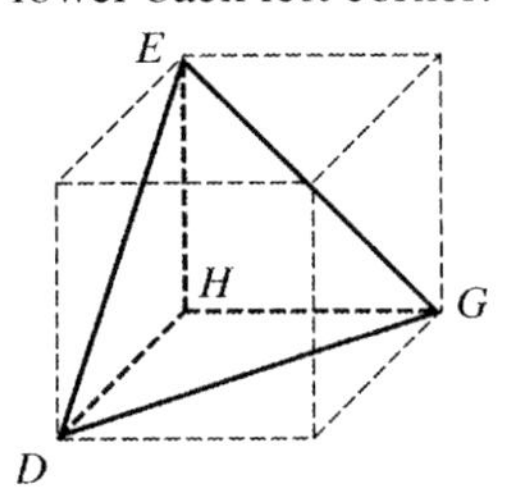

6. This problem may be solved by mental visualization or by actually constructing a model.

(a)

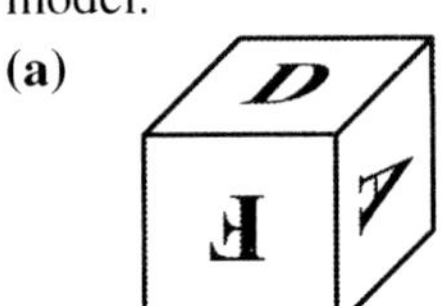

7. **(a)** 45°, since the dihedral angle between the adjacent sides of the cube is 90° and another pyramid would fit into the gap between the original pyramid and a vertical side of the cube. This can also be determined by visualizing a cross section half-way into the cube and parallel to the cube's front face. The pyramid forms a right isosceles triangle in this cross section.

(b) Filling the cube with six such pyramids, one sees that three pyramids sit around an edge formed by the lateral sides of a pyramid. Thus three copies of the dihedral angle give a full revolution of 360° around this edge, so the dihedral angle measures 120°.

14. **(a)** $F = 10$, $V = 7$, $E = 15$, so $V + F = E + 2$ since $7 + 10 = 15 + 2$

15. **(a)** Create the net by "unfolding" the figure.

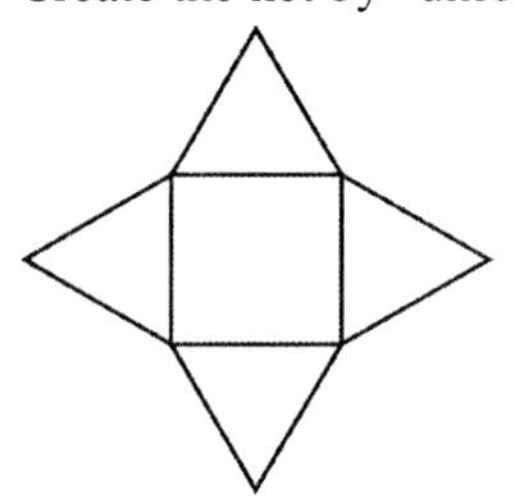

16. **(a)** $V = 10$, $F = 12$, $E = 20$
$V + F = 22 = E + 2$, so Euler's formula holds.

25. **(a)** Suppose the faces of a polyhedron consist of a p-gon, a q-gon, an r-gon, and so on. Since each edge of the polyhedron borders two faces, the sum $p + q + r + \cdots$ is twice the number of edges. That is, $p + q + r + \cdots = 2E$. Since there are F faces and $p, q, r, \ldots$ are all 3 or greater, we have $2E \geq 3 + 3 + \cdots = 3F$.

(b) Each of the V vertices of a polyhedron is the endpoint of 3 or more edges that meet at the vertex. Thus, $3V$ is less than or equal to the total number of ends of the edges. But each of the E edges has 2 ends, so there are $2E$ ends of edges. We see that $3V \leq 2E$.

(c) Adding $3V \le 2E$ and $3F \le 2E$ shows that $3V + 3F \le 4E$. But $V + F = E + 2$ (Euler's formula), so $3V + 3F = 3E + 6$. Comparing this to the inequality, we see that $3E + 6 \le 4E$. Subtracting $3E$ from both sides shows that $6 \le E$.

(d) Suppose $E = 7$. Since $3F \le 2E = 14$, we see that F is no larger than 4 ($F \ge 5$ would give $3F \ge 15$). Similarly, $3V \le 2E = 14$ means that $V \le 4$. Since both $V \le 4$ and $F \le 4$, then $V + F \le 8$.
But $V + F = E + 2$ (Euler), and $E = 7$, so $V + F = 9$. This contradicts $V + F \le 8$, so our assumption $E = 7$ is not possible.

(e) A pyramid with a base of 3, 4, 5, ..., n, ... sides has 6, 8, 10, ..., $2n$, ... edges, respectively. Slicing off a tiny corner at one vertex somewhere on the base of the pyramid adds three new edges, giving us polyhedra with 9, 11, 13, ..., $2n + 3$, ... edges. All together, the pyramids and pyramids with a truncated base corner give us polyhedra with 6, 8, 9, 10, 11, ... edges.

27. (a) The pyramid must have a base and five other sides, so the base is a pentagon.

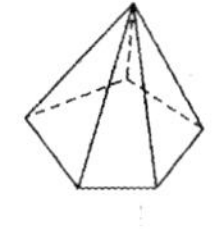

(b) The double pyramid must have three faces above and three faces below the center polygon, so the center polygon is a triangle.

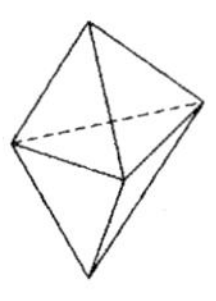

28. (a)

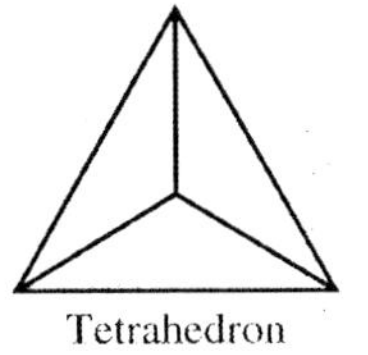

Tetrahedron

(b)

Octahedron

(c)

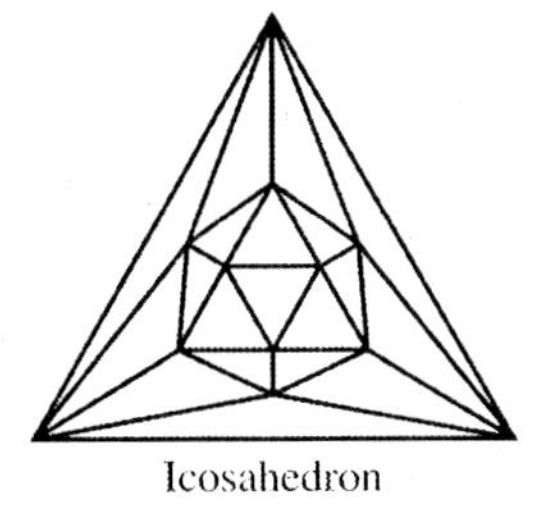

Icosahedron

32. (a) When folded, points A, C, E, and G all correspond to the apex of the pyramid. Thus, for example, $\overline{AB}$ and $\overline{BC}$ represent the same edge from the pyramid and so $AB = BC$. Likewise, $CD = DE$, $EF = FG$, and $GH = HA$. To see why $\overline{AP}$ is perpendicular to $\overline{BH}$, let R be the plane that is perpendicular to $\overline{BH}$ and contains A, and consider what happens to point A as $\triangle ABH$ is folded upward to form the pyramid. As the figure is folded along $\overline{BH}$, A moves along a circle in plane R. This circle contains the apex of the pyramid and it also contains the point that would be the apex if the pyramid were turned upside down. Since P is on the line segment that connects these points (indeed, it is the midpoint), we conclude that P is in plane R. Therefore, in the net, $\overline{AP}$ is in plane R. Since R is perpendicular to $\overline{BH}$, we conclude that $\overline{AP}$ is perpendicular to $\overline{BH}$. Likewise, $\overline{CP}$, $\overline{EP}$, and $\overline{GP}$ are perpendicular to $\overline{BD}$, $\overline{DF}$, and $\overline{FH}$, respectively.

33. The axis of all the hinges on a door must be along the intersection of the planes of the wall and the plane of the opened door. Since planes meet in a line, the axes of the hinges must be along a single line.

Problem Set 9.4

1. The networks I and III each represent the information correctly, since there is an edge between vertices when, and only when, the individuals have met. Network II does not represent the information given; for example, it incorrectly indicates that Coralee has met Bianca.

2. (a) Yes, it is traversable because it has no odd vertices. *AHGFEDCBHFDBA* is one Euler path.

(c) No, it is not traversable because it has more than two odd vertices (A, B, C, D, E, and F).

(e) Yes, it is traversable because it has only two odd vertices. The odd vertices, C and D, must be the endpoints of any Euler path. One Euler path is *CEAEFBFDBACD*, as shown below. (Note that there are two distinct edges joining A and E, and two distinct edges joining B and F.)

4. (a)

Figure	D = total degree	E = number of edges
(a)	$2+4+2+4+2+4+2+4=24$	12
(b)	$2+3+4+2+2+3+4+2=22$	11
(c)	$3+3+3+3+3+3+6=24$	12
(d)	$2+4+2+4+2+4+4+4+4=30$	15
(e)	$4+4+3+3+4+4=22$	11
(f)	$4+3+4+5+2+5+4+3=30$	15

5. (a) Yes, since exactly two vertices are odd and the network is connected.

(b) $D=2+2+4+8+3+6+6+1=32$. Since $D=2E$, there are 16 edges.

7. (a) Draw a vertex for each land mass and an edge for each bridge.

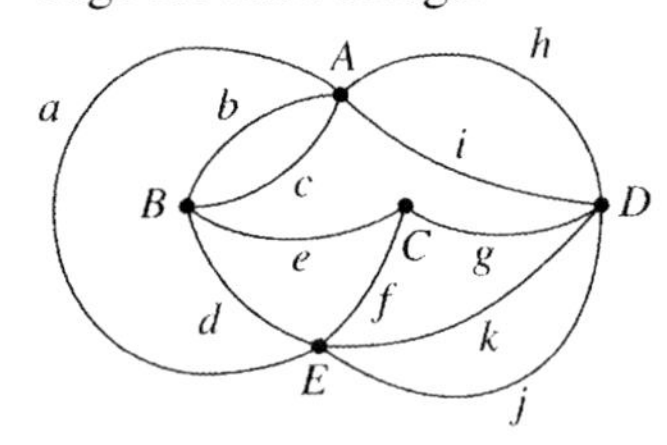

8. (a)

9. (a) There are 6 vertices, 7 regions (including the exterior region), and 11 edges. Therefore, $V=6$, $R=7$, and $E=11$.

12. (a) The triangle in the network shows that at least three different times must be scheduled, say science at 2:00, math at 3:00, and basketball at 4:00. Then band can meet at 2:00 and history can meet at 3:00 with no student having a conflict.

(b) Answers will vary.

13. (a) The two paths can be combined to give a new path that goes from one of the two vertices to the other and back again. This would be a closed path with distinct edges.

14. Since all trees are connected, a traversable tree is one that has zero or two odd vertices. But every tree with two or more vertices has at least two vertices of degree 1. So a nontrivial traversable tree must have exactly two vertices of degree 1 (and no other odd vertices). The only way this is possible is if the tree consists of a number of vertices connected in a row, or a single vertex with no edges.

17. Consider a network with a vertex for each person and an edge corresponding to each handshake. The number of edges joining any two people would be the number of times they shook hands. By the result stated in Problem 16, the number of odd vertices in this network is even. Therefore, there are an even number of people who have shaken hands an odd number of times.

18. (a) Several ways are possible. The endpoints must be the odd vertices in the network, that is, A, B, E, H, I, and J. One possibility is AB, $JFBGJIDAEI$, $EFGHCKH$.

(b) Temporarily add $m-1$ edges between $m-1$ distinct pairs of the odd vertices. The new edges make these into even vertices. Only two odd vertices remain, so the network is traversable. Following one of the added edges is equivalent to lifting the pencil. The $m-1$ lifts mean there are m strokes.

21. (a)

23. In the original configuration, current can flow from any pin to any other pin. This just means that the network is connected. Thus, any 7 of the original 12 wires that produce a connected network will produce the desired result. An example is HA, AB, BG, GF, FD, DC, DE.

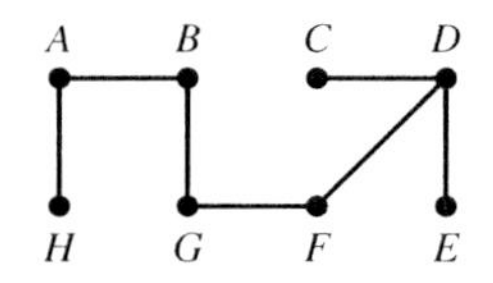

Chapter 9 Review Exercises

1. **(a)** $\overrightarrow{AC}$

(b) $\overline{BD}$

(c) AD

(d) $\angle ABC$ or $\angle CBA$ (or simply $\angle B$). Note that all other angles shown, when considered as a union of two rays, contain D.

(e) $m(\angle BCD)$, $m(\angle DCB)$, or $m(\angle C)$, or 90°

(f) $\overrightarrow{DC}$

2. **(a)** $\angle BAD$, since its measure appears to be less than 90°.

(b) $\angle BCD$, since its measure is marked as 90°.

(c) $\angle ABC$, $\angle ADC$, since their measures appear to be greater than 90°.

3. **(a)** $180° - 37° = 143°$

(b) $90° - 37° = 53°$

4. Since $\angle p$ and a 125° angle are adjacent supplementary angles, $p = 180° - 125° = 55°$. Since l and m are parallel, the small triangle is a right triangle, and so $q = 180° - 90° - p = 35°$. Since r corresponds to p along a transversal of parallel lines, $r = p = 55°$. Since $\angle r$ and $\angle s$ are adjacent supplementary angles, $s = 180° - 55° = 125°$. In summary, $p = 55°$, $q = 35°$, $r = 55°$, and $s = 125°$.

5. Since $\angle x$ and a 135° angle are adjacent supplementary angles, $x = 180° - 135° = 45°$. Since the sum of the measures of the angles in a triangle is 180°, $y = 180° - 102° - x = 180° - 102° - 45° = 33°$. Since $\angle y$ and $\angle z$ are adjacent supplementary angles, $z = 180° - y = 180° - 33° = 147°$. In summary, $x = 45°$, $y = 33°$, and $z = 147°$.

6. **(a)** (iv) **(b)** (i)

(c) (vi) **(d)** (v)

(e) (ii) **(f)** (iii)

7. **(a)** No, because obtuse angles have measure greater than 90° and the sum of the three interior angles of a triangle is 180°. Two obtuse angles would produce a sum greater than 180°.

(b) Yes, try angles of 100°, 100°, 100°, and 60°.

(c) No, because acute angles have measure less than 90°, so if all the interior angles are acute the sum would be less than $4 \cdot 90° = 360°$. But the sum of the interior angles of a quadrilateral must be 360°.

8. The figure is a hexagon, so the interior angles add up to $(6 - 2)(180°) = 720°$. Therefore, $3x + 3x + 3x + x + 5x + x = 720°$, which means $16x = 720°$, or $x = 45°$. The angles are 135°, 135°, 135°, 45°, 225°, and 45°.

9. Use the total-turn theorem. He turns through 360°.

10. **(a)** 6, one for each face of the cube.

(b) $\overline{CD}$, $\overline{EF}$, $\overline{GH}$

(c) $\overline{CG}$, $\overline{DH}$, $\overline{EH}$, $\overline{FG}$

(d) 45°, the same as $m(\angle DAH)$, since $\overrightarrow{AD}$ and $\overrightarrow{AH}$ each lie in one of the two planes and are perpendicular to $\overleftrightarrow{AB}$, the line of intersection of the two planes.

11. Square right prism; triangular pyramid (or tetrahedron); oblique circular cylinder; sphere; hexagonal right prism

12. **(a)**

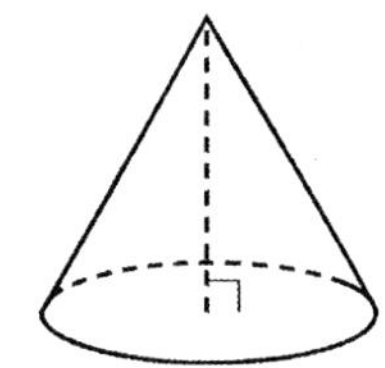

(b)

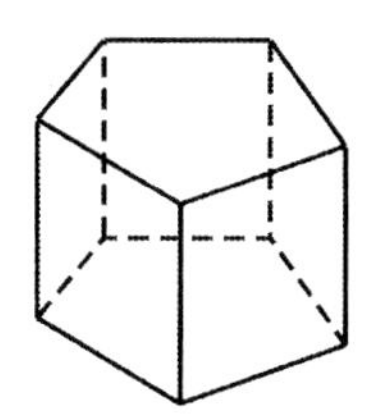

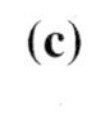

(c)

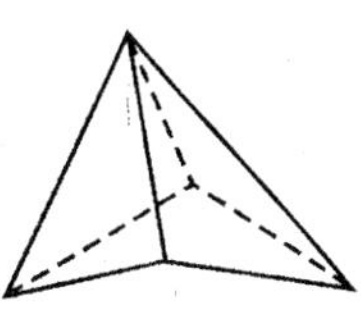

13. **(a)**

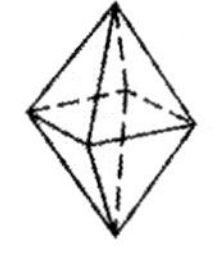

(b) There are 6 vertices, 8 faces, and 12 edges, so $V = 6$, $F = 8$, and $E = 12$. Thus $V + F = 14$ and $E + 2 = 14$. Since these quantities are equal, Euler's formula holds.

14. Note that $F = 14$ and $E = 24$. Then Euler's formula, $V + F = E + 2$, means that $V + 14 = 24 + 2$, so $V = 12$. There are 12 vertices.

15. **(a)** It has four odd vertices (A, B, C, and E).

(b) An edge should be added between any two of the vertices A, B, C, and E. Then the Euler path must begin and end at the remaining two of these vertices. If an edge is added that joins A and C, then one Euler path is $BADACFCBFEDBE$, as shown below. (Note that there are two edges joining A and D, and two edges joining C and F.) Many other Euler paths are possible.

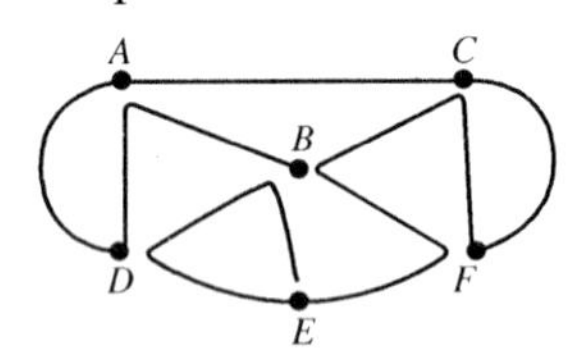

16. Construct a network with vertices A, B, C, D, and E and edges corresponding to bridges. Since just two vertices, A and D, have odd degree, there is an Euler path. The Euler path corresponds to a walking path which crosses each bridge exactly once.

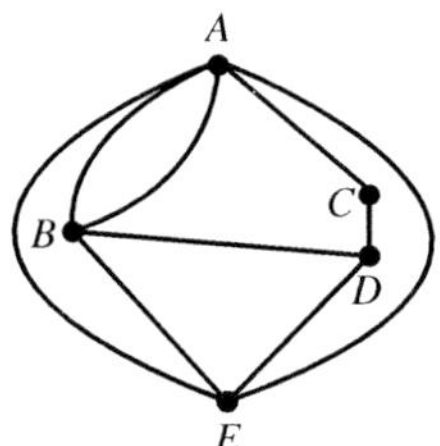

17. There are 11 vertices, 7 regions (including the exterior region), and 16 edges. Therefore, $V = 11$, $R = 7$, and $E = 16$. Then $V + R = 18$ and $E + 2 = 18$. Since these quantities are equal, it is true that $V + R = E + 2$, so Euler's formula holds.

Chapter 9 Test

1. The average interior angle measure for an n-gon is $\frac{(n-2)(180°)}{n}$. We want this value to be 144°, so $\frac{180(n-2)}{n} = 144$. This gives $180n - 360 = 144n$, which implies $n = 10$. The polygon has 10 sides.

2. Many examples are possible.

(a) A simple closed curve:

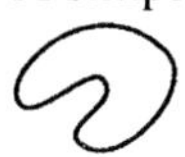

(b) A convex heptagon:

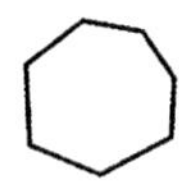

(c) A nonclosed simple polygon curve:

(d) A closed nonsimple polygonal curve:

3. **(a)** 24. Each edge borders a square face and a triangular face. So, just count the edges of all the square faces (that is, $6 \cdot 4 = 24$) or the edges of all the triangular faces (that is, $8 \cdot 3 = 24$).

(b) 12. Either count in the diagram (4 on the top, 4 on the bottom, and 4 in between) or use Euler's formula, $V + F = E + 2$.

4.

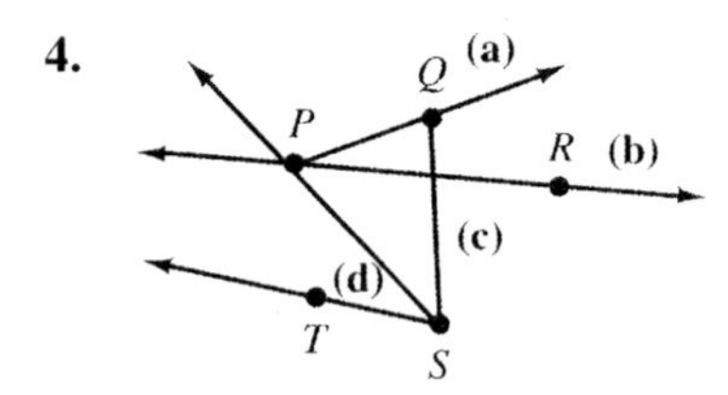

5. **(a)** Since $E = 11$ and $R = 7$, Euler's formula gives $V + R = E + 2$, or $V + 7 = 11 + 2$. Therefore, $V = 6$. There are 6 vertices.

(b) Many connected planar networks with 11 edges and 7 regions can be drawn. One example is shown:

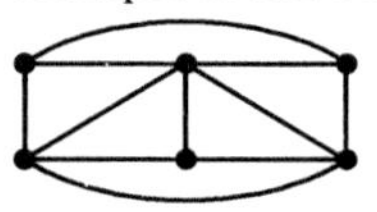

6. **(a)** The interior angles appear acute ($<90°$) at A and C

(b) The interior angles appears to be right ($= 90°$) at F.

(c) The interior angle appears to be obtuse ($> 90°$ and $< 180°$) at D.

(d) The interior angle appears to be straight ($= 180°$) at E.

(e) The interior angle appears to be reflex ($> 180°$ and $< 360°$) at B.

7. (a) True. The sides of a square are all the same length.

(b) False. The sum of the measures of a right angle, an obtuse angle, and a third angle would be greater than 180°.

(c) True. Any triangle with at least two congruent sides is isosceles.

(d) True. The sides of a square can be separated into two pairs of congruent adjacent sides.

8. The total turn angle is 1080°, so the turn at each of the eight vertices is $\frac{1080°}{8} = 135°$.

This means that the interior angle at each point is $180° - 135° = 45°$.

9. (a) Diagrams will vary. One example is shown.

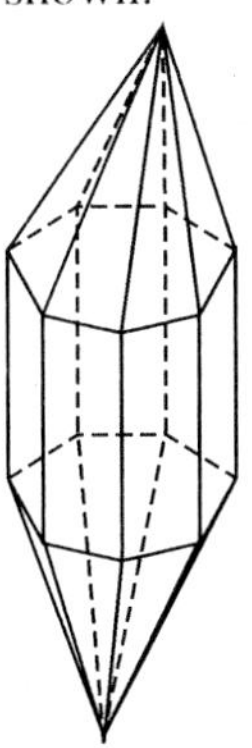

(b) $V = 16$, $F = 21$, $E = 35$

(c) $V + F = 16 + 21 = 37$, and $E + 2 = 35 + 2 = 37$. Since these quantities are equal, Euler's formula holds.

10. The same as an interior angle of a regular pentagon, which is $\frac{3 \cdot 180°}{5} = 108°$.

11. The small triangle on the right has angles 40°, $180° - 140° = 40°$, and $180° - r$, so $40° + 40° + (180° - r) = 180°$, which gives $r = 80°$. Since $\angle s$ corresponds to the supplement of $\angle r$ along a transversal of parallel lines, $s = 180° - r = 180° - 80° = 100°$. Since $\angle t$ corresponds to the supplement of a 140° angle along a transversal of parallel lines, $t = 180° - 140° = 40°$. In summary, $r = 80°$, $s = 100°$ and $t = 40°$.

12. Note that (i) has 2 odd vertices, (ii) has 2 odd vertices, (iii) has 4 odd vertices, and (iv) has no odd vertices.

(a) (iii), as this is the only network with 4 or more odd vertices.

(b) (iv), as this is the only network with no odd vertices.

(c) (i) and (ii) are traversable at some but not every vertex, since these networks have two odd vertices.

(d) (iii) requires an additional edge for traversability. The edge can join any two of the four odd vertices (that is, the three vertices on the inner circle and the vertex at the top of the diagram).

Chapter 10 Measurement: Length, Area, and Volume

Problem Set 10.1

1. **(a)** Height, length, thickness, area, diagonal, weight

(c) Height, width, depth

4. **(a)** Answers will vary anywhere from 40 to 54.

(b) Pens are a difficult unit of area because the circular portions of area don't fit well together, but leave gaps between them.

6. **(a)** $1 \text{ acre} = \frac{1}{640} \text{mi}^2$

$$= \left(\frac{1}{640} \text{mi}^2\right)\left(\frac{5280 \text{ ft}}{1 \text{ mi}}\right)^2$$

$$= 43{,}560 \text{ ft}^2$$

8. **(a)** $33 \text{ cL} = 33 \times 10^{-2} \text{L} = 330 \times 10^{-3} \text{L} = 330 \text{ mL}$

(b) Not quite; 1 L = 1000 mL and three bottles have a volume of 3×33 cL = 990 mL.

9. **(a)** $58{,}728 \text{ g} = 58.728 \times 10^3 \text{ g} = 58.728 \text{ kg}$

(c) $0.23 \text{ kg} = 0.23 \times 10^3 \text{ g} = 230 \text{ g}$

10. **(a)** 3.5 kg

12. **(a)** About 28 cm by 22 cm

(c) About 2 cm

14. **(a)** one centipede

(b) one microphone

(c) two kilo mockingbirds (*To Kill a Mockingbird*)

(d) one decacards (deck of cards)

(e) a nano (nanny) goat

19. **(a)** The correct answer is 0.00005 km.

(b) Answers may vary, but Brent needs to be shown why getting a decimal is a possible correct answer. He needs to see what 5 cm look like and then compare 5 cm to one decimeter. Next, show him that 5 cm = 0.5 dm and explain why. His teacher can then show him 5 cm compared to meters and show that 5 cm = 0.05 m and why. Seeing this comparison may help Brent realize why there can be a decimal when converting. It is important to discuss the reasons why the number looks so different when converting from centimeters to kilometers.

21. **(a)** 1 jill = 2 jacks = 4 jiggers = 8 mouthfuls; 1 cup = 2 jills = 16 mouthfuls; 1 pint = 2 cups = 32 mouthfuls

24. $\left(\frac{100 \text{ km}}{9 \text{ L}}\right)\left(\frac{3.7854 \text{ L}}{1 \text{ gal}}\right)\left(\frac{1 \text{ mi}}{1.6 \text{ km}}\right) \approx 26.3$ mi/gal

26. **(a)** $(5 \text{ gal})\left(\frac{4 \text{ qt}}{1 \text{ gal}}\right)\left(\frac{32 \text{ oz}}{1 \text{ qt}}\right) = 640 \text{ oz}$

Since $\frac{640}{80} = 8$, add 8 liquid ounces of concentrate.

(b) $80 \times 65 \text{ mL} = 5200 \text{ mL} = 5.2 \text{ L}$, so add 5.2 liters of water.

28. **(a)** $\frac{22{,}300 \text{ kg}}{1.77 \text{ kg/L}} - 7682 \text{ L} \approx 4917 \text{ L}$

4917 liters were added.

29. $1 \text{ ha} \approx \left(10{,}000 \text{ m}^2\right)\left(\frac{1 \text{ km}}{1000 \text{ m}}\right)^2 \cdot \left(\frac{1 \text{ mi}}{1.6 \text{ km}}\right)^2\left(\frac{640 \text{ acres}}{1 \text{ mi}^2}\right) = 2.5 \text{ acres}$

31. $\left(\frac{25 \text{ in}}{1 \text{ min}}\right)\left(\frac{60 \text{ min}}{1 \text{ hr}}\right)\left(\frac{24 \text{ hr}}{1 \text{ day}}\right)\left(\frac{14 \text{ day}}{1 \text{ fortnight}}\right) \cdot \left(\frac{1 \text{ ft}}{12 \text{ in}}\right)\left(\frac{1 \text{ furlong}}{660 \text{ ft}}\right) \approx 63.6 \frac{\text{furlong}}{\text{fortnight}}$

Problem Set 10.2

3. The dodecagon and square have the same area, equal to the sum of the areas of the same subregions in their dissections.

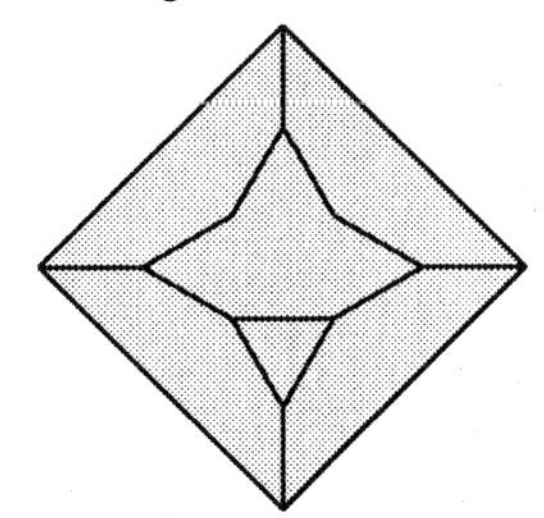

5. (a) 12 units

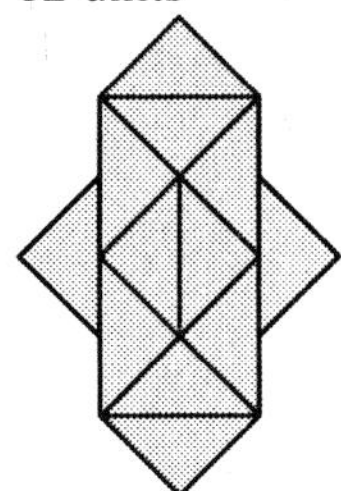

7. (a) $\frac{1}{2}\times(4\text{ mm}+6\text{ mm})\times 2\text{ mm}$

$+\frac{1}{2}\times 6\text{ mm}\times 5\text{ mm} = 25\text{ mm}^2$

10. (a) $36\text{ cm}^2 \div 3\text{ cm} = 12\text{ cm}$

11. (a) 1 cm by 24 cm; 2 cm by 12 cm; 3 cm by 8 cm; 4 cm by 6 cm. The dimensions can also be given in opposite order.

12. (a) $A = (16.2\text{ ft})(5\text{ ft}) = 81\text{ ft}^2$;
$P = 2(16.2\text{ ft}) + 2(6.4\text{ ft}) = 45.2\text{ ft}$

13. (a) $A = \frac{1}{2}(81.2\text{ m})(41.0\text{ m}) = 1664.6\text{ m}^2$;

$P = 49.0\text{ m} + 68.1\text{ m} + 81.2\text{ m}$
$= 198.3\text{ m}$

14. (a) $A = \frac{1}{2}(19+25)10 = 220$ square units

16. (a) $\triangle ABC$; all triangles have the same base and $\triangle ABC$ has the smallest height.

(b) $\triangle ABF$; it has the largest height.

(c) $\triangle ABD$ and $\triangle ABE$; they have equal heights and the same base.

18. (a) 9 square units

19. (a) $100\text{ m} + 100\text{ m} + 2\pi \cdot 25\text{ m}$
$= (200 + 50\pi)\text{m} \approx 357\text{ m}$

(b) $(50\text{ m})(100\text{ m}) + \pi(25\text{ m})^2$
$= (5000 + 625\pi)\text{m}^2 \approx 6963\text{ m}^2$

21. (a) $\pi(2)^2 - \pi(1)^2 = 3\pi$ square units ≈ 9.4 square units

24. (b) Yes, the perimeter will become infinitely large as either w approaches zero or as w becomes very large itself. Thus, long, skinny rectangles can all have the same area. This is a different result from problem 23.

25. Along large semicircle: $\frac{1}{2}(2\cdot\pi\cdot 8\text{ m}) = 8\pi\text{ m}$

Along the two smaller semicircles:

$\frac{1}{2}(2\cdot\pi\cdot 3\text{ m}) + \frac{1}{2}(2\cdot\pi\cdot 5\text{ m}) = 8\pi\text{ m}$

The distances are the same.

26. (a) Since the distance from the North Pole to the equator is one-quarter the circumference, the circumference is (4)(10,000,000)(1 m) = 40,000,000 m.

(b) $12{,}755\pi\text{ km} \approx 40{,}071\text{ km} = 40{,}071{,}000\text{ m}$

(c) The equator is larger since the earth bulges slightly at the equator and is slightly flattened at the poles.

33. The common overlap reduces the area of both regions by the same amount, so the difference in area is unchanged. Thus the area is 20 cm^2.

37. (a)

The area of the unshaded region in the above figure is $1^2 - \frac{1}{4}\pi(1^2) = 1 - \frac{\pi}{4}$ square units. There are two such unshaded regions in the figure below.

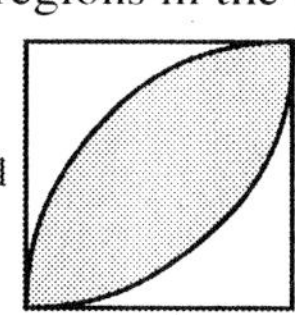

(*continued on next page*)

(*continued*)

The area of the shaded region in the figure is $1-\left(1-\frac{\pi}{4}\right)-\left(1-\frac{\pi}{4}\right)=\frac{\pi}{2}-1$ square units.

38. The areas of the rectangular portions of sidewalk total (60 ft + 70 ft + 40 ft + 80 ft + 50 ft)(8 ft) = 2400 ft^2. The pieces formed with circular areas have total turning of 360°, so when placed together would form a circle with radius 8 ft, and area of $\pi(8\text{ ft})^2 = 64\pi\text{ ft}^2$. Total area is $(2400+64\pi)\text{ ft}^2 \approx 2601\text{ ft}^2$.

39. **(a)** Erin walks $2\pi R$ and Nerd walks $2\pi(R + L)$, so Nerd walks $2\pi L$ farther than Erin.

42. Draw $\overline{AP}, \overline{BP}$, and $\overline{CP}$.

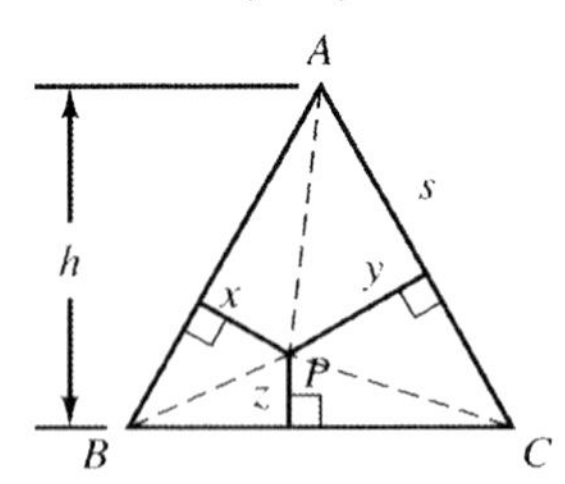

Then $\text{area}(\triangle ABC)=\frac{1}{2}sh$

$= \text{area}(\triangle ABP) + \text{area}(\triangle BPC) + \text{area}(\triangle CPA)$

$=\frac{1}{2}sx+\frac{1}{2}sz+\frac{1}{2}sy=\frac{1}{2}s(x+y+z)$.

Therefore, $\frac{1}{2}sh=\frac{1}{2}s(x+y+z)$, so $h = x + y + z$.

Alternate visual proof :

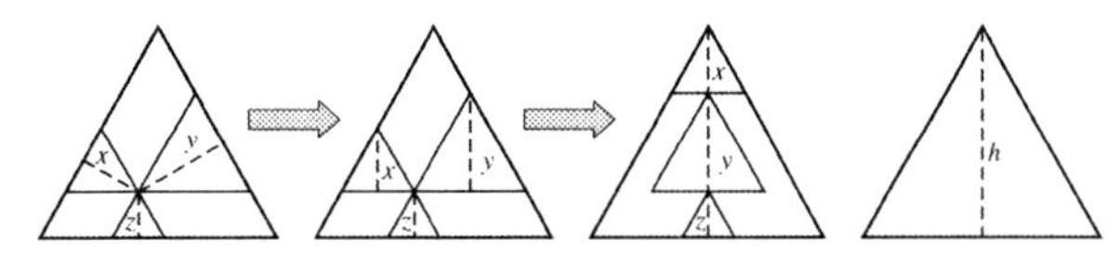

50. The lawn has an area $75\text{ ft}\times 125\text{ ft} = 9375\text{ ft}^2$. Since 21 in. = $\frac{7}{4}$ ft, the lawn area is equivalent to a rectangle 21 in. wide and $9375\div\frac{7}{4}=5357.14....$ That is, Kelly will walk about 5357 feet, a bit over a mile (1 mile = 5280 ft).

52. Consider the carpet as a 6-ft by 4-ft rectangle with two semicircular ends of radius 2 ft. Then the carpet's area is $(6\text{ ft})(4\text{ ft})+\pi(2\text{ ft})^2 \approx 36.57\text{ ft}^2 \approx 5266\text{ in}^2$, so about 5266 in. of braid, or about 439 ft is required.

54. Since each tile measures $\frac{8}{12}=\frac{2}{3}$ foot on a side, the dimensions of the kitchen in tile units are $10\div\frac{2}{3}=15$ and $12\div\frac{2}{3}=18$. Therefore, the number of tiles needed is $15 \times 18 = 270$.

56. View one side of length 300 ft as a base. The altitude of the triangle is greatest if the angle is 90°. Thus $m(\angle A) = 90°$.

58. The circumscribed circle has four times the area of the inscribed circle.

Problem Set 10.3

1. **(a)** By the Pythagorean theorem, $x^2 = 7^2 + 24^2 = 49 + 576 = 625$. Therefore, $x=\sqrt{625}=25$.

(c) By the Pythagorean theorem, $x^2 + 5^2 = 22^2$ or $x^2 = 22^2 - 5^2 = 484 - 25 = 459$. Therefore, $x=\sqrt{459}=3\sqrt{51}$.

(e) By the Pythagorean theorem, $x^2 = 1^2 + 1^2 = 1 + 1 = 2$. Therefore $x=\sqrt{2}$.

2. **(a)** By the Pythagorean theorem, $x^2 + (2x)^2 = (25)^2$ or $5x^2 = 625$. Therefore, $x=\sqrt{125}=5\sqrt{5}$.

4. **(a)** $x^2 = 10^2 + 15^2 = 325$, so $x=\sqrt{325}$; $y^2 = x^2 + 7^2 = 325 + 49 = 374$, so $y=\sqrt{374}$.

5. **(a)**
$$x^2 + 12^2 = 13^2$$
$$x^2 = 13^2 - 12^2$$
$$x=\sqrt{13^2-12^2}$$
$$x=\sqrt{169-144}=\sqrt{25}=5$$

6. **(a)** height $= \sqrt{15^2 - 9^2} = 12$, so area $= (20)(12) = 240$ square units

8. The smaller circle has radius 1 and area $\pi(1^2) = \pi$ square units. By the Pythagorean theorem, the larger circle has radius $\sqrt{1^2 + 1^2} = \sqrt{2}$ and area $\pi\left(\sqrt{2}\right)^2 = 2\pi$ square units. Thus, the area between the two circles is $2\pi - \pi = \pi$ square units. Therefore, the areas are equal.

10. $AC = \sqrt{4+1} = \sqrt{5};\ AD = \sqrt{5+1} = \sqrt{6};$
$AE = \sqrt{6+1} = \sqrt{7};\ AF = \sqrt{7+1} = \sqrt{8};$
$AG = \sqrt{8+1} = \sqrt{9} = 3$

12. Consider the following figure.

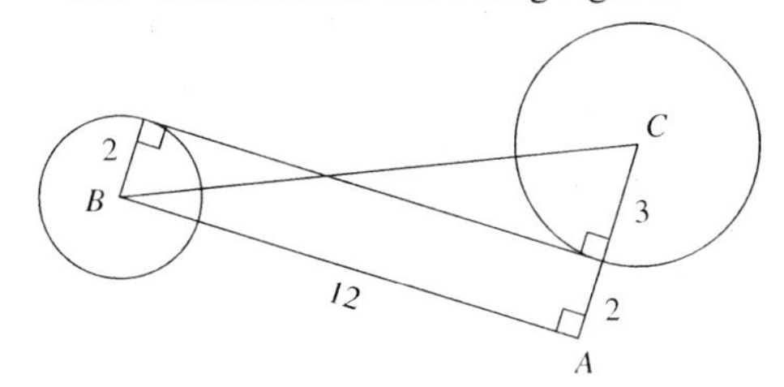

The distance between centers, B and C, is $\sqrt{12^2 + 5^2} = 13$.

14. **(a)** $(21)^2 + (28)^2 = 1225 = (35)^2$; yes

(c) $(12)^2 + (35)^2 = 1369 = (37)^2$; yes.

(e) $\left(7\sqrt{2}\right)^2 + \left(4\sqrt{7}\right)^2 = 210 \neq 308$
$= \left(2\sqrt{77}\right)^2$; no.

16. **(a)** Consider the diagram of the right triangle with semicircles drawn on the three sides shown.

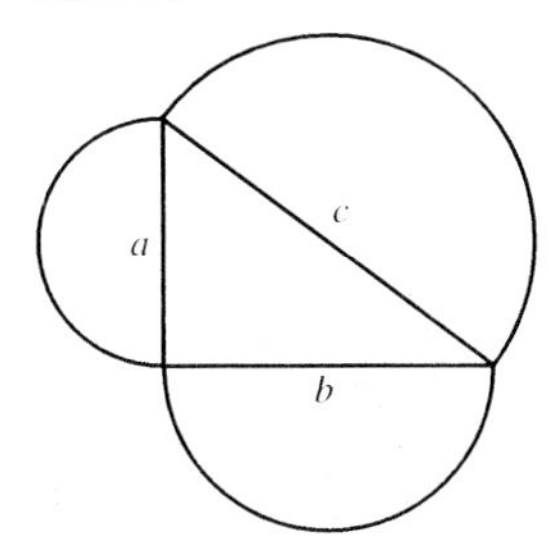

The areas of the three semicircles are

$$\frac{1}{2}\pi\left(\frac{a}{2}\right)^2, \frac{1}{2}\pi\left(\frac{b}{2}\right)^2, \text{ and } \frac{1}{2}\pi\left(\frac{c}{2}\right)^2.$$

Since the triangle is a right triangle, $a^2 + b^2 = c^2$. Multiplying through by $\frac{\pi}{2^3} = \frac{\pi}{8}$, we have

$$\frac{1}{2}\pi\left(\frac{a}{2}\right)^2 + \frac{1}{2}\pi\left(\frac{b}{2}\right)^2 = \frac{1}{2}\pi\left(\frac{c}{2}\right)^2.$$

This shows that the short answer to Sean's questions is *yes*.

22. **(a)**

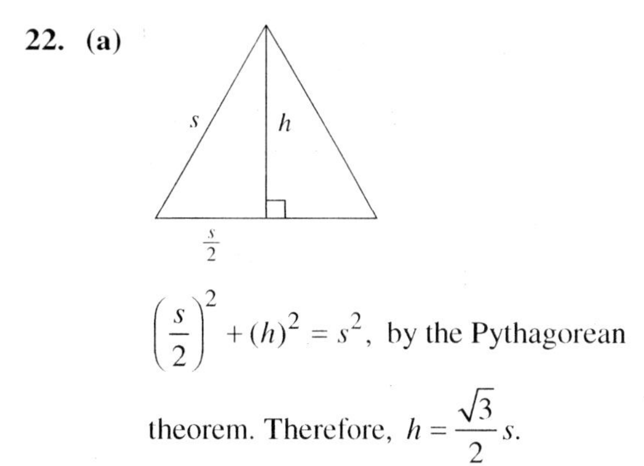

$\left(\frac{s}{2}\right)^2 + (h)^2 = s^2$, by the Pythagorean theorem. Therefore, $h = \frac{\sqrt{3}}{2}s$.

24. Flattening the top of the box suggested gives the diagram shown. By the Pythagorean theorem, $AC' = \sqrt{12^2 + 5^2} = \sqrt{169} = 13$ and $AC'' = \sqrt{9^2 + 8^2} = \sqrt{145} = 12.04...$ The shortest distance is therefore about 12 inches crossing over the 9 inch edge.

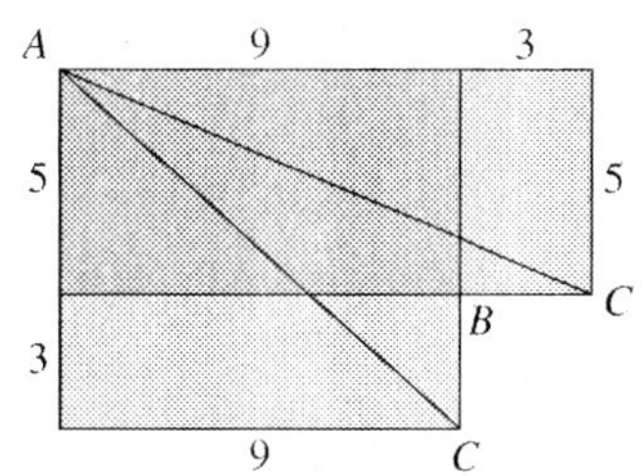

28. **(a)**

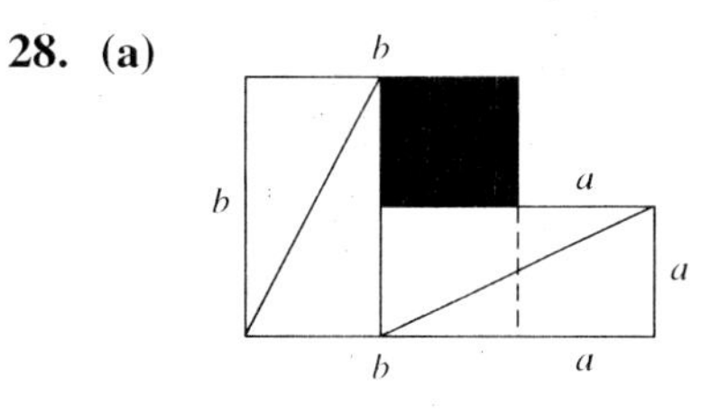

(b) Since the square and double square are tiled by the same five shapes, their areas are equal. The respective areas are c^2 and $a^2 + b^2$, so $c^2 = a^2 + b^2$.

30.

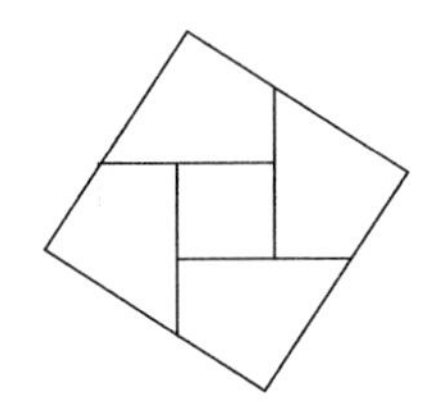

32. Answers will vary. The ladder cannot be vertical, so the height is less than 24 ft. (The ladder can be vertical if it is bolted against a vertical wall.) If the base is 7 feet from the wall the top of the ladder is still nearly 23 feet off the ground.

34. Let d be the depth of the pond. The stem length is $d + 2$ (in feet). With the stem held to the side, a right triangle is formed with legs of length 6 and d and hypotenuse of length $(d + 2)$. Then $6^2 + d^2 = (d+2)^2$, or

$36 + d^2 = d^2 + 4d + 4$, so $d = 8$.

38. **(a)** $d \approx 1.2\sqrt{100} = 1.2(10) = 12$ miles

(b) $d \approx 1.2\sqrt{1353} \approx 1.2(36.78) = 44.1$ miles

(c)
$$d \approx 1.2\sqrt{h}$$
$$40 \approx 1.2\sqrt{h}$$
$$\frac{40}{1.2} \approx \sqrt{h}$$
$$\left(\frac{40}{1.2}\right)^2 \approx h \Rightarrow 1111 \approx h$$

Yertle is approximately 1111 feet high.

39. The sum of the areas of the equilateral triangles on the legs equals the area of the equilateral triangle on the hypotenuse.

Problem Set 10.4

2. **(a)** $V = Bh = (7\text{ cm})(4\text{ cm})(3\text{ cm}) = 84\text{ cm}^3$

(c) $V = Bh = \pi \times (10\text{ m})^2 \times 4\text{ m}$
$= 400\pi\text{ m}^3 \approx 1257\text{ m}^3$

3. **(a)** $V = \frac{1}{3}Bh = \frac{1}{3}(12\text{ ft})(8\text{ ft})(10\text{ ft}) = 320\text{ ft}^3$

(c) $V = \frac{1}{3}Bh = \frac{1}{3} \times \pi(5\text{ cm})^2 \times (12\text{ cm})$
$= 100\pi\text{ cm}^3 \approx 314\text{ cm}^3$

4. The volume of the solid is the sum of the volume of the right circular cylinder and the volume of the hemisphere.

$V_{\text{cyl}} = \pi(3^2)(7) = 63\pi\text{ cm}^3$

$V_{\text{hemi}} = \dfrac{\frac{4}{3}\pi r^3}{2} = \dfrac{\frac{4}{3}\pi \cdot 3^3}{2} = 18\pi\text{ cm}^3$

$V_{\text{solid}} = 63\pi + 18\pi = 81\pi\text{ cm}^3$

6. The volume of the solid with radius r is
$\frac{2}{3}\pi r^3 + \pi r^2(7)$.

(a) If $r = 1$, the formula yields
$\frac{2}{3}\pi + 7\pi = \frac{23\pi}{3}\text{ in}^3$.

7. Have Eno carefully fill an oddly shaped container with an open top completely and exactly full of rice. Then have him pour the rice into a suitable box and smooth it out so that the top is as flat as possible. He can then measure the dimensions of the bottom of the box and also the height of the rice in the box and so determine the volume of rice (and hence of the oddly shaped container) in *cubic* units by using the formula $V = lwh$ for the volume of a rectangular solid.

16. **(a)** The circumference of the cone is
$\frac{3}{4}2 \cdot \pi(4\text{ in}) = 6\pi\text{ in}$, so the radius is
$\frac{6\pi\text{ in}}{2\pi} = 3\text{ in}$.

19. The volume of a box is
$(4\text{ in})(5\text{ in})(8\text{ in}) = 160\text{ in}^3$, so the volume of two boxes is 320 in^3. The volume of a "tub" is
$\pi(3\text{ in})^2(10\text{ in}) \approx 283\text{ in}^3$, so two boxes is a better buy.

21. **(a)** The scale factor is $\frac{5}{13}$, so the weight is

$\left(\frac{5}{13}\right)^3(106.75) \approx 6.07$ pounds.

Problem Set 10.5

2. **(a)** $SA = 2 \cdot \frac{1}{2}(20+15)(12)\text{ cm}$
$+ (2)(13+15+12+20)\text{ cm}$
$= 540\text{ cm}^2$

(c) $SA = 2 \cdot \pi(15 \text{ ft})^2 + (12 \text{ ft})2\pi(15 \text{ ft})$
$= 810\pi \text{ ft}^2 \approx 2545 \text{ ft}^2$

3. (a) slant height $= \sqrt{(40 \text{ m})^2 + (30 \text{ m})^2} = 50 \text{ m}$

$SA = (60 \text{ m})^2 + 4 \cdot \frac{1}{2}(60 \text{ m})(50 \text{ m})$
$= 9600 \text{ m}^2$

(c) $SA = \pi(6 \text{ in.})^2 + \pi(6 \text{ in.})(15 \text{ in.})$
$= 126\pi \text{ in}^2 \approx 396 \text{ in}^2$

4. (a) $SA = 4\pi(2200 \text{ km})^2 = 19{,}360{,}000\pi \text{ km}^2$
$\approx 6.08 \times 10^7 \text{ km}^2$

$V = \frac{4}{3}\pi(2200 \text{ km})^3 \approx 4.46 \times 10^{10} \text{ km}^3$

(c) $SA = 4\pi(4 \text{ ft})^2 + (20 \text{ ft})(2\pi)(4 \text{ ft})$
$= 224\pi \text{ ft}^2 \approx 704 \text{ ft}^2$

$V = \frac{4}{3}\pi(4 \text{ ft})^3 + \pi(4 \text{ ft})^2(20 \text{ ft}) \approx 1273 \text{ ft}^3$

5. $SA_{\text{cyl}} = B + ph = 9\pi + 6\pi(7) = 51\pi$
$SA_{\text{hemi}} = 2\pi r^2 = 2\pi(9) = 18\pi$
$SA_{\text{solid}} = 51\pi + 18\pi = 69\pi$

8. Let r be the radius of the sphere, so r is the radius of the cylinder and $2r$ is the height of the cylinder.

area(sphere) $= 4\pi r^2$ square units;

area(cylinder) $= 2 \cdot \pi r^2 + (2\pi r)(2r)$
$= 6\pi r^2$ square units

Thus, $\frac{\text{area(sphere)}}{\text{area(cylinder)}} = \frac{4\pi r^2}{6\pi r^2} = \frac{2}{3}$.

9. (a) Slant height $= \sqrt{(20 \text{ cm})^2 - (12 \text{ cm})^2}$
$= 16 \text{ cm}$

10. (a) Since the diameter of the 16″ pizza is 2 times that of the 8″ pizza, the area of the 16″ pizza is $2^2 = 4$ times that of the 8″ pizza by the similarity principle. Therefore, it will feed 4 people.

(b) Since 1.4^2 is about 2.0, the area of one 14″ pizza is nearly 2 times the area of one 10″ pizza by the similarity principle. Hence, one 14″ pizza is nearly the same amount of pizza as two 10″ pizzas, but costs two dollars less. It is better to buy one 14″ pizza.

12. (a) Since each small circle has $\frac{1}{2}$ the diameter of the large circle, each small circle has $\frac{1}{4}$ the area. Thus the shaded region has $1 - \frac{1}{4} - \frac{1}{4} = \frac{1}{2}$ the area of the large circle.

14. (a) Doubling the radius increases the volume by a factor of 4, while halving the height halves the volume, so the volume is doubled. The new volume is 200 mL.

15. (a) Doubling the sides increases the volume by a factor of 8, so the new cube holds 8 liters.

16. If r is the radius and the slant height is equal to the radius, then the surface area is $\pi r^2 + \pi r^2 = 2\pi r^2$. Since $2\pi r^2 = 16$, $r = \sqrt{\frac{8}{\pi}}$ cm.

25. (a) Since $\frac{2}{16} = \frac{1}{8} = \left(\frac{1}{2}\right)^3$, the scale factor is $\frac{1}{2}$, and so the height is $\frac{1}{2}(6 \text{ in}) = 3$ in.

26. (a)

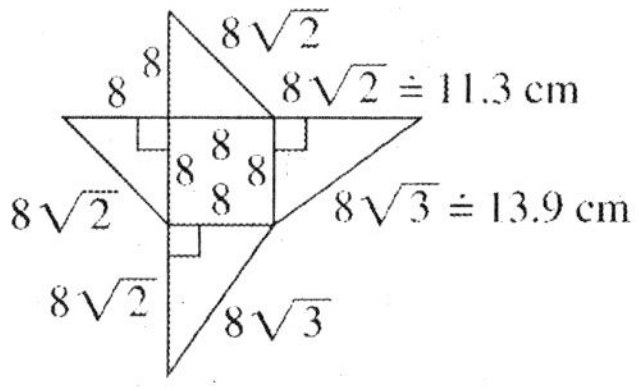

$8\sqrt{2}$, 8, 8, $8\sqrt{2} \doteq 11.3$ cm, 8, 8, 8, 8, $8\sqrt{2}$, $8\sqrt{3} \doteq 13.9$ cm, $8\sqrt{2}$, $8\sqrt{3}$

27. Let s be the radius of the semicircle. Then the slant height of the cone is s. Let d be the diameter of the cone. Then the circumference of the cone is $\pi d = \frac{1}{2} 2\pi s$, so $d = s$.

28. The diameter at the top is 3 times the diameter of the cylinder. Thus, the cross-sectional area is greater by a factor of $3^2 = 9$, so 9 inches of water in the cylinder correspond to 1 inch of rainfall.

Chapter 10 Review Exercises

1. (a) Centimeters (b) Millimeters
   (c) Kilometers (d) Meters
   (e) Hectares (f) Square kilometers
   (g) Milliliters (h) Liters

2. (a) 4 L (b) 190 cm
   (c) 200 m^2

3. (60 cm)(40 cm)(35 cm) = 84,000 cm^3 = 84 L

4. $\frac{300 \text{ ft}}{3 \text{ sec}} \cdot \frac{1 \text{ mile}}{5280 \text{ ft}} \cdot \frac{60 \text{ sec}}{1 \text{ min}} \cdot \frac{60 \text{ min}}{1 \text{ hr}} \approx 68 \text{ mph}$

5. Dissect the trapezoid by a horizontal line through M and rotate the bottom half by 180° as shown in the figure.

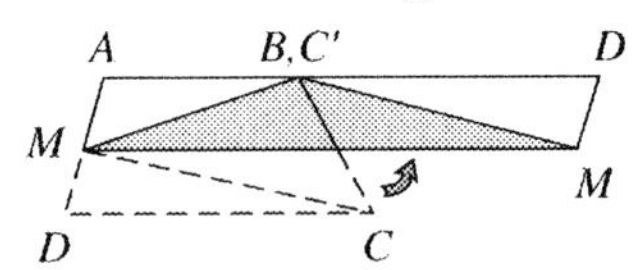

The triangle has half the area of the parallelogram $AD'M'M$ since they both have the same base and height, so it also has half the area of the trapezoid $ABCD$.

6. (a) A line from the center to the corner dissects the piece into two triangles of altitude 10, one with base 9 inches and the other with base 7 inches. The area of the piece is
$\frac{1}{2}\cdot 7\cdot 10+\frac{1}{2}\cdot 9\cdot 10=\frac{1}{2}\cdot 16\cdot 10=80 \text{ in}^2$,
which is 1/5th of the total area
$20^2 = 400 \text{ in}^2$ of the cake. So, yes, it is a fair piece.

   (b) The reasoning shown in part (a) shows that any piece with 16 inches along the perimeter has area 80 in^2. Therefore, the cuts are arranged as shown in the figure.

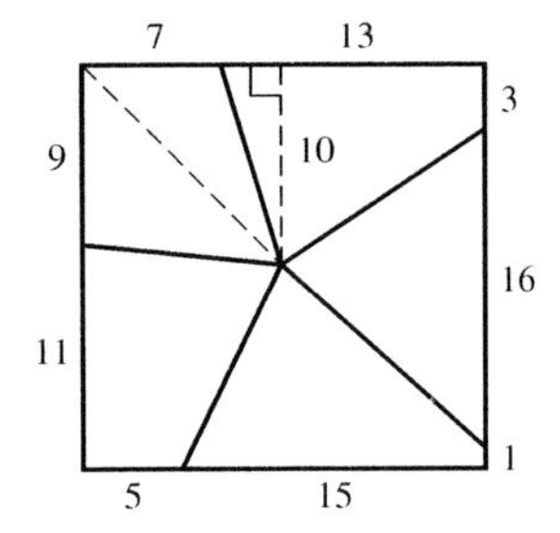

7. (a) 1 ft 4 in = 16 in, 4 ft = 48 in, so the area is
$(16 \text{ in})(48 \text{ in}) = 768 \text{ in}^2 = 5\frac{1}{3} \text{ ft}^2$.

   (b) Dissect the figure with a diagonal from the bottom left vertex to the top right vertex to form two triangles. The area is
$\frac{1}{2}(8 \text{ m})(9 \text{ m})+\frac{1}{2}(3 \text{ m})(6 \text{ m}) = 45 \text{ m}^2$.

   (c) $\frac{1}{2}(5 \text{ cm} + 7\text{cm})(3 \text{ cm}) = 18 \text{ cm}^2$

8. (a) 11 square units (b) 9 square units
   (c) $9\frac{1}{2}$ square units

9. (a) $A = (3 \text{ ft})(4 \text{ ft}) + \frac{1}{2}\cdot \pi \cdot (1.5 \text{ ft})^2$
$\approx 15.5 \text{ ft}^2$;
$P = 4 \text{ ft} + 3 \text{ ft} + 4 \text{ ft} + \frac{1}{2}\cdot 2\pi(1.5 \text{ ft})$
$= 11 \text{ ft} + \pi(1.5 \text{ ft}) \approx 15.7 \text{ ft}$

   (b) $A = \frac{3}{4}\cdot \pi(3 \text{ m})^2 = \frac{27}{4}\pi \text{ m}^2 \approx 21.2 \text{ m}^2$;
$P = 3 \text{ m} + 3 \text{ m} + \frac{3}{4}\cdot 2\pi(3 \text{ m})$
$= 6 \text{ m} + \frac{9}{2}\pi \text{ m} \approx 20.1 \text{ m}$

10. By the Pythagorean theorem,
$x^2 = \left(\sqrt{6}\right)^2 + \left(\sqrt{30}\right)^2 = 36$, so $x = 6$. Next compute the area two ways to find two equations involving x and y:
$A = \frac{1}{2}\left(\sqrt{6}\right)\left(\sqrt{30}\right) = 3\sqrt{5}$ and $A = \frac{1}{2}xy$,
$\frac{1}{2}xy = 3\sqrt{5}$ or $xy = 6\sqrt{5}$. Since $x = 6$,
$y = \sqrt{5}$.

11. The height of the cone is
$\sqrt{(35 \text{ cm})^2 - (10 \text{ cm})^2} = \sqrt{1125} \text{ cm}$
$\approx 33.5 \text{ cm}$

12. $\sqrt{(4 \text{ in})^2 + (10 \text{ in})^2} = \sqrt{116} \text{ in.} \approx 2\sqrt{29} \text{ in.}$

$\sqrt{(4 \text{ in})^2 + (12 \text{ in})^2} = \sqrt{160} \text{ in.} \approx 4\sqrt{10} \text{ in.}$

$\sqrt{(10 \text{ in})^2 + (12 \text{ in})^2} = \sqrt{244} \text{ in.} \approx 2\sqrt{61} \text{ in.}$

$\sqrt{(4 \text{ in})^2 + (10 \text{ in})^2 + (12 \text{ in})^2} = \sqrt{260} \text{ in.} \approx 4\sqrt{65} \text{ in.}$

13. $P = 3 + \sqrt{2} + \sqrt{10} + \sqrt{5} + 1 + 5 = 9 + \sqrt{2} + \sqrt{10} + \sqrt{5} \approx 15.8 \text{ units}$

14. **(a)** $V = \left[(10)(20) + \frac{1}{2}(8+20)(8)\right](30) \text{ ft} = 9360 \text{ ft}^3$

$SA = 2 \cdot \frac{1}{2}(8 \text{ ft} + 20\text{ft})(8 \text{ ft}) + 2 \cdot (10 \text{ ft})(20 \text{ ft}) + 2 \cdot (10 \text{ ft})(30 \text{ ft}) + 2 \cdot (10 \text{ ft})(30 \text{ ft}) + (8 \text{ ft})(30 \text{ ft}) + (20 \text{ ft})(30 \text{ ft}) = 2664 \text{ ft}^2$

(b) $V = \frac{1}{2} \cdot \frac{4}{3}\pi(7 \text{ m})^3 + \pi(7 \text{ m})^2(18 \text{ m}) = 1110\frac{2}{3}\pi \text{ m}^3 \approx 3489 \text{ m}^3$

$SA = \frac{1}{2} \cdot 4\pi(7)^2 + 2\pi(7)(18) + \pi(7)^2 = 399\pi \text{ m}^2 \approx 1253 \text{ m}^3$

(c) $V = \frac{1}{3}\pi(5 \text{ cm})^2(8 \text{ cm}) + \frac{1}{2} \cdot \frac{4}{3}\pi(5 \text{ cm})^3 = 150\pi \text{ cm}^3 \approx 471 \text{ cm}^3$

$SA = \frac{1}{2} \cdot 4\pi(5 \text{ cm})^2 + \pi(5 \text{ cm})\sqrt{5^2 + 8^2} = \left(50 + 5\sqrt{89}\right)\pi \text{ cm}^2 \approx 305 \text{ cm}^2$

15. $V(\text{sphere}) = \frac{4}{3}\pi(10 \text{ m})^3$ and $V(\text{four cubes}) = 4(10 \text{ m})^3$. Since $\pi > 3$, then $\frac{4}{3}\pi > 4$, showing that the sphere has the larger volume. Also, a sphere of radius 10 m can be inscribed in a cube with sides of length 20 m, which is the same as eight cubes with sides of length 10 m put together.

16. **(a)** The scale factor is $\frac{75 \text{ ft}}{50 \text{ ft}} = 1.5$: Heather needs $1.5(180 \text{ ft}) = 270$ ft of fencing.

(b) Since Johan's garden area is $\left(\frac{1}{1.5}\right)^2$ times that of Heather's, he needs $\left(\frac{1}{1.5}\right)^2 45 \text{ pounds} = 20 \text{ pounds}$ of fertilizer.

Chapter Test

1. **(a)** 8 cm^2 **(b)** 8 cm^2

(c) 8.5 cm^2

2. Each figure contains four full units of area. Figure A contains an additional eight half-units of area, and Figure B contains an additional four half-units of area. Jim should conclude that Figure A is four half-units, or two units, of area larger than Figure B.

3. **(a)** The hypotenuse of the triangle has length $\sqrt{(7 \text{ m})^2 + (24 \text{ m})^2} = 25 \text{ m}$.

$SA = 2 \cdot \frac{1}{2}(7)(24) + (7 + 24 + 25)(5); = 448 \text{ m}^2$

$V = \frac{1}{2}(7 \text{ m})(24 \text{ m})(5 \text{ m}) = 420 \text{ m}^3$

(b) $SA = 2 \cdot \frac{1}{2}\pi(4)^2 + \frac{1}{2} \cdot 2\pi(4)(6) + (6)(8) \approx 173.7 \text{ in}^2$

$V = \frac{1}{2}\pi(4 \text{ in})^2(6 \text{ in}) \approx 150.8 \text{ in}^3$

(c) The slant height is $\sqrt{(8 \text{ ft})^2 + (6 \text{ ft})^2} = 10 \text{ ft}$.

$SA = 4 \cdot \frac{1}{2}(12 \text{ ft})(10 \text{ ft}) + (12 \text{ ft})^2 = 384 \text{ ft}^2$

$V = \frac{1}{3}(12 \text{ ft})^2(8 \text{ ft}) = 384 \text{ ft}^3$

(d) The slant height is $\sqrt{(12 \text{ cm})^2 + (5 \text{ cm})^2} = 13 \text{ cm}$.

$SA = \pi(5 \text{ cm})^2 + \frac{2\pi(5 \text{ cm})}{2\pi(13 \text{ cm})} \cdot \pi(13 \text{ cm})^2 = 90\pi \text{ cm}^2 \approx 283 \text{ cm}^2$

$V = \frac{1}{3}\pi(5 \text{ cm})^2(12 \text{ cm}) = 100\pi \text{ cm}^3 \approx 314 \text{cm}^3$

4. **(a)** millimeters **(b)** meters

(c) meters **(d)** kilometers

(e) milliliters **(f)** liters

5. Cross-sectional view:

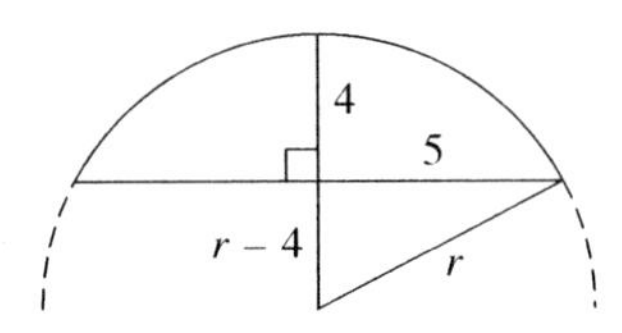

$(r-4)^2 + (5)^2 = r^2$, or

$r^2 - 8r + 16 + 25 = r^2$. Then

$r = \dfrac{16+25}{8\text{ mm}} = 5.125$ mm.

6. **(a)** $P = \sqrt{52} + \sqrt{13} + \sqrt{65} \approx 18.9$ units

(b) Yes, it is a right triangle since $\left(\sqrt{52}\right)^2 + \left(\sqrt{13}\right)^2 = 65 = \left(\sqrt{65}\right)^2$.

7. **(a)** $2161\text{ mm} = 2161 \times 10^{-3}\text{m}$
$= 216.1 \times 10^{-2}\text{m} = 216.1\text{ cm}$

(b) $1.682\text{ km} = 1.682 \times 10^{3}\text{m}$
$= 168,200 \times 10^{-2}\text{m}$
$= 168,200\text{ cm}$

(c) $0.5\text{ m}^2 = 0.5(100\text{ cm})^2 = 5000\text{ cm}^2$

(d) $1\text{ ha} = 10,000\text{ m}^2$

(e) $4719\text{ mL} = 4719 \times 10^{-3}\text{L} = 4.719\text{ L}$

(f) $3.2\text{ L} = 3.2 \times 1000\text{ cm}^3 = 3200\text{ cm}^3$

8. Let the radius of the circle be r. Then the circumscribed square has sides of length $2r$ and the inscribed square has sides of length $\sqrt{2}r$.

$$\frac{\text{area(inscribed)}}{\text{area(circumscribed)}} = \frac{\left(\sqrt{2}r\right)^2}{(2r)^2} = \frac{2r^2}{4r^2} = \frac{1}{2}$$

Alternate solution: The scale factor of the large to the small square is $\dfrac{1}{\sqrt{2}}$, so the small square has $\left(\dfrac{1}{\sqrt{2}}\right)^2 = \dfrac{1}{2}$ the area of the large square.

9. By the Pythagorean theorem, the ladder reaches
$\sqrt{(15\text{ ft})^2 - (6\text{ ft})^2} = \sqrt{189}\text{ ft} \approx 13.7\text{ ft}$.

10. **(a)** $(1147\text{ in})\left(\dfrac{1\text{ yd}}{36\text{ in}}\right) \approx 31.86\text{ yd}$

(b) $(7942\text{ ft})\left(\dfrac{1\text{ mi}}{5280\text{ ft}}\right) \approx 1.5\text{ mi}$

(c) $(32.4\text{ yd}^2)\left(\dfrac{3\text{ ft}}{1\text{ yd}}\right)^2 = 291.6\text{ ft}^2$

(d) $(9402\text{ acres})\left(\dfrac{1\text{ mi}^2}{640\text{ acres}}\right) \approx 14.69\text{ mi}^2$

(e) $(7.6\text{ yd}^3)\left(\dfrac{3\text{ ft}}{1\text{ yd}}\right)^3 = 205.2\text{ ft}^3$

(f) $(5961\text{ in}^3)\left(\dfrac{1\text{ ft}}{12\text{ in}}\right)^3 \approx 3.45\text{ ft}^3$

11. The scale factor of Papa Bear to Mama Bear is $\dfrac{4}{5}$ and the scale factor of Mama Bear to Baby Bear is $\dfrac{1}{2}$.

	Papa Bear	Mama Bear	Baby Bear
Length of Suspenders	50	40	20 in
Weight	468.75	240	30 lb
Number of fleas	6000	3840	960

Note that weight is based on volume, and the number of fleas is based on surface area.

12. $A = \dfrac{1}{2}(9\text{ ft})(12\text{ ft}) + (8\text{ ft})(12\text{ ft}) - \dfrac{1}{2}\pi(4\text{ ft})^2$
$\approx 125\text{ ft}^2$

$P = 15\text{ ft} + 9\text{ ft} + 8\text{ ft} + 12\text{ ft} + \dfrac{1}{2} \cdot 2\pi(4\text{ ft})$
$\approx 56.6\text{ ft}$

13. (a) $\frac{2}{3}\pi(5 \text{ in})^2 \approx 52.4 \text{ in}^2$

(b) $\frac{1}{2}(2.6 \text{ m} + 1.4 \text{ m})(3 \text{ m}) = 6.0 \text{ m}^2$

(c) $\frac{1}{12} \cdot \pi(24 \text{ cm})^2 \approx 151 \text{ cm}^2$

14. $A = \frac{1}{2} \cdot (6 \text{ cm})(16 \text{ cm}) + \frac{1}{2}(15 \text{ cm})(16 \text{ cm})$

$= 168 \text{ cm}^2$

Sides are 10 cm and 17 cm by the Pythagorean theorem, so $P = 54$ cm.

15.

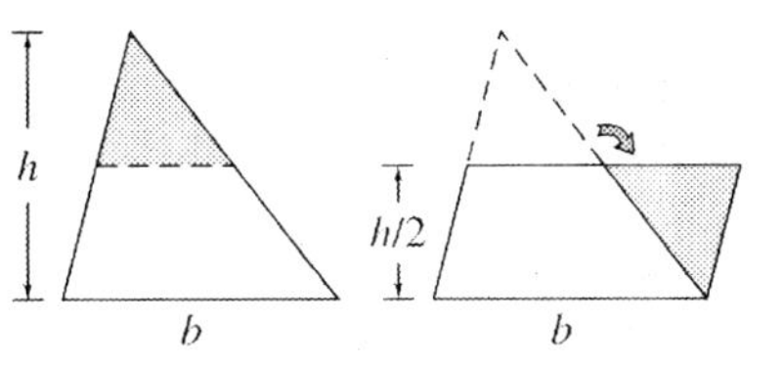

area(triangle) = area(parallelogram)

$= \left(\frac{b}{2}\right)(h) = \frac{1}{2}bh$

16. (a) $SA = 4\pi(10 \text{ m})^2 = 400\pi \text{ m}^2 \approx 1257 \text{ m}^2$

$V = \frac{4}{3}\pi(10 \text{ m})^3 \approx 4189 \text{ m}^3$

(b) $SA = \pi(5)^2 + 2\pi(5)(6) + \frac{1}{2} \cdot 4\pi(5)^2$

$= 135\pi \text{ cm}^2 \approx 424 \text{ cm}^2$

$V = \pi(5 \text{ cm})^2(6 \text{ m}) + \frac{1}{2} \cdot \frac{4}{3}\pi(5 \text{ cm})^3$

$\approx 733 \text{ cm}^3$

17. $V(\text{peel}) = \frac{4}{3}\pi(2.5 \text{ m})^3 - \frac{4}{3}\pi(1.75 \text{ in})^3$

$\approx 43.0 \text{ in}^3$, and $V(\text{grapefruit}) = \frac{4}{3}\pi(2.5 \text{ in})^3$

$\approx 65.4 \text{ in}^3$. About 66% is peel. Alternate solution: The scale factor is $\frac{2.5 - 0.75}{2.5} = 0.7$.

Since $(0.7)^3 = 0.343$, it follows that 34.3% of the grapefruit is not peel, and 65.7% is peel.

Chapter 11 Transformations, Symmetries, and Tilings

Problem Set 11.1

1. (a) Not a rigid motion; distances between particular cards will change.

(c) No, distances between particular pieces almost certainly will have changed.

2. (a) The figure is moved 3 units to the right.

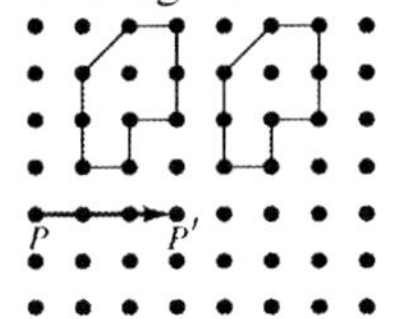

4. (a) $360° - 60° = 300°$

(c) 43°, since the remainder after dividing 3643 by 360 is 43.

5. (a)

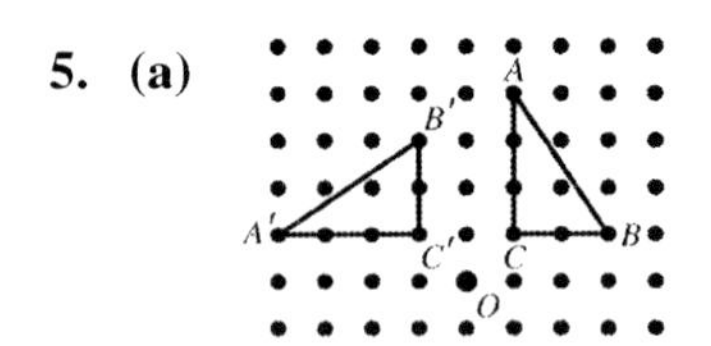

6. (a) The center of rotation is the same distance from A as from A', so it is on the perpendicular bisector of $\overline{AA'}$. (Similarly, it is on the perpendicular bisector of $\overline{BB'}$, but this turns out to be the same line.) By using trial and error among the points on the perpendicular bisector, we find that the only point that can produce the desired transformation of both points simultaneously is point O.

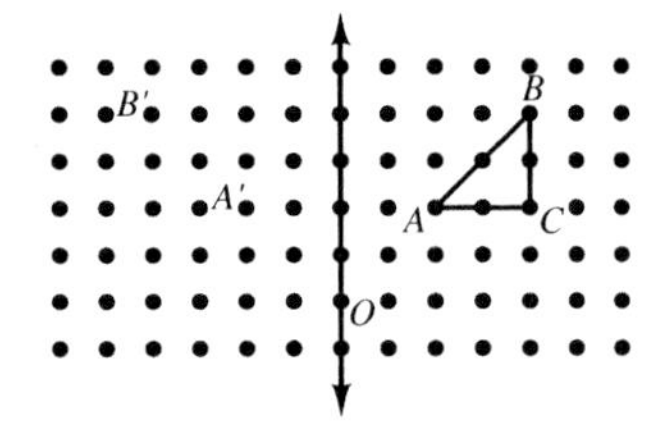

7. The nonequivalent rigid motions that move the triangle to point downward and leave point O fixed are rotations about O through 60°, 180°, or 300°, or a reflection across one of the three mirror lines through O parallel to one of the sides of the triangle.

11. (a) Draw the perpendicular bisector of $\overline{PP'}$. Since $\overline{PP'}$ is horizontal, the line of reflection is vertical.

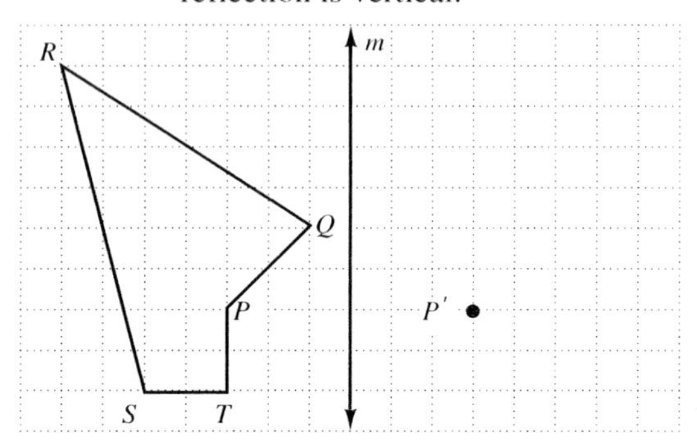

13. (a), (b) For part (a), notice that the glide arrow is 7 units long, so A, B, C, D, E is 7 units to the right of $ABCDE$. For part (b), reflect A, B, C, D, E, across the line of reflection to find the image of $ABCDE$ under the glide reflection.

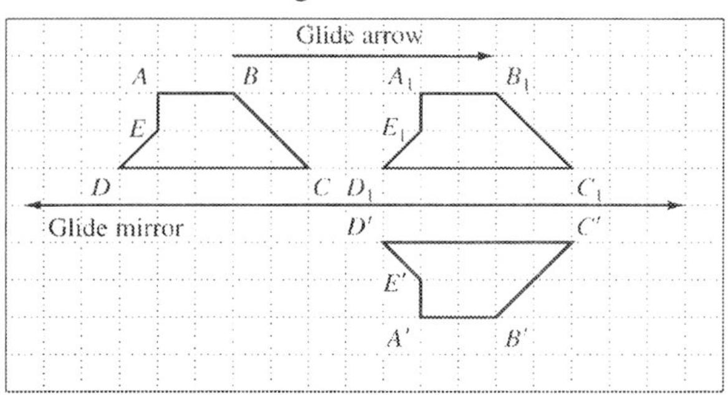

15. (a) The distance from m_1 to m_2 must be 3 units, or half the length of the slide arrow.

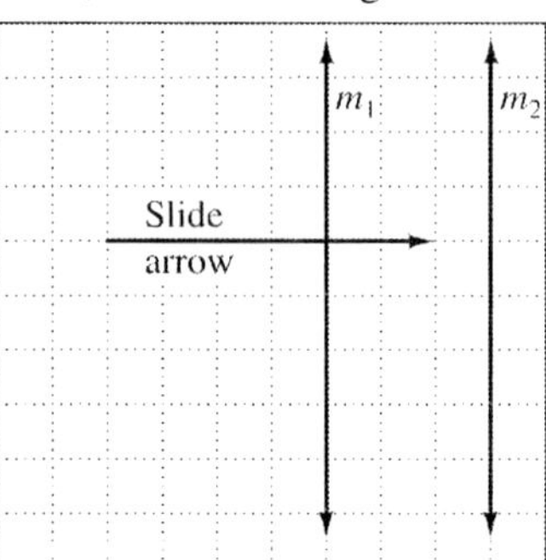

17. (a) A' is the point on $\overrightarrow{OA}$ such that $OA' = 2 \cdot OA$. B' is the point on $\overrightarrow{OB}$ such that $OB' = 2 \cdot OB$. C' is the point on $\overrightarrow{OC}$ such that $OC' = 2 \cdot OC$.

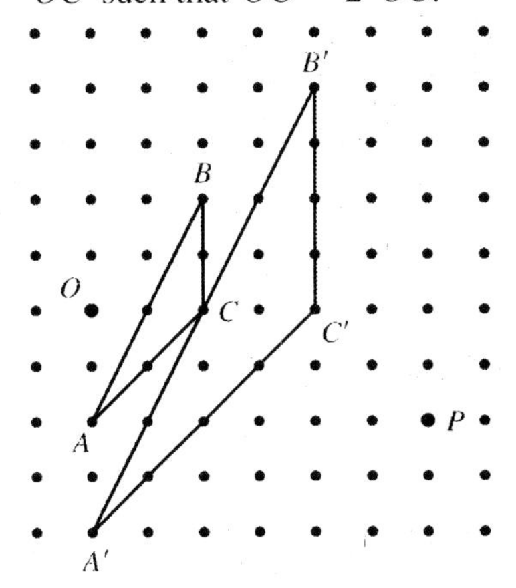

(b) A' is the point on $\overrightarrow{PA}$ such that

$PA' = \frac{1}{2} \cdot PA.$

B' is the point on $\overrightarrow{PB}$ such that

$PB' = \frac{1}{2} \cdot PB.$

C' is the point on $\overrightarrow{PC}$ such that

$PC' = \frac{1}{2} \cdot PC.$

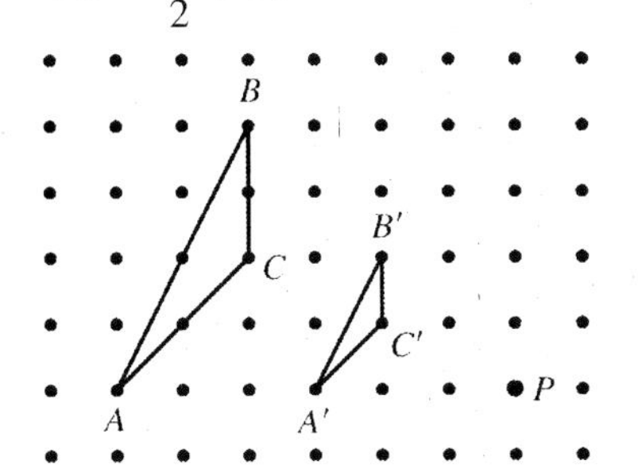

18. (a) The center is at P, the intersection of $\overleftrightarrow{DG}$ and $\overleftrightarrow{EH}$ The scale factor is $\frac{GI}{DF} = \frac{6}{4} = \frac{3}{2}$.

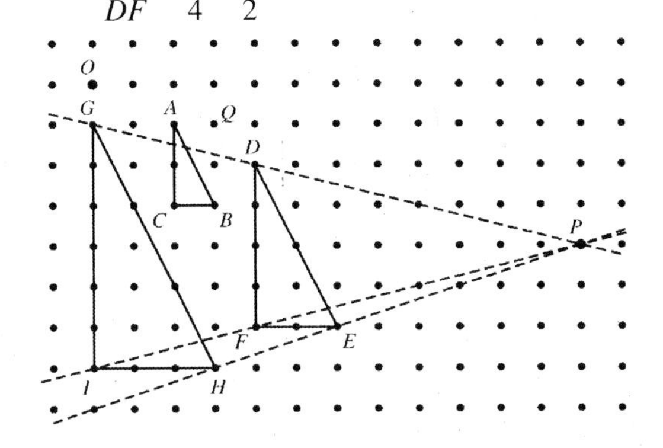

20. We need to rotate the figure 90° counterclockwise so that, for example, $\overline{A_1B_1}$ will be parallel to $\overline{A'B'}$. Then we can perform a size transformation whose center is found by intersecting two lines such as $\overleftrightarrow{B_1B'}$ and $\overleftrightarrow{D_1D'}$ Answers will vary; one possibility is the following. Rotate 90° counterclockwise about B. Then perform a size transformation centered at P with scale factor 2.

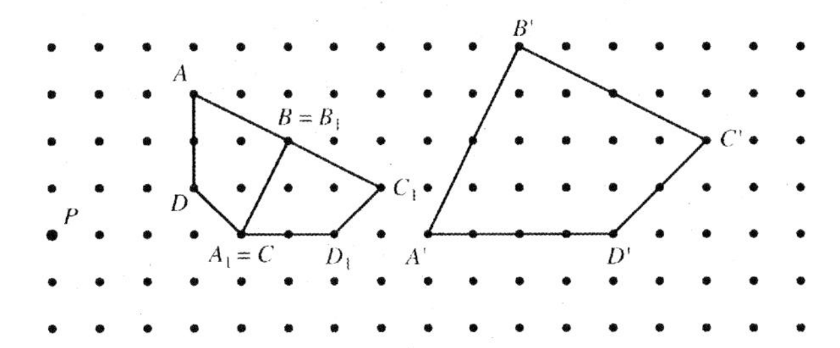

25. (a) Since $A'P' = AP = 2$ cm, P' is some point on the circle of radius 2 cm centered at A'.

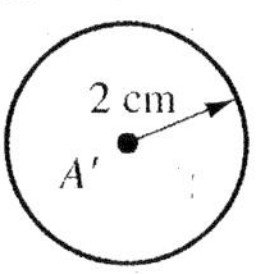

26. Since each of the turns preserves orientation, the combined transformation must preserve orientation, so it is either a translation or a rotation. But there is an overall turn of 180°, so it remains only to determine the center of this rotation. On a 1 cm square grid, the two 90° rotations take O_1 to O_1' (and O_2 to O_2') as shown. Since the rotation center of a half turn is half-way between any point and its image, it must be the point O. The motion is equivalent to a 180° rotation (half turn) about the point O.

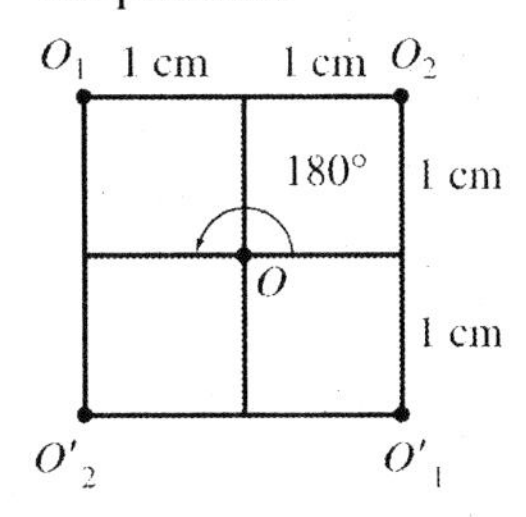

28. (a), (b) Reflection of $\triangle ABC$ across m_1 gives $\triangle A_1B_1C_1$. Reflection of $\triangle A_1B_1C_1$ across m_2 gives $\triangle A_2B_2C_2$. Finally, reflection $\triangle A_2B_2C_2$ across m_3 gives $\triangle A'B'C'$. For part (b), l is the line that contains the midpoints of $\overline{AA'}$, $\overline{BB'}$, and $\overline{CC'}$. (It is also the perpendicular bisector of any one of these.)

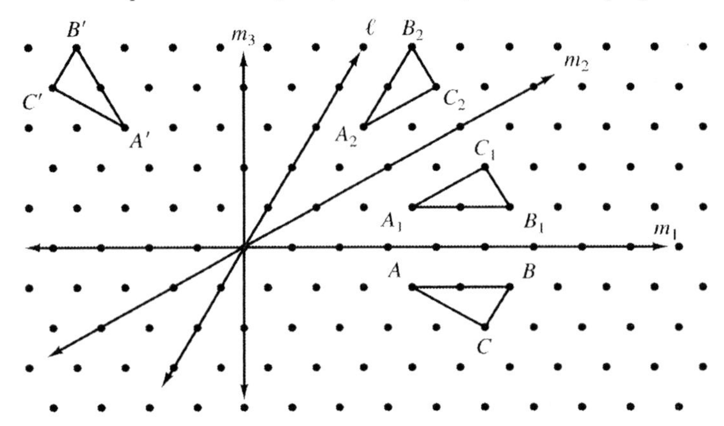

30. (a), (b) m_1 is the perpendicular bisector of $\overline{AA'}$. m_2 is the perpendicular bisector of $\overline{BB'}$. The image of C_1 across m_2 is C'.

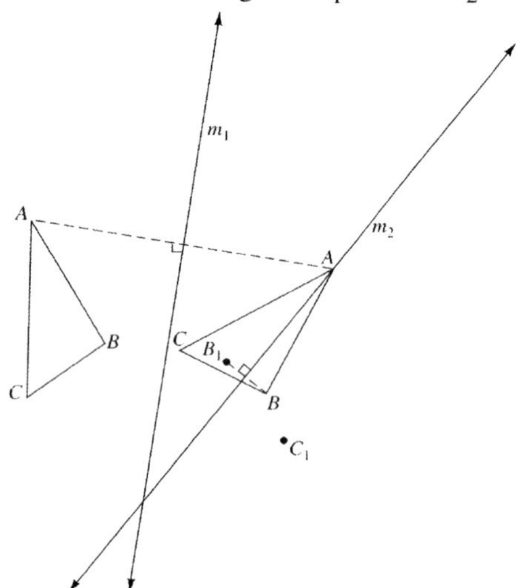

(c) The rigid motion that takes $\triangle ABC$ to $\triangle A'B'C'$ is a reflection across m_1 followed by a reflection across m_2. Since two consecutive reflections across intersecting lines are equivalent to a rotation, the basic rigid motion is a rotation about the point P of intersection of m_1 and m_2, through an angle twice the measure x of the directed angle from line m_1 toward line m_2.

32. (a) A translation: Six reflections give an orientation preserving rigid motion, so it is either a rotation or a translation. Since a rotation has a fixed point (namely the rotation center), the motion is a translation.

34. (a) Since A and B have opposite orientations (which is most easily observed by looking at the tail) and are not reflections of each other, the motion is a glide reflection.

35. (a) The line $\overleftrightarrow{PP'}$ passes through O, so O is the intersection of $\overleftrightarrow{PP'}$ and line l.

(b) Since $\overline{PQ}$ and $\overline{P'Q'}$ are parallel, Q' lies on the line through P' that is parallel to $\overline{PQ}$. Since Q' also lies on $\overrightarrow{OQ}$ (which is part of line l), Q' is the intersection of line l and the line through P' that is parallel to $\overline{PQ}$.

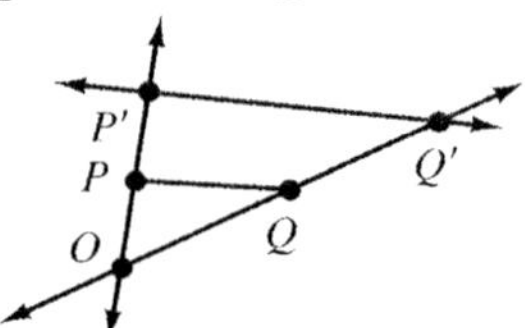

38. (a) $\overline{RS'}$ is the reflection of $\overline{RS}$ across m, so $RS' = RS$ since a reflection preserves all distances. Similarly, $QS' = QS$.

40. A single mirror reverses orientation, so the double reflection seen in a corner mirror preserves orientation. The corner mirror reflection of your right hand will appear as a right hand.

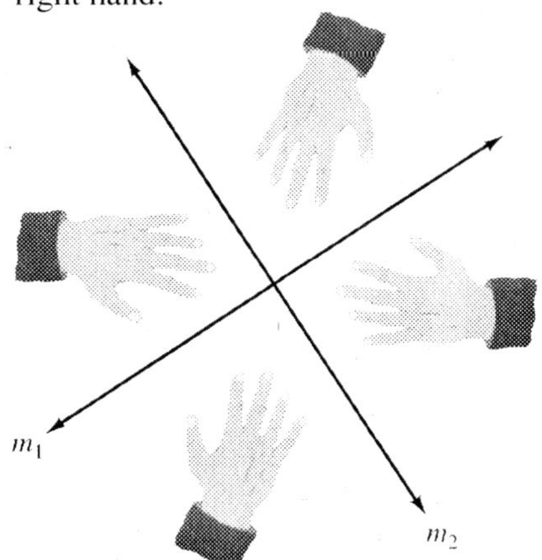

47. **(a)** Measurements show triangle XYZ is an equilateral triangle.

(b) The centers of the rotated equilateral triangles form an equilateral triangular grid.

Problem Set 11.2

1. **(a)** 6 lines of symmetry

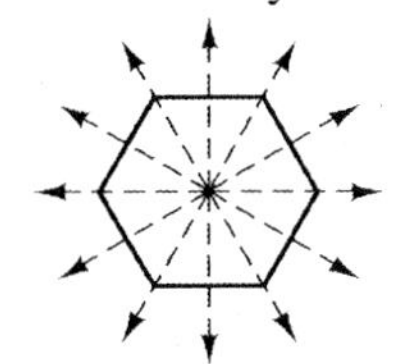

(c) No lines of symmetry

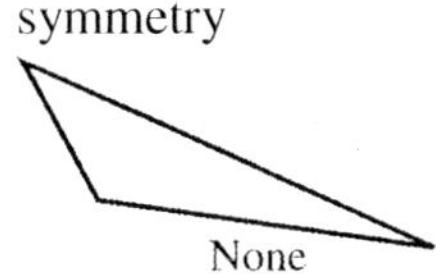

3. **(a)** Answers will vary. One possibility is an isosceles triangle.

4. **(a)** a kite **(b)** an isosceles trapezoid

5. **(a)**

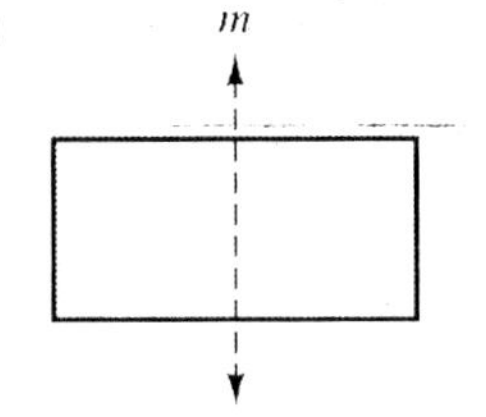

(c)

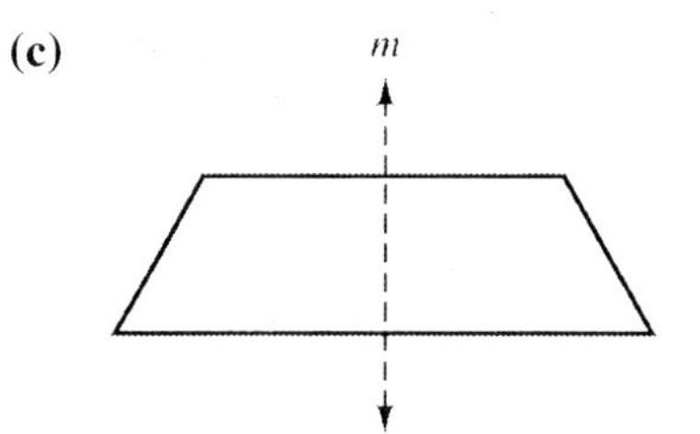

8. **(a)** Unfolding the horizontal fold produces the figure on the left, below. Unfolding the vertical fold then produces the figure on the right.

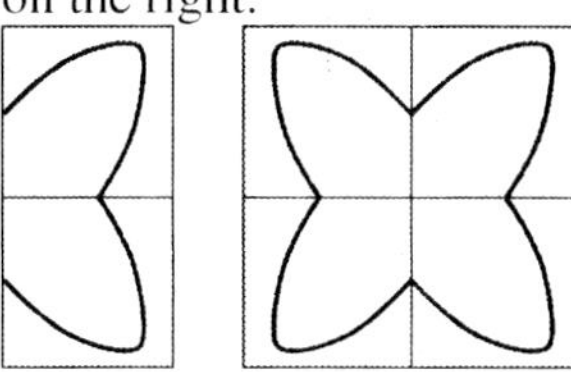

9. **(a)** One vertical line of symmetry

10. Note that some of the patterns shown represent light reflections rather than actual patterns in the wheel covers themselves. These should be ignored when determining symmetry.

(a) Five lines of symmetry and 72° rotation symmetry

(b) 72° rotation symmetry

11. **(a)**

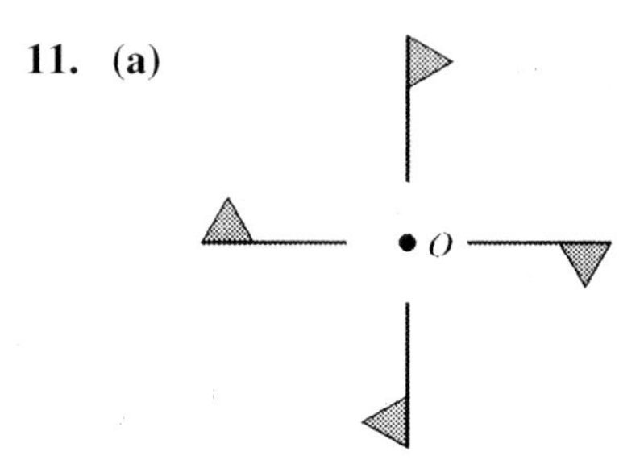

12. Note that any regular n-gon has n lines of symmetry. If n is odd, there is a line of symmetry passing through each vertex and the midpoint of the opposite side. If n is even, there are $\frac{n}{2}$ lines of symmetry through pairs of opposite vertices, and $\frac{n}{2}$ lines of symmetry through the midpoints of opposite sides, for a total of n lines of symmetry.

(a) Equilateral triangle

(b) Square

13. **(a)** 0, 8 **(b)** 0, 3, 8

(c) 0, 8 **(d)** 0, 8

16. **(a)** Translation symmetry and vertical line symmetry: *m*1

17. **(a)** Translation symmetry, vertical line symmetry, horizontal line symmetry, glide reflection symmetry, and half-turn symmetry: *mm*

26. **(a)** No letter is the same as another letter written backwards. That is, no letter or digit reflects vertically into a different letter or digit, so it must reflect into itself.

28. The pattern would be the same after reflecting across the vertical line and rotating 180°, and this pair of transformations is equivalent to a reflection across a horizontal line. Therefore, the pattern must also have a horizontal line of symmetry, so it would be an *mm* pattern.

29. **(a)** The pattern has no vertical line of symmetry and, disregarding the color scheme, it has horizontal glide symmetry. Its type is 1*g*.

(c) The pattern has vertical line symmetry and horizontal glide symmetry. Its type is *mg*.

31. **(a)** The motion is a translation. The pattern does not have vertical line symmetry, horizontal glide reflection, or half-turn symmetry. Its type is 11.

p	p	p	p	p	p

(c) The motion is a half-turn. The pattern has half-turn symmetry only. Its type is 12.

p	d	p	d	p	d

32. **(a)** Three directions of reflection symmetry (including vertical); the same three directions of glide symmetry; 120° rotation symmetry

36. **(a)** As left-handed people know well, not all scissors are symmetric.

(c) A dress shirt is not symmetric due to buttons and pockets.

(e) Tennis rackets have two planes of bilateral symmetry (ignoring such details as how they are strung and wrapping around the handle).

37. **(a)** There is symmetry across the main diagonal line of entries in the sense that the same numbers appear on both sides of this diagonal. In other words, addition is commutative.

Problem Set 11.3

1. Many different tilings can be formed. For example:

(a) The pattern shown consists of squares within squares—all formed from identical triangles.

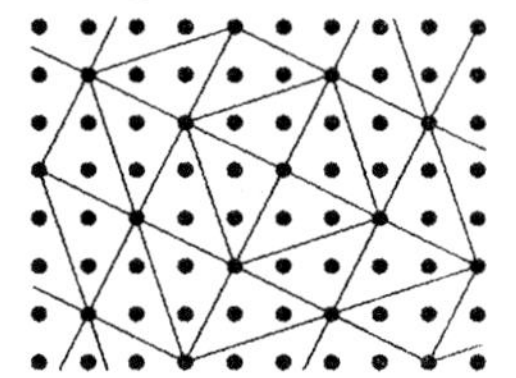

2. **(a)** As shown in Example 11.12, any quadrilateral can tile the plane.

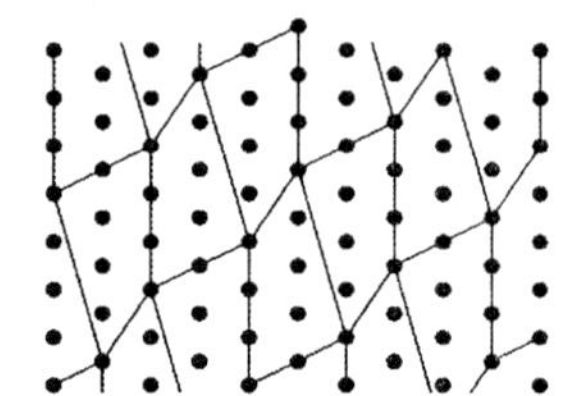

17. **(a)** Squares, pentagons, hexagons, heptagons, octagons

(b) Actually, it is a "real" tiling by "fake" regular polygons. Many vertex figures that appear in the tiling cannot correspond to regular polygons. For example, a pentagon, hexagon, and octagon meet at each of many vertices. If these were regular figures, they would have interior angles of 108°, 120°, and 135°, respectively. But $108° + 120° + 135° = 363° \neq 360°$, so these cannot be all regular polygons.

24. Here's one way to make a tiling 12-gon, by modifying a square tile:

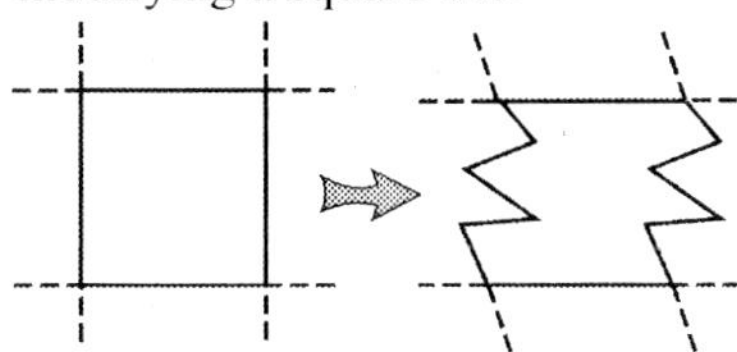

Similarly, modifying an equilateral triangle by a midpoint modification with opposite parallel congruent sides will form an 11-gon that tiles:

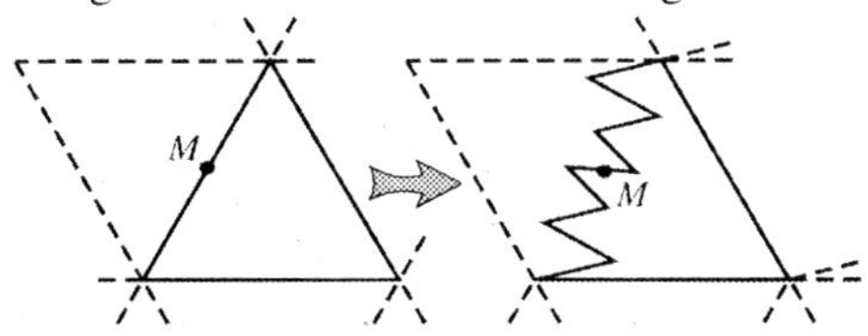

The same idea can be used for any $n \geq 6$, using a modified square for even n, and a modified pentagon for odd n.

26. **(a)** Any four identical triangles can be arranged to form a new triangle that has the same shape, but is twice as large.

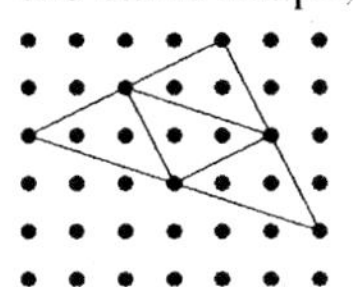

(c)

28. **(a)** Four Sphinxes can be arranged to create a larger sphinx.

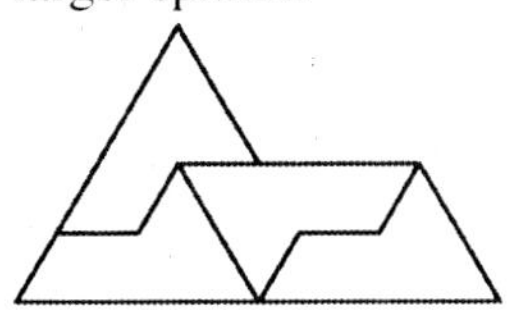

Chapter 11 Review Exercises

1. Move each point 3 units right and 1 unit down.

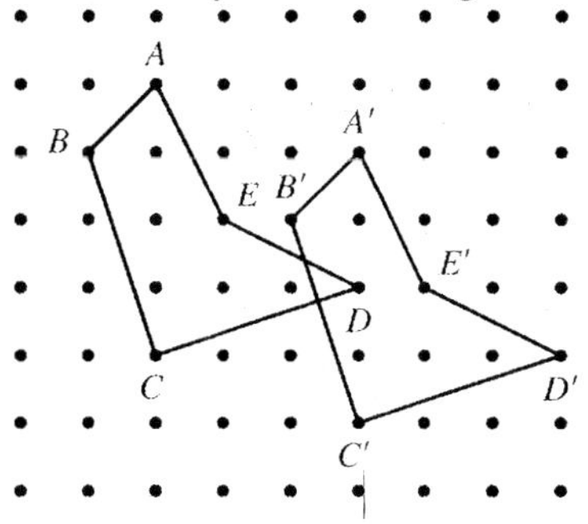

2. Find the perpendicular bisectors of $\overline{AA'}$ and $\overline{BB'}$. Their intersection is the turn center O, and the measure of $\angle AOA'$ is the turn angle.

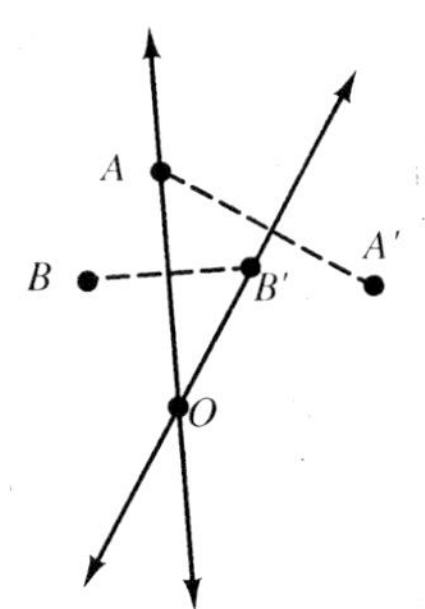

3. Reflection across line l where line l is determined by any two of the midpoints of $\overline{AA'}$, $\overline{BB'}$, and $\overline{CC'}$. It is also the perpendicular bisector of each of these segments.

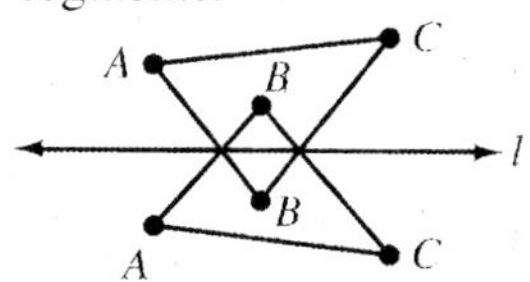

4. Draw any two vertical lines two inches apart. There are then three ways to choose the order of the successive lines of reflection.

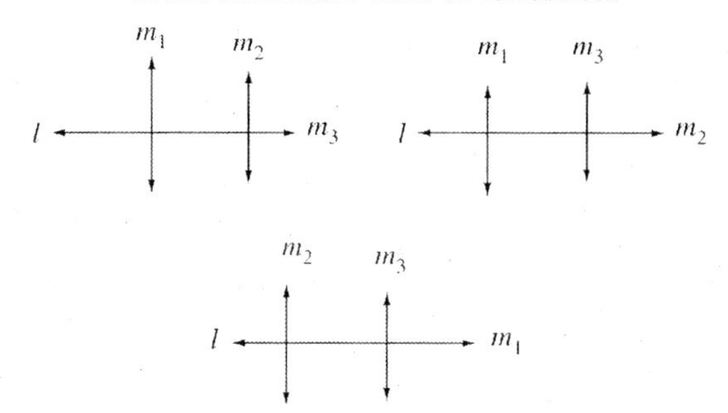

5.

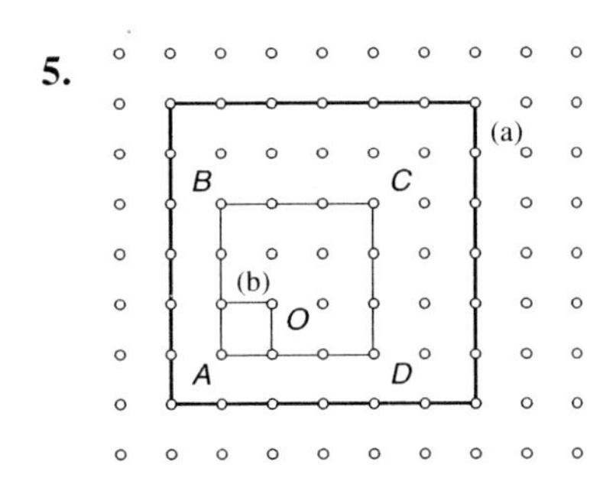

6. By the Pythagorean theorem, each diagonal of $ABCD$ has measure $\sqrt{2}AB$. Each diagonal of $JKLM$ has measure AB, so a size transformation about the center P of $ABCD$ with scale factor $\frac{1}{\sqrt{2}} = \frac{\sqrt{2}}{2}$ should be accomplished, along with a 45° rotation about P. These transformations can be done in either order.

7. (a) Note that $A = A_1 = A_2$ and $C = C_1$.

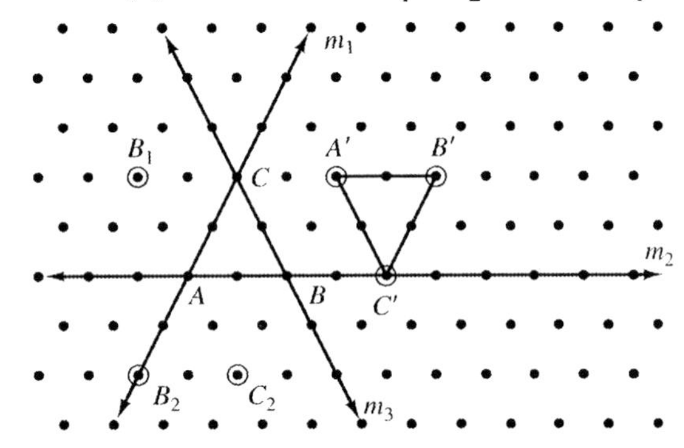

(b) Glide reflection, 3 units right and reflect across the line l parallel to $\overline{AB}$ and midway between C and $\overline{AB}$.

8. (a) 1, about 20° counterclockwise from horizontal

(b) 2, horizontal and vertical

(c) None, due to pattern interweaving

(d) None, due to pattern interweaving

(e) 3, including vertical

(f) Infinitely many (all lines through the center point, since the figure has circular symmetry)

9. The letters in the upper row have either rotational or mirror symmetry, unlike those in the second row.

10. (a) scalene triangle

(b) isosceles but not equilateral triangle

(c) equilateral triangle

(d) rhombus

(e) regular hexagon

11. (a) equilateral triangle

(b) parallelogram

(c) a regular 9-gon (enneagon, or nonagon)

12. (a) None

(b) 180°

(c) 180°

(d) 72°, 144°, 216°, 288°

(e) 120°, 240°

(f) Any angle

13. (a) Translation symmetry and vertical line symmetry (type $m1$)

(b) Translation symmetry and horizontal line symmetry (type $1m$)

14. The angles are 60° for the triangle, 90° for the square, and 120° for the hexagon. The angles of the four polygons must add up to 360°, so the fourth angle is 90°, and therefore the fourth polygon is a square.

15. Answers will vary. Two examples are shown.

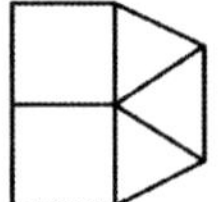

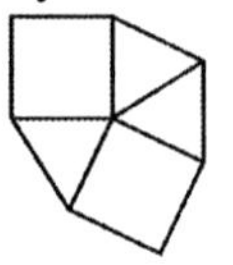

16. Use 180° rotations of the tile about the midpoint of the sides.

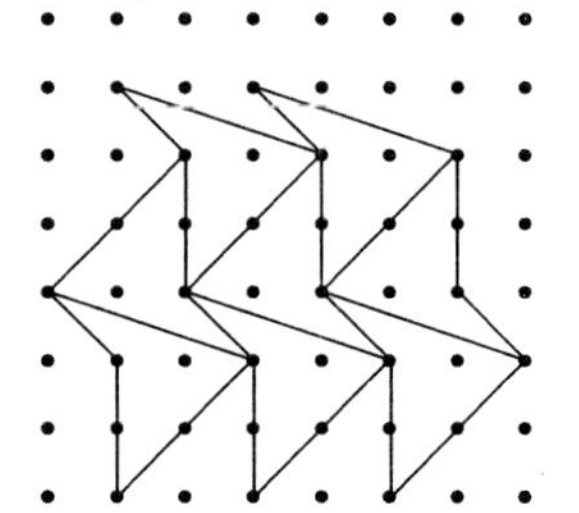

Chapter 11 Test

1. (a), (b), (c), (d), and (e) will tile the plane because any triangle, quadrilateral, pentagon with two pairs of parallel sides, hexagon with one pair or three pairs of opposite parallel sides of the same length will tile the plane. The plane cannot be tiled with any convex polygon having seven or more sides.

2. Various pairs of lines are possible. The distance between the lines must be half the distance between P and P'. In addition, the lines must be perpendicular to $\overline{PP'}$ and m_2 must be "above" m_1.

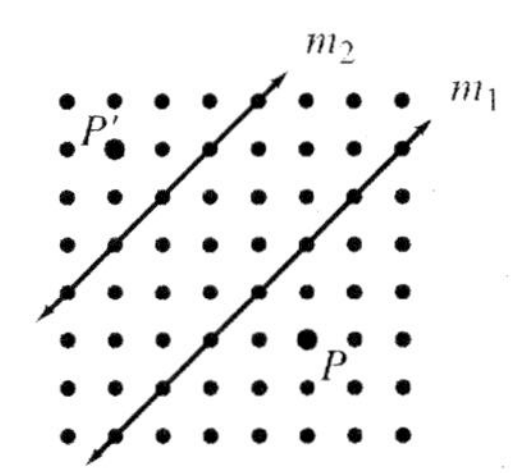

3. Move C 3 units right and 2 units down to find C' (2 units right of B').

Move B' 3 units left and 2 units up to find B (2 units left of C).

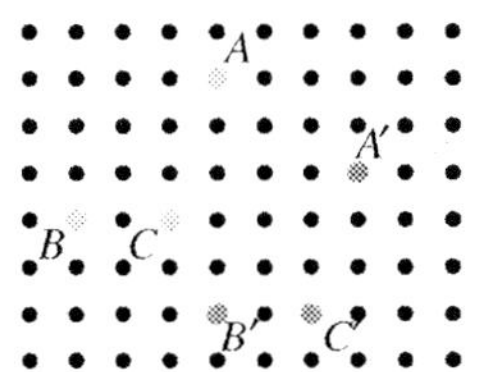

4. **(a)** Rotation (half-turn)

(b) Rotation

5. **(a)** Glide-reflection, since the first two reflections accomplish a translation in a direction parallel to the third line of reflection.

(b) Reflection, since the 3 lines of reflection are parallel.

(c) Reflection, since the 3 lines of reflection are concurrent.

(d) Glide-reflection, since the 3 lines of reflection are neither parallel nor concurrent.

6. **(a)** Translation symmetry and glide reflection symmetry (type 1*g*)

(b) Translation symmetry, vertical line symmetry, horizontal reflection (and glide reflection) symmetry, and half-turn symmetry (type *mm*)

(c) Translation symmetry and half-turn symmetry (type 12)

(d) Translation symmetry, vertical line symmetry, horizontal glide reflection symmetry, and half-turn symmetry (type *mg*)

7. Many transformations are possible. Here is one sequence: Translate the square so A is taken to A'; rotate about A' by 45°; perform a dilation about A' with scale factor $\frac{3\sqrt{2}}{2}$ (since $A'B' = 3\sqrt{2}$ and $AB = 2$).

8. **(a)** The center is P, the intersection of the perpendicular bisectors of $\overline{AA'}$ and $\overline{BB'}$. The rotation angle is 90°.

(b)

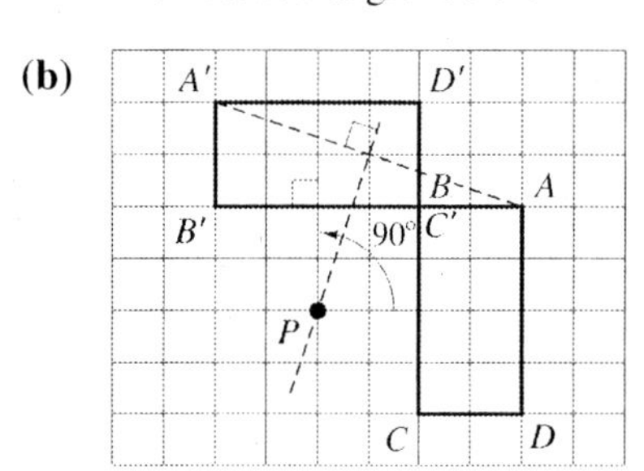

9. A vertex figure must use two octagons and a square, since 135° + 135° + 90° is the only combination that uses both 135° and 90° angles and adds up to 360°.

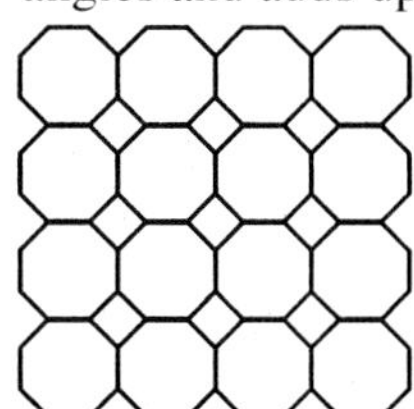

10. Construct the perpendicular bisector of AA'. This is the line of reflection. Label B and C, the points of intersection of the line of reflection with the rays of $\angle A$. Since these points are unchanged by the reflection, the desired angle is $\angle BA'C$.

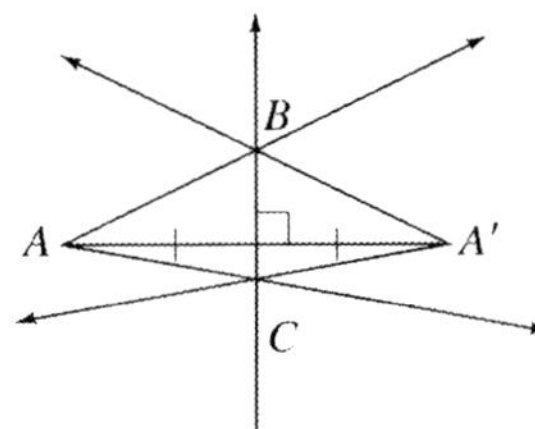

11. **(a)** Since A is unmoved, the center is A.

(b) The scale factor is $\frac{AV}{AB} = \frac{8}{8+4} = \frac{2}{3}$.

(c) $VW = \frac{2}{3} \cdot 20 = \frac{40}{3}$

(d) $ZF = \frac{3}{2} \cdot 18 - 18 = 27 - 18 = 9$

12. Many pairs of lines are possible. The two lines need to intersect at point O and the directed angle between the lines should be half the measure of $\angle QOQ'$.

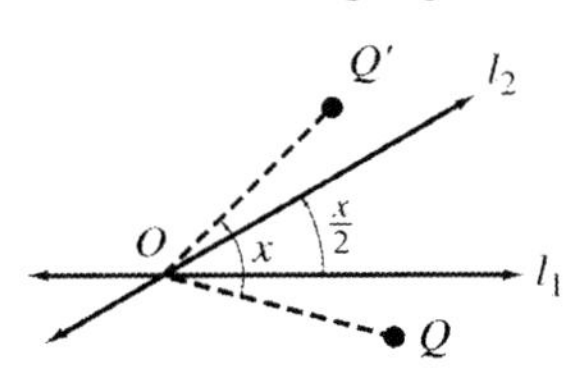

13. **(a)** Point (180° rotation) symmetry, and two diagonal lines of symmetry.

(b) Four lines of symmetry, and 90° rotation symmetry.

14.

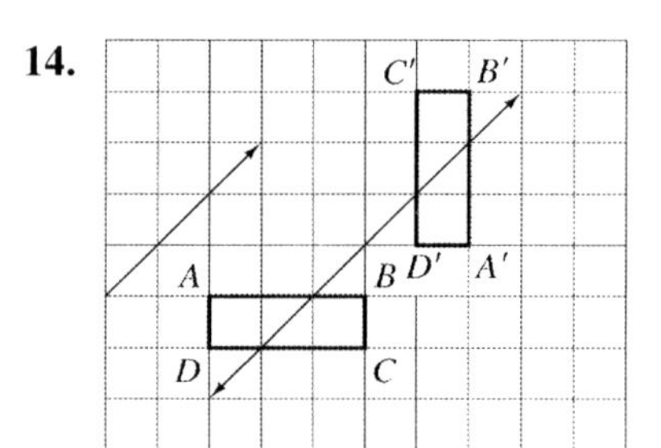

15. **(a)**

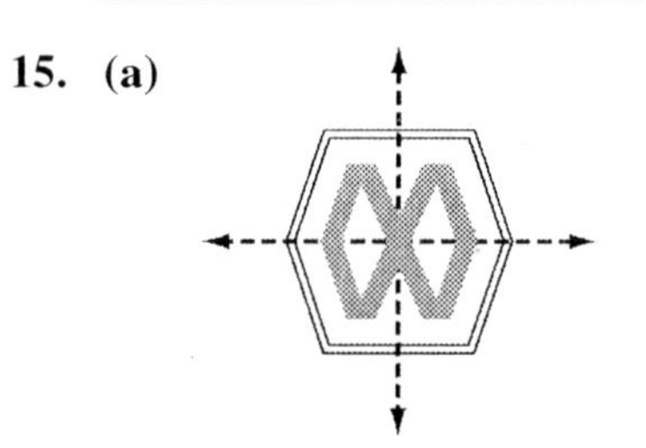

(b)

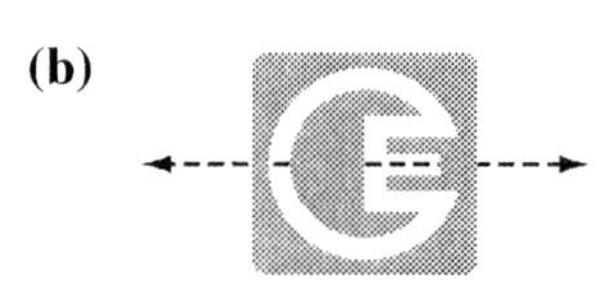

(c)

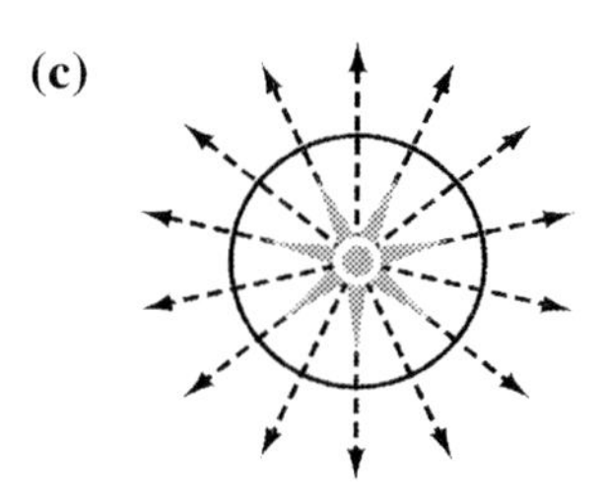

16. **(a)** 180°

(b) None

(c) $\frac{360°}{7}$

Chapter 12 Congruence, Constructions, and Similarity

Problem Set 12.1

1. **(a)** $L \leftrightarrow K, H \leftrightarrow W, S \leftrightarrow T$

(b) $\overline{LH} \leftrightarrow \overline{KW}, \overline{HS} \leftrightarrow \overline{WT}, \overline{SL} \leftrightarrow \overline{TK}$

(c) $\angle L \leftrightarrow \angle K, \angle H \leftrightarrow \angle W, \angle S \leftrightarrow \angle T$

(d) $\triangle LHS \cong \triangle KWT$

3. **(a)** Draw a line segment and mark a point E. Set the compass to AB and determine a point G on the line segment with an arc centered at E. Set the compass to CD and draw an arc centered at G, away from E. This determines a point F on the line segment. Thus, $EF = x + y$.

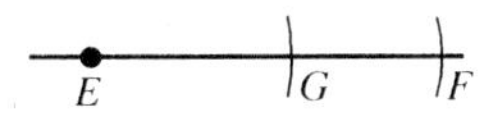

(b) Draw a line segment and mark a point E. Set the compass to AB and determine a point G on the line segment with an arc centered at E. Set the compass to CD and draw an arc centered at G, towards E. This determines a point F on the line segment. Thus $EF = x - y$.

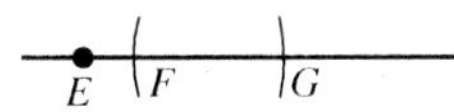

7. **(a)** Use the ruler to draw segment $\overline{AB}$ of length 5 cm. Use the protractor to construct $\angle A$ with measure 28°. Along the terminal side, use the ruler to locate a point C that is 5 cm away from A. Draw segment $\overline{BC}$ to complete the triangle.

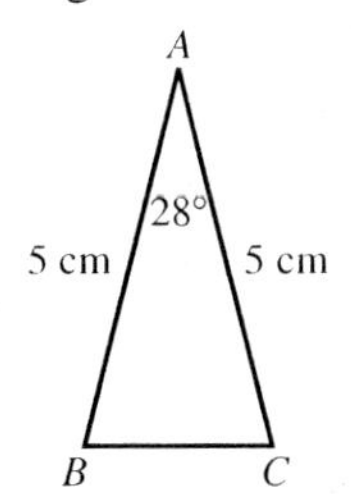

There is only one possible triangle.

(c) This is impossible by the triangle inequality since $2 + 5 < 8$.

(e) Use the protractor to draw a right angle at B. Use the ruler to draw segment $\overline{AB}$ of length 6 cm along one side of the angle and then draw segment $\overline{BC}$ of length 4 cm along the other side. Draw segment $\overline{AC}$ to complete the triangle.

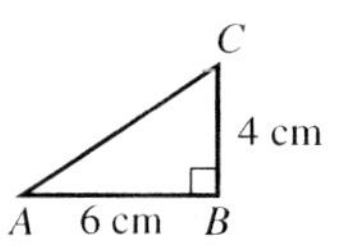

There is only one possible triangle.

8. **(a)** $\overline{BD} \cong \overline{BD}$ and $\overline{AD} \cong \overline{CD}$. $\angle BDA \cong \angle BDC$ since they are right angles. $\triangle ABD \cong \triangle CBD$ by the SAS property.

(c) $\overline{AB} \cong \overline{EF}, \overline{BC} \cong \overline{FD}$, and $\overline{AC} \cong \overline{ED}$. $\triangle ABC \cong \triangle EFD$ by the SSS property.

(e) $\angle BAD \cong \angle CAD, \angle BDA \cong \angle CDA$, and $\overline{AD} \cong \overline{AD}$. $\triangle ABD \cong \triangle ACD$ by the ASA property.

(g) $\angle DAC \cong \angle BCA, \overline{AC} \cong \overline{CA}$, and $\overline{CD} \cong \overline{AB}$, but no conclusion is possible since there is no "SSA" property.

(i) $\overline{AB} \cong \overline{DE}, \overline{AC} \cong \overline{DC}$, and $\angle ACB \cong \angle DCE$, but no conclusion is possible since there is no "SSA" property.

9. Let $\triangle ABC$ be equilateral. Since $AB = BC$, it follows from the isosceles triangle theorem that $\angle A \cong \angle C$. In the same way, since $BC = CA$ it follows that $\angle B \cong \angle A$. Thus all three angles are congruent.

12. **(a)** $\Delta QPT \cong \Delta SPT$ by the SSS property.

13. **(a)** Since alternate interior angles between parallel lines are congruent, $\angle ABD \cong \angle CDB$ and $\angle ADB \cong \angle CBD$. $\overline{DB} \cong \overline{BD}$, thus, $\triangle ABD \cong \triangle CDB$ by the ASA property.

17. Place a corner C of the rectangular sheet of paper and lay the ruler across the plate to meet the points A and B on the edge of the plate crossed by the sides of the sheet of paper, as shown in the figure below.

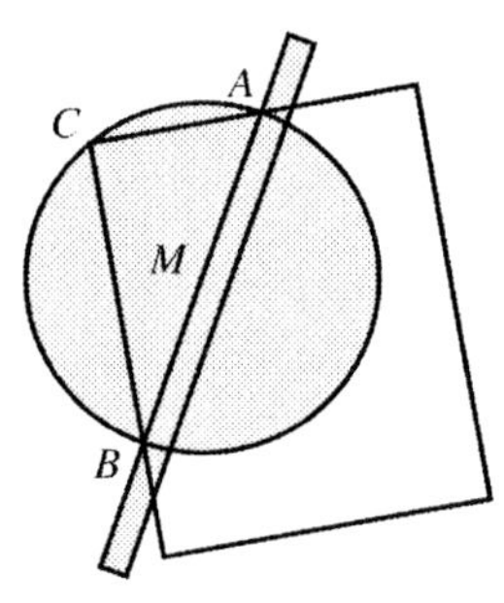

By the converse of Thale's theorem, $\overline{AB}$ is a diameter of the plate. Repeating the procedure at a second point C' will allow you to construct a second diameter $\overline{A'B'}$. The center of the circle is where the two diameters intersect.

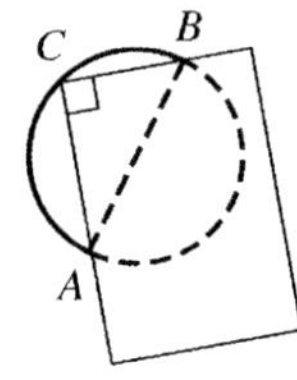

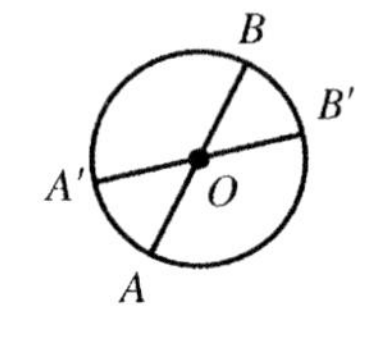

18. Let a be the measure of the third side in centimeters. By the triangle inequality, $a < 4 + 9$ or $a < 13$. Also by the triangle inequality, $9 < a + 4$, so $5 < a$. Thus, the third side is longer than 5 cm and shorter than 13 cm.

19. **(a)** The sum of the three sides is 14 cm, so the fourth side must be less than 14 cm: $0 < s < 14$ cm, where s is the length of the fourth side.

24. **(a)** Since $AB = AE$, $\triangle ABE$ is isosceles. Therefore, $\angle B \cong \angle E$ by the isosceles triangle theorem.

(b) Since $AC = AD$, $\triangle ACD$ is isosceles. Therefore, $\angle ACD \cong \angle ADC$ by the isosceles triangle theorem.

(c) $\angle ACB$ and $\angle ACD$ are supplementary, as are $\angle ADC$ and $\angle ADE$. Since $\angle ACD \cong \angle ADC$,
$m(\angle ACB) = 180° - m(\angle ACD)$
$= 180° - m(\angle ADC) = m(\angle ADE)$, so $\angle ACB \cong \angle ADE$. Using $\angle B \cong \angle E$ from part (a) and $AC = AD$, $\triangle ABC \cong \triangle AED$ by the AAS property.

(d) Using the result of part (c), $\overline{BC} \cong \overline{ED}$ because corresponding parts of congruent triangles are congruent. Thus $BC = DE$.

26. **(a)** Using $\angle A \cong \angle A$ and the given conditions, $\triangle ABC \cong \triangle ADE$ by the ASA property. The conditions are sufficient.

(c) Using $\angle A \cong \angle A$ and the given conditions, $\triangle ABC \cong \triangle ADE$ by the SAS property.

27. If $AB = CD = a$, $BC = AD = b$, and $AC = BD = c$, then each face of the tetrahedron is a triangle with sides of length a, b, and c. By the SSS property, the faces of the tetrahedron are congruent to one another.

29. **(a)** By the triangle inequality, $QP + QT > TP$. Therefore, $QP + QT + TR > TP + TR$. But $QT + TR = QR$, so $QP + QR > TP + TR$.

30. **(a)** $EA + ED > DA$ by the triangle inequality.

32. **(a)** The angles at the vertices of a quadrilateral can change even though the lengths of the sides are fixed. In the case of the rack, the bolts only fix the lengths AB, BC, CD, and AD. (There is no "SSSS" congruence property for quadrilaterals.)

33. **(a)** The framework forms a parallelogram, but not necessarily a rectangle.

36. **(a, b)** The measure of $\angle APB$ is always a constant, always satisfying the equation $m(\angle AOB) = 2 \cdot m(\angle APB)$.

(c) Draw the diameter $\overline{PQ}$. There are two cases to consider.
Case 1:
A and B lie on opposite sides of $\overline{PQ}$. We see that $\triangle POA$ is isosceles, so the base angles are congruent. That is, $m(\angle APO) = m(\angle OAP) = x$. Therefore, $m(\angle AOQ) = 2x = 2m(\angle APO)$. By the same reasoning, $m(\angle BOQ) = 2y = 2m(\angle BPO)$. Thus $m(\angle AOB) = 2x + 2y = 2m(\angle APB)$.

(*continued on next page*)

(*continued*)

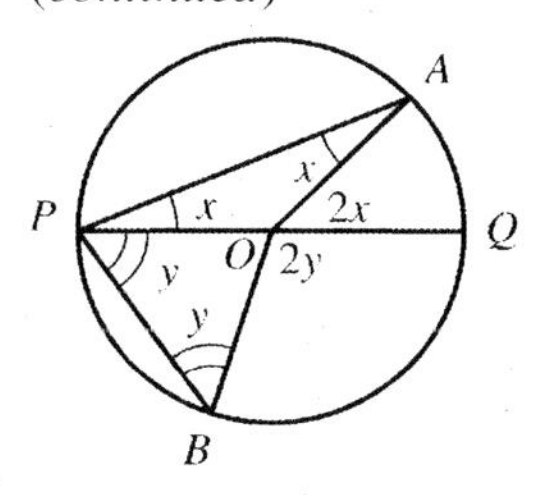

Case 1

Case 2

Case 2:

A and B lie on the same side of $\overline{PQ}$.

Again, $\triangle POA$ is isosceles, so $m(\angle APO) = m(\angle OAP) = x$. Therefore, $m(\angle AOQ) = 2x$. By the same reasoning, $m(BOQ) = 2y$. Thus, $m(\angle AOB) = 2x - 2y = 2m(\angle APB)$.

Problem Set 12.2

1. **(a)** Step 1: Draw a line through P that intersects l. Label the intersection point A.
Step 2: Draw arcs of equal radius centered at A and P. Label as B the intersection point of the arc at A with $\overleftrightarrow{AP}$. Label as C the intersection point of the arc at A with l. Label as D the intersection point of the arc at P with $\overleftrightarrow{AP}$.
Step 3: Set the compass to radius BC, and draw the arc centered at D. Label as E the intersection with the arc drawn at P.
Step 4: Construct the line k through P and E.

(b) The construction gives the congruence of the corresponding angles, $\angle BAC \cong \angle DPE$ Therefore $k \parallel l$ by the corresponding angles property.

3. **(a)** The corresponding angles property guarantees that m is parallel to l.

(b) Align the ruler with the line; slide the drafting triangle, with one leg of the right triangle on the ruler, until the second leg of the triangle meets point P.

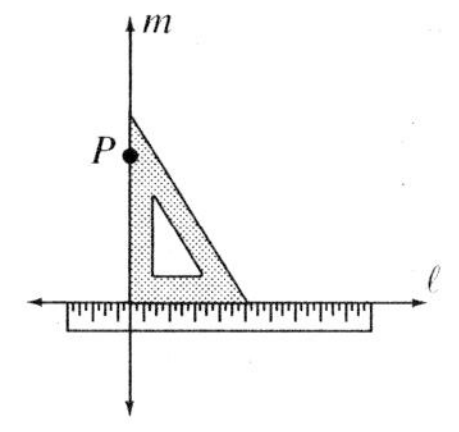

7. The jar lid will also give the circumscribing circle through A, B, C. When it is drawn, each of the four points A, B, C, and P is the intersection of three of the four circles.

8. Construct the perpendicular to side $\overrightarrow{BA}$ at T and the angle bisector. Let O be their point of intersection. The circle centered at O passing through T is the desired circle.

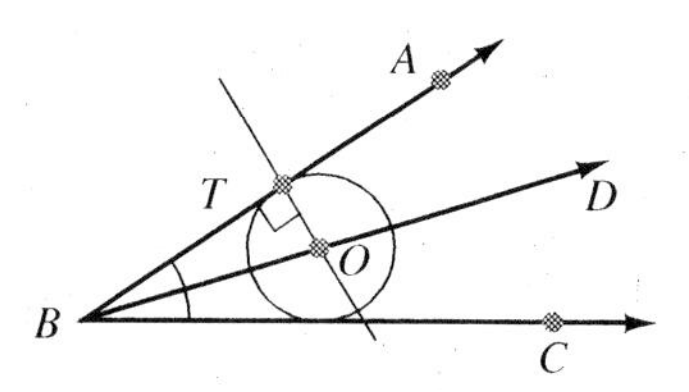

9. **(a)** The circumcenter will be inside the acute triangle

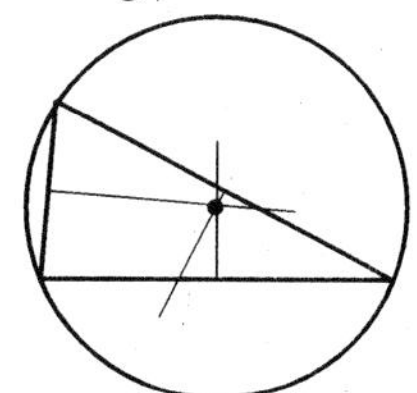

(b) The circumcenter will be at the midpoint of the hypotenuse of the right triangle.

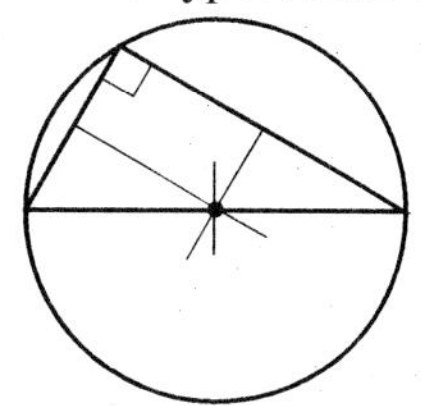

(c) The circumcenter will be outside the obtuse triangle.

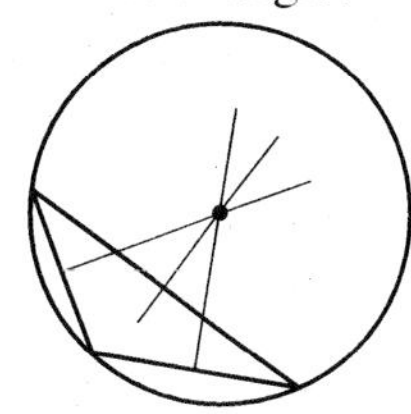

(d) The circumcenter is inside, on, or outside a triangle if, and only if, the triangle is acute, right, or obtuse, respectively.

11. The three circles intersect at a single point, S, as shown in the figure.

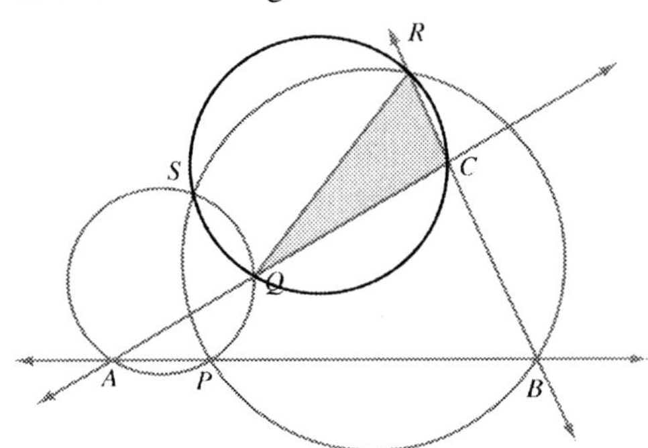

13. Since $\triangle PQS$ is inscribed in a circle with diameter $\overline{PQ}$, it has a right angle at point S by Thale's theorem. Thus $\overline{PS} \perp \overline{SQ}$. Similarly, $\triangle PQT$ is inscribed in a semi-circle with diameter $\overline{PQ}$, so $\overline{PT} \perp \overline{TQ}$. Hence $\overline{PS}$ and $\overline{PT}$ are tangent to the circle at Q.

15. By Thale's theorem, $\angle ADB$ is a right angle. $\triangle ODB$ is an equilateral triangle since all sides have the length of the radius of the circles, so $\triangle ODB$ is equiangular by problem 8 of Problem Set 12.1. Thus, $m(\angle ODB) = 60°$.

Moreover, $ODBE$ is a rhombus, so $\overline{DE}$ is a bisector of $\angle ODB$, so

$$m(\angle ADE) = m(\angle ADB) - m(\angle EDB) = 90° - 30° = 60°.$$

Similarly, $m(\angle AED) = 60°$. Therefore, all angles of $\triangle ADE$ have measure 60°, so $\triangle ADE$ is equilateral.

17. Since $m(\angle 1) + m(\angle 2) + m(\angle 3) + m(\angle 4) = 180°$, $m(\angle 1) = m(\angle 2)$, and $m(\angle 3) = m(\angle 4)$, it follows that $2m(\angle 2) + 2m(\angle 3) = 180°$. Thus, $m(\angle 2) + m(\angle 3) = 90°$. Hence, m and n are perpendicular.

18. (a) Suppose the perpendicular bisector of chord $\overline{AB}$ intersects the circle at a point C. Then the circle is the circumscribing circle of $\triangle ABC$. The center of the circumscribing circle is the point of concurrence of the perpendicular bisectors of all three sides of $\triangle ABC$. In particular, it is on the perpendicular bisector of side $\overline{AB}$ contains the center of the circle.

19. Constructions may vary.

(a) Extend $\overline{AB}$, and construct the line at A that is perpendicular to $\overline{AB}$. Set the compass to radius AB and mark off this distance on the perpendicular line to determine a point C for which $AC = AB$. Similarly, construct a line through B perpendicular to $\overline{AB}$, and determine a point D (on the same side of $\overline{AB}$ as C) on this perpendicular so $BD = AB$. $ABDC$ is a square with given side $\overline{AB}$.

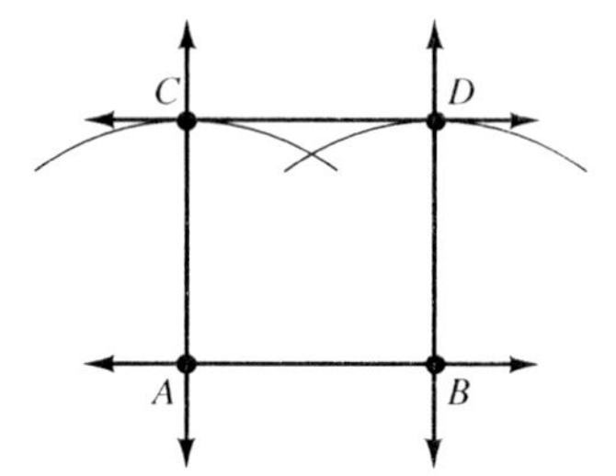

20. Constructions may vary.

(a) Extend $\overline{AB}$. Construct perpendicular rays to $\overline{AB}$ at both A and B, to the same side of $\overline{AB}$. Bisect the right angle at A. Let the bisector's intersection with the ray at B determine C. Construct the line through C perpendicular to $\overline{BC}$. Let D be the intersection with the ray constructed at A. Thus $ABCD$ is a square with the given side $\overline{AB}$.

(b) Construct the equilateral triangle $\triangle ABO$ with the given side $\overline{AB}$. The reflection of A across $\overline{OB}$ determines the point C. Repeat reflections to determine the remaining vertices D, E, and F to construct the regular hexagon $ABCDEF$.

23. By the Gauss-Wantzel Constructibility Theorem, the constructible regular polygons have n sides where $n = 2^r \cdot 4$ (where r is a whole number), $2^r \cdot p$ (where p is a fermat prime and r is a whole number), or $n = 2^r \cdot p_1 \cdot \ldots\ p_m$ (where $p_1, \ldots\ p_m$ are distinct fermat primes and r is a whole number). The only Fermat primes we need be concerned with are 3, 5, and 17, since the remaining Fermat primes are greater than 100.

(*continued on next page*)

(*continued*)

Therefore, the polygon is constructible for the following values of n:

$2^r \cdot 4$: 4, 8, 16, 32, 64, 128, ...

$2^r \cdot 3$: 3, 6, 12, 24, 48, 96, 192, ...

$2^r \cdot 5$: 5, 10, 20, 40, 80, 160, ...

$2^r \cdot 17$: 17, 34, 68, 136, ...

$2^r \cdot 3 \cdot 5$: 15, 30, 60, 120, ...

$2^r \cdot 3 \cdot 17$: 51, 102, ...

$2^r \cdot 5 \cdot 17$: 85, 170, ...

$2^r \cdot 3 \cdot 5 \cdot 17$: 255, 510, ...

In summary we have the following list.
Constructible: 3, 4, 5, 6, 8, 10, 12, 15, 16, 17, 20, 24, 30, 32, 34, 40, 48, 51, 60, 64, 68, 80, 85, 96
Nonconstructible: 7, 9, 11, 13, 14, 18, 19, 21, 22, 23, 25, 26, 27, 28, 29, 31, 33, 35, 36, 37, 38, 39, 41, 42, 43, 44, 45, 46, 47, 49, 50, 52, 53, 54, 55, 56, 57, 58, 59, 61, 62, 63, 65, 66, 67, 69, 70, 71, 72, 73, 74, 75, 76, 77, 78, 79, 81, 82, 83, 84, 86, 87, 88, 89 90, 91, 92, 93, 94, 95, 97, 98, 99, 100

29. **(a)** There are five points on C on line l for which triangle ABC is isosceles. As shown below, they are constructed by the circle at A through B, the circle at B through A, and the perpendicular bisector of $\overline{AB}$.

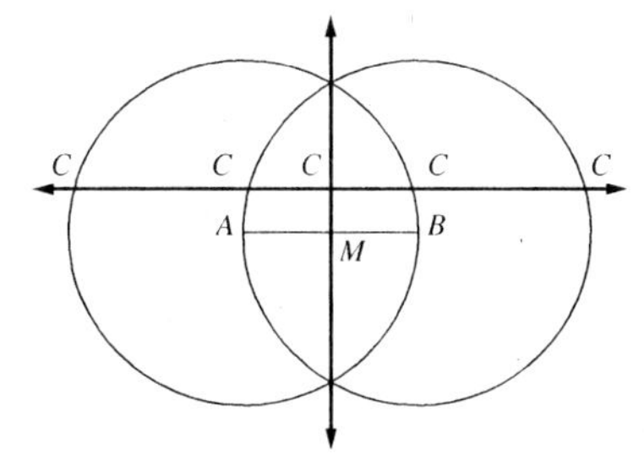

32. **(a)** The three altitudes of a triangle are concurrent (pass through a single point).

(b) The perpendicular bisectors of ΔPQR are known to be concurrent, since they intersect at the center of the circle that circumscribes ΔPQR. Since the perpendicular bisectors of ΔPQR are also the altitudes of ΔABC, the three altitudes are concurrent.

33. $m(\angle NOQ) = m(\angle PON) - m(\angle POQ)$
$= 144° - 120° = 24°$ since a central angle for a regular pentagon is 72° and a central angle for an equilateral triangle is 120°. A regular 15-gon has central angle
$\dfrac{360°}{15} = 24°$, so laying off segments of length QN would give 15

39. **(a)** Use the laws of exponents and the distributive property of multiplication over addition.

$$\begin{aligned}
F_5 &= 2^{2^5} + 1 \\
&= 2^{32} + 1 \\
&= 2^{12+20} + 1 \\
&= 2^{12} \cdot 2^{20} + 1 \\
&= (4096) \cdot (1,048,576) + 1 \\
&= 4 \cdot (1,048,576) \cdot 10^3 \\
&\qquad + 96 \cdot (10^6 + 48,576) + 1 \\
&= 4,194,304,000 + 96,000,000 \\
&\qquad + 4,663,296 + 1 \\
&= 4,294,967,297
\end{aligned}$$

40. **(a)** G, H, and P are collinear. The Euler line passes through G, H, and P.

(b) $\dfrac{GH}{PG} = 2$. Thus G is one third of the distance from P to H along the Euler line.

(c) The circle intersects all sides of $\triangle ABC$ at their midpoints.

(d) The circle bisects each of the segments $\overline{AH}$, $\overline{BH}$, and $\overline{CH}$.

42. $\triangle TRI$ is equilateral.

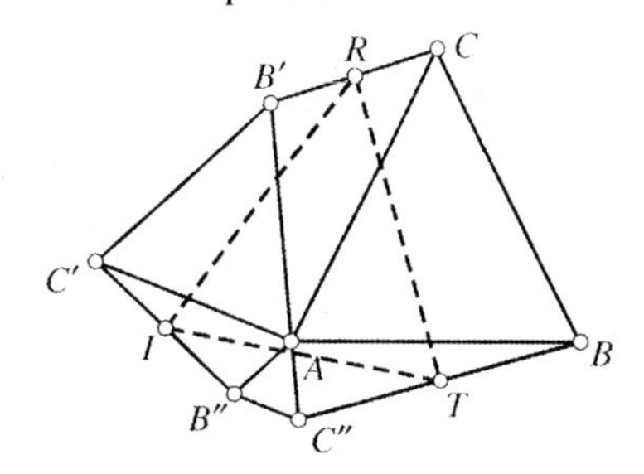

Problem Set 12.3

1. (a) $m(\angle A) = 180° - m(\angle B) - m(\angle C)$
$= 180° - 60° - 90° = 30°$.
$\triangle ABC \sim \triangle PNO$ by the AA similarity property. The scale factor from $\triangle ABC$ to $\triangle PNO$ is $\frac{PN}{AB} = \frac{12}{8} = \frac{3}{2}$.

(c) $\angle HIG \cong \angle UIT$ since they are vertical angles. $\triangle GHI \sim \triangle TUI$ by the AA similarity property. The scale factor is $\frac{UT}{HG} = \frac{8}{5}$.

2. (a) Yes; since all angles measure 60°, equilateral triangles are similar by the AA similarity property.

(c) Yes; since any two such triangles have angles of measure 36° and 90°, they are similar by the AA similarity property.

(e) Yes; any two congruent triangles are similar by the AA similarity property, the SSS similarity property, or the SAS similarity property. The scale factor is 1.

3. (a) $\frac{15}{12} = \frac{a}{8}$; $a = \frac{15}{12} \cdot 8 = 10$

(c) $\frac{15+3}{15} = \frac{c+2}{c}$; $18c = 15c + 30$; $c = 10$

4. (a) By the Pythagorean theorem, the hypotenuse of triangle XYZ is $\sqrt{5^2 + 12^2} = \sqrt{169} = 13$, so its perimeter is 5 + 12 + 13 = 30. Since the perimeter of triangle ABC is 3000 feet, the scale factor is 100. Thus, triangle ABC is a right triangle with legs of length 500 and 1200, and a hypotenuse of length 1300.

6. (a) No; for example, a square and a nonsquare rectangle are convex quadrilaterals with congruent angles, yet they are not similar.

8. (a) $\overline{AB} \parallel \overline{CD}$ so $\angle ABE \cong \angle CDE$ by the alternate interior angles theorem. $\angle AEB \cong \angle CED$ since they are vertical angles. $\triangle ABE \sim \triangle CDE$ by the AA similarity property.

(b) By similarity, $\frac{x}{36} = \frac{17}{51}$, so $x = \frac{17}{51} \cdot 36 = 12$. Also, $\frac{26}{y} = \frac{17}{51}$, so $y = 26 \cdot \frac{51}{17} = 78$.

10. (a) $\angle CAD \cong \angle BAC$, since they are the same angle and $m(\angle ADC) = m(\angle ACB) = 90°$. By the AA similarity property, $\triangle ADC \sim \triangle ACB$. Likewise, $\triangle CDB \sim \triangle ACB$. Thus $\triangle ADC \sim \triangle CDB$.

14. (a) Draw an arc of large enough radius so that point B is on the seventh line above the line with point A.

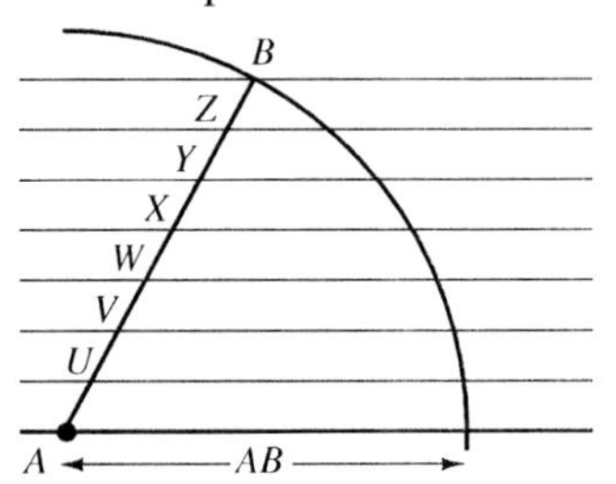

20. (a) $\angle ACB \cong \angle CDA$ since they are right angles. $\angle BAC \cong \angle CAD$ since they are the same angle. $\triangle ABC \sim \triangle ACD$ by the AA similarity property.
$\angle BCA \cong \angle BDC$ since they are right angles. $\angle ABC \cong \angle CBD$ since they are the same angle. $\triangle ABC \sim \triangle CBD$ by the AA similarity property.

(b) Using $\triangle ACD \sim \triangle ABC$, then $\frac{AD}{AC} = \frac{AC}{AB}$ or $\frac{x}{b} = \frac{b}{c}$. Similarly, $\triangle CBD \sim \triangle ABC$ gives $\frac{BD}{BC} = \frac{CB}{AB}$ or $\frac{y}{a} = \frac{a}{c}$.

(c) $x = \frac{b^2}{c}$ and $y = \frac{a^2}{c}$. Also, $x + y = c$, so $c = \frac{b^2}{c} + \frac{a^2}{c}$, or $c^2 = a^2 + b^2$.

22. (a) $DJ = \sqrt{1^2 + \left(\frac{1}{2}\right)^2} = \sqrt{\frac{5}{4}} = \frac{1}{2}\sqrt{5}$

(b) By similarity, $\dfrac{PS}{AD} = \dfrac{TS}{JD}$. Therefore, using part (a),

$$PS = AD \cdot \frac{TS}{JD} = 1 \cdot \frac{\frac{1}{2}}{\frac{\sqrt{5}}{2}} = \frac{1}{\sqrt{5}} = \frac{\sqrt{5}}{5}.$$

(c) Area $(PQRS) = (PS)^2 = \dfrac{1}{5}$. That is the inner small square has 20% of the area of the large square $ABCD$.

25. The midpoints W, X, Y, and Z are the vertices of the parallelogram $WXYZ$, and therefore are in the same plane. Thus, we know that the diagonals of any parallelogram intersect at their common midpoints.

27. Construct $\overline{AC}$. Let L denote the midpoint of $\overline{AC}$, which is also the midpoint of $\overline{BD}$. By Example 12.14, P is the centroid of $\triangle ABC$ and $BP = \dfrac{2}{3}BL$. Since $BL = \dfrac{1}{2}BD$, this shows $BP = \dfrac{2}{3} \cdot \dfrac{1}{2}BD = \dfrac{1}{3}BD$. By the same reasoning, Q is the centroid of $\triangle ADC$ and $QD = \dfrac{1}{3}BD$. Finally, $PQ = BD - BP - QD$ $= \left(1 - \dfrac{1}{3} - \dfrac{1}{3}\right)BD = \dfrac{1}{3}BD$. Therefore, $BP = PQ = QD$.

28. $\angle APC \cong \angle BPD$ since they are vertical angles. $\angle PDB \cong \angle PCA$ since they are right angles. $\triangle ACP \sim \triangle BDP$ by the AA similarity property. Since $\dfrac{AC}{BD} = \dfrac{4}{2} = 2$, the scale factor is 2. Therefore, $CP = 2DP$ and $AP = 2BP$. Since $CD = 4$ and $CD = CP + DP = 3DP$, $DP = \dfrac{4}{3}$, and $CP = \dfrac{8}{3}$. Using the Pythagorean theorem,

$$BP = \sqrt{2^2 + \left(\frac{4}{3}\right)^2} = \sqrt{4 + \frac{16}{9}} = \sqrt{\frac{52}{9}} = \frac{2}{3}\sqrt{13}$$

and $AP = 2BP = \dfrac{4}{3}\sqrt{13}$.

30. The right triangles also have a congruent angle at the vertex at the mirror, so the triangles are similar by the AA property. Assuming Mohini's eyes are 5" beneath the top of her head, this gives the proportion $\dfrac{h}{5'} = \dfrac{15'}{4'}$, making the pole $h = (5')\left(\dfrac{15'}{4'}\right) = 18'9''$ high.

Mohini must take into account how far from the ground her eyes are.

32. By similar triangles $\dfrac{6' - x}{6'} = \dfrac{5.25'}{18'}$. Therefore,

$$x = 6'\left(1 - \frac{5.25'}{18'}\right) = 4.25' = 4'3''.$$

36. **(a)** $\triangle BCP \sim \triangle DAP$

(b) Part (a) gives $\dfrac{PA}{PC} = \dfrac{PD}{PB}$ or $PA \cdot BP = PC \cdot PD$.

Chapter 12 Review Exercises

1. **(a)** True. $\angle A \cong \angle D$, $\angle B \cong \angle E$, $\angle C \cong \angle F$, $\overline{AB} \cong \overline{DF}$, and $\overline{BC} \cong \overline{DE}$.

(b) False. After pairing up the congruent angles, the sides with equal lengths are not corresponding sides in the triangles, so the two triangles are not congruent.

2. **(a)** $\overline{AD} \cong \overline{AB}$, $\overline{CD} \cong \overline{CB}$ and $\overline{AC} \cong \overline{AC}$. $\triangle ACD \cong \triangle ACB$ by SSS.

(b) $\overline{AC} \cong \overline{EC}$, $\overline{CD} \cong \overline{CB}$, and $\angle C \cong \angle C$. $\triangle ACD \cong \triangle ECB$ by SAS.

(c) $\angle A \cong \angle B$, $\angle F \cong \angle C$, and $\overline{FD} \cong \overline{CE}$. $\triangle ADF \cong \triangle BEC$ by AAS.

(d) $\overline{CA} \cong \overline{CE}$, $\overline{CD} \cong \overline{CB}$, and $\angle C \cong \angle C$. $\triangle ACD \cong \triangle ECB$ by SAS.

(e) $\triangle ABE$ is isosceles, so $\angle B \cong \angle E$ by the isosceles triangle theorem. $\angle BAC \cong \angle EAD$ and $\overline{BA} \cong \overline{EA}$. $\triangle ABC \cong \triangle AED$ by ASA. (It is also true that $\triangle ABD \cong \triangle AEC$ by ASA.)

(f) $\Delta ABC \cong \Delta DCB$ by SSS.

3. $m(\angle B) = m(\angle G) = 62°$, $AB = FG = 2.1$ cm, and $BC = GH = 3.2$ cm, so $\triangle ABC \cong \triangle FGH$ by SAS.

(a) $AC = FH = 2.9$ cm

(b) $m(\angle H) = 180° - m(\angle F) - m(\angle G)$
$= 180° - 78° - 62° = 40°$

(c) $m(\angle A) = m(\angle F) = 78°$

(d) $m(\angle C) = m(\angle H) = 40°$

4. $\triangle APB \cong \triangle DPC$ by the SSS property; $\triangle ABC \cong \triangle DCB$ by the SSS property since $AC = AP + PC = DP + PB = DB$, $AB = DC$, and $BC = CB$; $\triangle ADB \cong \triangle DAC$ by the SSS property since $AC = DB$, $DC = AB$, and $AD = DA$.

5. $\angle B \cong \angle C$ by the isosceles triangle theorem. By construction, $BF = CD$ and $BD = CE$. Therefore, $\triangle BDF \cong \triangle CED$ by SAS, so $DE = DF$.

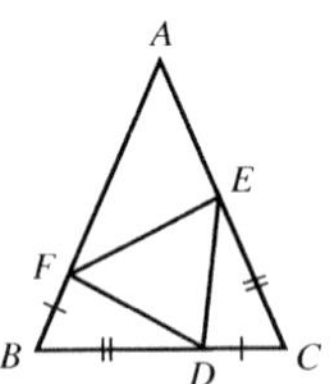

6. Let the diagonals intersect at P. By the triangle inequality applied to ΔAPB and ΔCPD, we have $AB < AP + PB$ and $CD < CP + PD$. Adding the inequalities shows that $AB + CD < AP + PC + BP + PD = AC + BD$. Similarly, $BC + DA < BD + CA$.

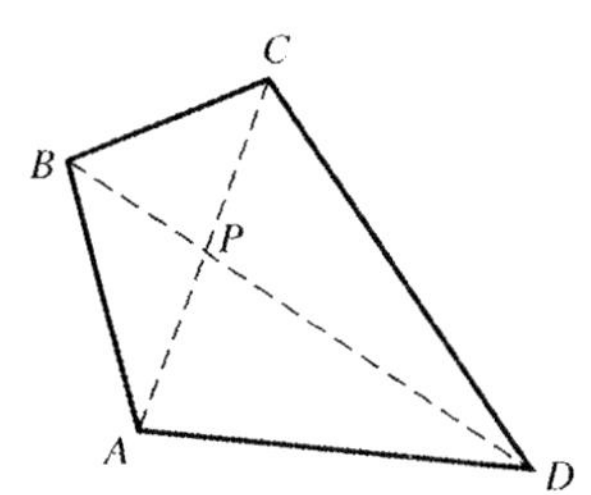

7. (a) See construction 7 of Section 12.2.

(b) See construction 4 of Section 12.2

(c) See construction 6 of Section 12.2.

(d) Draw a circle at any point A on line m, and let it intersect line l at B and C. Construct $\overline{AB}$ and $\overline{BC}$. Draw circles of the same radius at B and C to determine the respective midpoints M and N of $\overline{AB}$ and $\overline{BC}$. (See construction 6 of Section 14.2) Then $k = \overleftrightarrow{MN}$ is the desired line. Alternatively, construct a line perpendicular to l at a point on l. This determines a perpendicular segment between l and m. The perpendicular bisector of the segment is the desired line k.

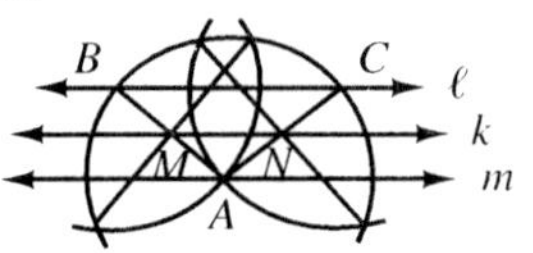

8. (a) Reflect one side of $\angle A$ to the other.

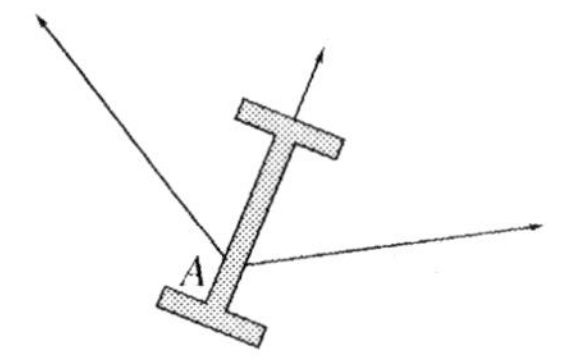

(b) Pivot the Mira about P until the line reflects onto itself.

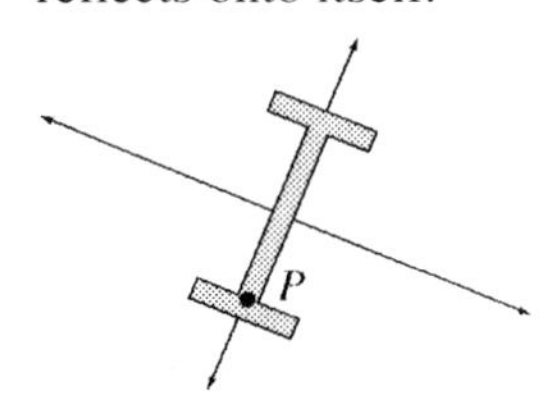

(c) Reflect A onto B.

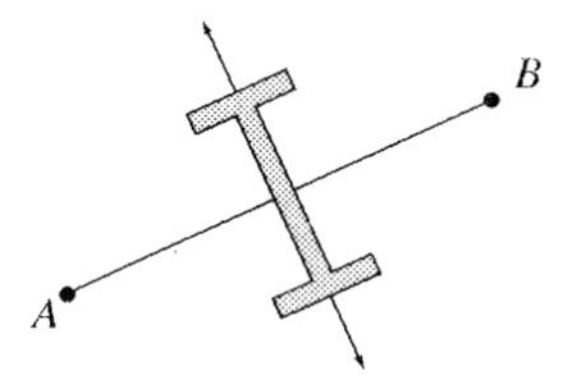

(d) Reflect m onto line ℓ.

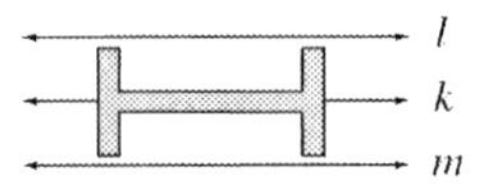

9. (a) Construct $\angle A$, mark off length AB, and draw a circle at B of radius BC. The circle intersects the other ray from A at two points, C_1 and C_2, giving two triangles $\triangle ABC_1$ and $\triangle ABC_2$.

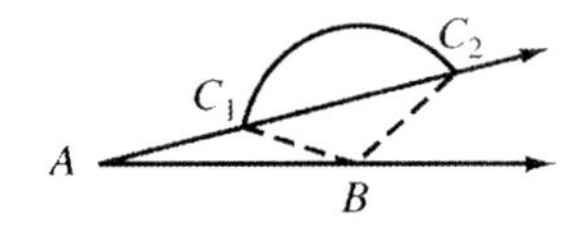

(b) Only $\triangle ABC_1$ has $\angle C = \angle C_1$ obtuse.

10. Find the midpoint M of $\overline{AD}$. Then draw circles of radius $\frac{1}{2}AD$ centered at A, D, and M. $AB = CD = DE = FA = \frac{1}{2}AD$ by the construction. Since $\triangle BCM$ and $\triangle EFM$ are equilateral, $BC = EF = \frac{1}{2}AD$. Thus, $ABCDEF$ is a regular hexagon.

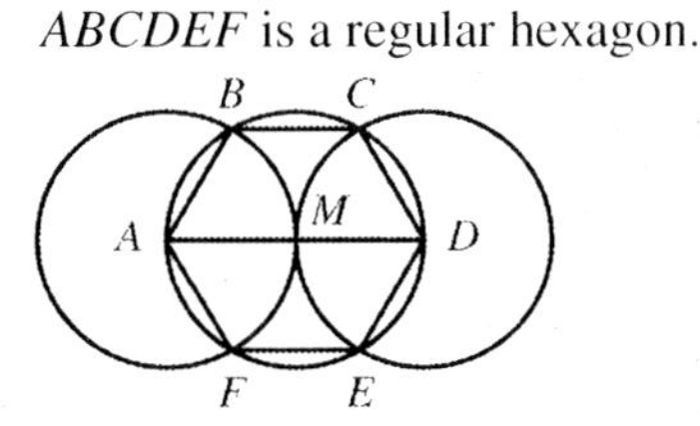

11. **(a)** Yes; measure the side lengths and see if the SSS similarity property applies.

(b) Yes; measure the angles and see if the AA similarity property applies.

12. Bisect sides $\overline{AB}$ and $\overline{AC}$ to determine their midpoints, M and N, respectively. Extend $\overline{AB}$ beyond B and draw the circle at B through M. Let E be the intersection with the extension. Similarly, extend $\overline{AC}$ beyond C, draw the circle at C through N, and let F be the intersection of this circle with the extension. Choose $D = A$, so $\angle A \cong \angle D$. Since

$AM = MB = BE = \frac{1}{2}AB$ and

$AN = NC = CF = \frac{1}{2}AC$,

$DE = AE = AM + MB + BE = \frac{3}{2}AB$ and

$DF = AF = AN + NC + CF = \frac{3}{2}AC$. Thus,

$\frac{DE}{AB} = \frac{DF}{AC} = \frac{3}{2}$. $\triangle ABC \sim \triangle DEF$ by the SAS similarity property.

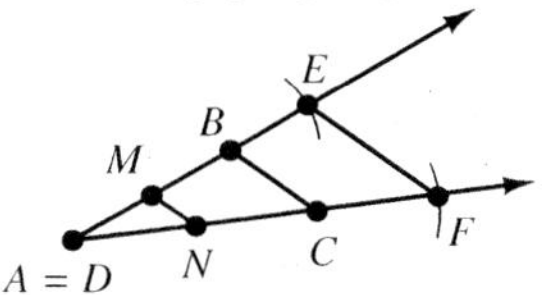

13. **(a)** $\angle A \cong \angle Q$ and $\frac{QP}{AB} = \frac{QR}{AC} = \frac{3}{2}$, so $\triangle ABC \sim \triangle QPR$ by the SAS similarity property. The scale factor is $\frac{3}{2}$.

(b) $\frac{XZ}{CB} = \frac{XY}{CA} = \frac{YZ}{AB} = 3$, so $\triangle ABC \sim \triangle YZX$ by the SSS similarity property. The scale factor is 3.

(c) $m(\angle F) = m(\angle H) = 70°$ by using the isosceles triangle theorem. $\triangle ABC \sim \triangle HGF$ by the AA similarity property. The scale factor is $\frac{HG}{AB} = \frac{4}{6} = \frac{2}{3}$.

(d) $\frac{BD}{AB} = \frac{DC}{BD} = \frac{BC}{AD} = \frac{1}{2}$. $\triangle ABD \sim \triangle BDC$ by the SSS similarity property. The scale factor is $\frac{1}{2}$.

14. Method 1: Draw additional line segments parallel to the given transversals. This creates similar triangle, from which it follows that $\frac{x}{9} = \frac{16}{12} \Rightarrow x = 9\left(\frac{16}{12}\right) = 12$ and $\frac{y}{12} = \frac{15}{9} \Rightarrow y = 12\left(\frac{15}{9}\right) = 20$.

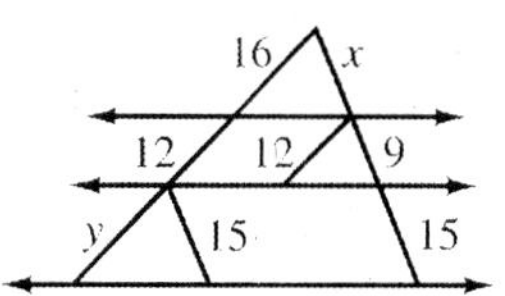

Method 2: Notice that, since k, l, and m are parallel, the small, medium, and large triangles are similar. $\frac{x}{x+9} = \frac{16}{16+12}$ or $28x = 16x + 144$. Thus $12x = 144$ or $x = 12$. $\frac{16}{16+12+y} = \frac{x}{x+9+15}$ or $\frac{16}{28+y} = \frac{12}{36}$. Thus $y + 28 = 48$ or $y = 20$.

15. Let K be the top of the stick and Y the top of the pyramid. Then $\triangle KS_1S_2 \sim \triangle YP_1P_2$ with scale factor $\frac{P_1P_2}{S_1S_2} = \frac{270}{2} = 135$. Therefore the height of the pyramid is 135×3 feet $= 405$ feet.

Chapter 12 Test

1. **(a)** Make the assumption that they form the same angle relative to the ground, namely the angle of elevation, since the distant sun casts shadow rays that are essentially parallel.

   (b) Make the assumption that the person and the tree stand at the same angle with the ground, for example, both vertical. Then $\triangle ABC \cong \triangle DEF$ by the AA similarity property.

   (c) $\frac{DE}{AB} = \frac{DF}{AC}$, so $DE = \frac{56'}{7'} \cdot 6' = 48'$. The tree is 48 ft tall.

2. Draw the circle with radius MC centered at M. The circle passes through A, B, and C. Thus, by Thales' theorem, $\triangle ACB$ is a right triangle because it is inscribed in a semicircle of diameter $\overline{AB}$.

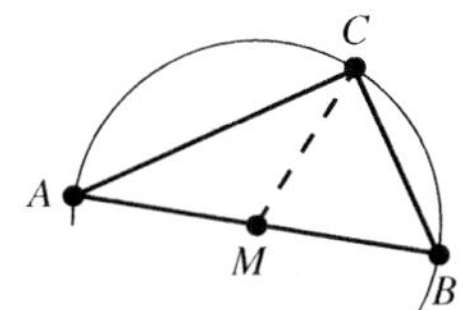

3. **(a)** $\triangle ADE \sim \triangle ACB$ by the AA similarity property, since both triangles contain $\angle A$ and a right angle.

   (b) $\triangle ABC \sim \triangle XYZ$ by the SAS similarity property since $\frac{XY}{AB} = \frac{YZ}{BC} = \frac{5}{4}$ and $\angle B \cong \angle Y$.

   (c) $\triangle DEG \sim \triangle EFG$ by the SSS similarity property, since $\frac{EF}{DE} = \frac{FG}{EG} = \frac{GE}{GD} = 2$.

   (d) $\triangle AEB \sim \triangle CED$ by the AA similarity property since $\angle AEB \cong \angle CED$ being vertical angles and $\angle EBA \cong \angle EDC$ being alternate interior angles between parallel lines.

4. **(a)** $\overline{AC} \cong \overline{AC}$, $\angle ACD \cong \angle ACB$, and $\angle D \cong \angle B$. $\triangle ACD \cong \triangle ACB$ by the AAS congruence property.

   (b) $\overline{AB} \cong \overline{AD}, \overline{AC} \cong \overline{AC}$, and $\angle BAC \cong \angle DAC$. $\triangle ABC \cong \triangle ADC$ by the SAS congruence property.

   (c) $m(\angle ADC) = m(\angle ADB) + m(\angle BDC) = m(\angle BCA) + m(\angle ACD) = m(\angle BCD)$, so $\angle ADC \cong \angle BCD$. $\angle ACD \cong \angle BDC$ and $\overline{DC} \cong \overline{CD}$. $\triangle ADC \cong \triangle BCD$ by the ASA congruence property.

   (d) $\angle ABE \cong \angle CBD$ since they are vertical angles. $\angle A \cong \angle C$ and $\overline{AE} \cong \overline{CD}$. $\triangle ABE \cong \triangle CBD$ by the AAS congruence property.

   (e) $\angle BDC \cong \angle FDE$ since they are vertical angles. $\angle DBC \cong \angle DFE$ since they are right angles. $\overline{BD} \cong \overline{FD}$. $\triangle BCD \cong \triangle FED$ by the ASA congruence property. (Note that $\triangle ACF \cong \triangle AEB$ also.)

   (f) $\overline{AB} \cong \overline{AD}, \overline{BC} \cong \overline{DC}$, and $\overline{AC} \cong \overline{AC}$. $\triangle ABC \cong \triangle ADC$ by the SSS congruence property.

5. $\angle A \cong \angle A$ and $\angle ADE \cong \angle ACB$ since they are right angles. $\triangle ADE \sim \triangle ACB$ by the AA similarity property. Thus
$\frac{AE}{AB} = \frac{AD}{AC} = \frac{AD}{AD + DC} = \frac{2DC}{2DC + DC} = \frac{2}{3}$, so
$AE = \frac{2}{3}AB = \frac{2}{3} \cdot 12 = 8$.
$EB = AB - AE = 12 - 8 = 4$.

6. **(a)** $m(\angle W) = m(\angle M) = 53^\circ$

   (b) Scale factor $= \frac{WU}{MK} = \frac{20}{16} = \frac{5}{4}$

   (c) $\frac{UV}{KL} = \frac{5}{4}$, so $UV = \frac{5}{4}KL = \frac{5}{4} \cdot 20 = 25$.

7. Let x be the length of third side in feet. Use the triangle inequality. $x < 10 + 16$, so $x < 26$. $16 < 10 + x$, so $6 < x$. Thus the third side is greater than 6 feet and less than 26 feet.

8. Since $ABCDE$ is a regular pentagon, $\angle A \cong \angle B \cong \angle C \cong \angle D \cong \angle E$ and $AB = BC = CD = DE = EA$. Since $AP = BQ = CR = DS = ET$ is given, $PB = QC = RD = SE = TA$. Thus $\triangle APT \cong \triangle BQP \cong \triangle CRQ \cong \triangle DSR \cong \triangle ETS$ by the SAS property, so $PT = QP = RQ = SR = TS$. Therefore, $PQRST$ is equilateral. If $m(\angle APT) = x$ and $m(\angle ATP) = y$, then the measure of each interior angle of $PQRST$ is $180^\circ - x - y$. Thus, $PQRST$ is regular.

9. (a) Draw circles of radius PQ, one centered at P and one centered at Q. The circles intersect at points R and S for which $\triangle PQR$ and $\triangle PQS$ are equilateral.

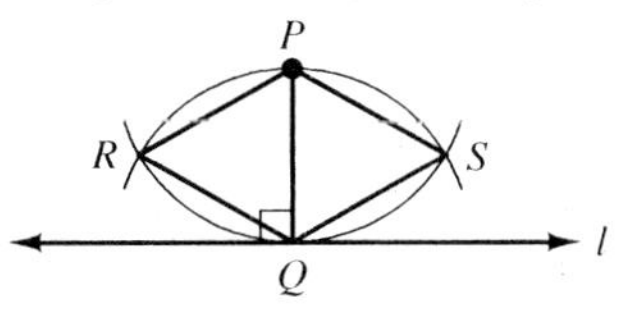

(b) Construct the line l that is perpendicular to $\overline{PQ}$ at Q. Then, construct the angle bisectors of both $\angle QPR$ and $\angle QPS$, and denote their intersections with line l as T and U. Since both $\angle QPR$ and $\angle QPS$ measure 60°, then $\angle TPQ$ and $\angle UPQ$ each measure 30°. Thus $\angle PTQ$, $\angle PUQ$, and $\angle TPU$ each measure 60°. Since $\triangle PTU$ has now shown to be equiangular, it is also equilateral.

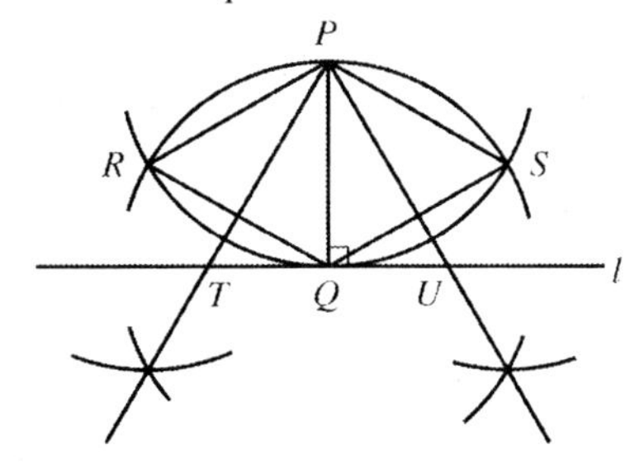

10. Construct a segment $\overline{DE}$ that is congruent to $\overline{AB}$. Construct rays at D and E to the same side of $\overline{DE}$ to form angles that are respectively congruent to $\angle A$ and $\angle B$. Let F be a point of intersection of the rays. Then $\triangle DEF \cong \triangle ABC$.

11. Since $\angle F \cong \angle G$, $FH = GH$ by the isosceles triangle theorem. Since $FG = FH$, $\triangle FGH$ is equilateral.

12. (a) $\triangle ABC$ and $\triangle DEF$ are right triangles. Using the Pythagorean theorem, we have $AB = 12 = DE$ and $DF = 9 = AC$, so $\triangle ABC \cong \triangle DEF$ by SSS or SAS.

(b) They are not congruent. The hypotenuse of $\triangle ABC$ is 5. Since $EF = 5$, the hypotenuse of $\triangle DEF$ must have length greater than 5, so the hypotenuses cannot correspond.

(c) $\triangle ABD$ and $\triangle CBD$ are right triangles. Using the Pythagorean theorem, we have $AB = 3 = CB$ and $CD = 4 = AD$. $\triangle ABD \cong \triangle CBD$ by SSS or SAS.

(d) $\angle ABE \cong \angle DBC$ since they are right triangles. $AB = DB$ and $BE = BC$. $\triangle ABE \cong \triangle DBC$ by the SAS property.

Chapter 13 Statistics: The Interpretation of Data

Problem Set 13.1

3. Count the number of scores in each of the intervals to obtain the frequency and thus the height of the bar above the interval. For example, the first interval (20–29) has no scores so the height is 0. The second interval (30–39) has 1 score so the height above the interval is 1.

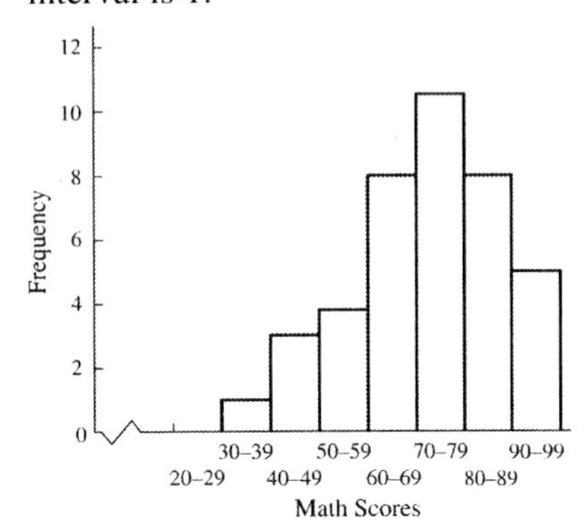

4. (a)

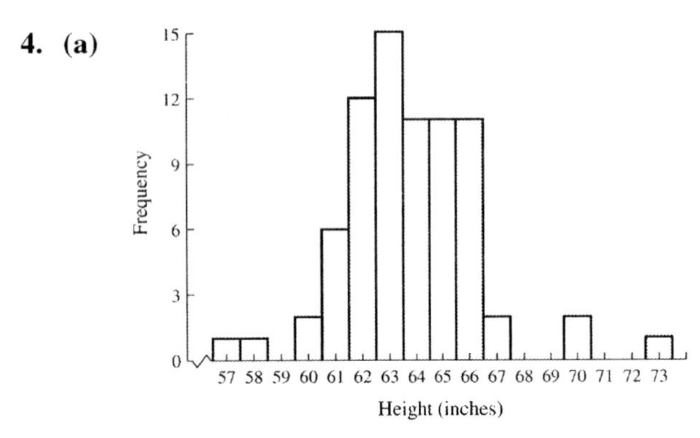

7. Find the central angle to the nearest degree, for each expense.

Taxes: $\frac{21,000}{64,000} \cdot 360° \approx 118°$

Rent: $\frac{10,800}{64,000} \cdot 360° \approx 61°$

Food: $\frac{5,000}{64,000} \cdot 360° \approx 28°$

Clothes: $\frac{2,000}{64,000} \cdot 360° \approx 11°$

Car payments: $\frac{4,800}{64,000} \cdot 360° = 27°$

Insurance: $\frac{5,200}{64,000} \cdot 360° \approx 29°$

Charity: $\frac{7,000}{64,000} \cdot 360° \approx 39°$

Savings: $\frac{6,000}{64,000} \cdot 360° \approx 34°$

Misc: $\frac{2,200}{64,000} \cdot 360° \approx 12°$

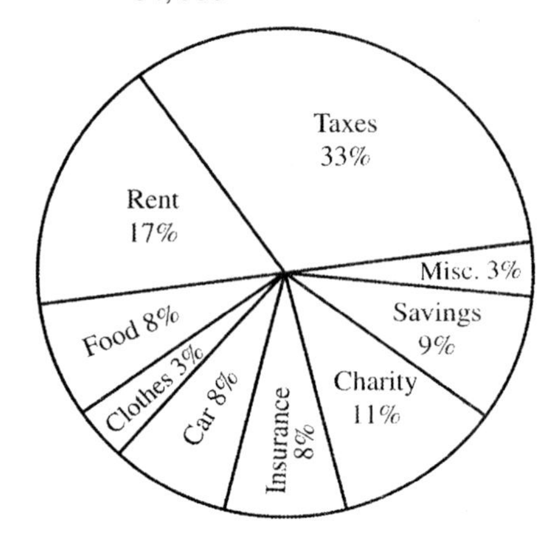

9. (a) 12 students received a C.

20. (a) Histogram (A) emphasizes the changes in the daily Dow Jones average by its choice of vertical scale. The changes appear to be large in histogram (A) thus exaggerating the report of stock activity on the evening news. Histogram (B) makes it clear that the changes are minimal.

(b) $\frac{10086 - 10050}{10086} \cdot 100\% = \frac{36}{10086} \cdot 100\%$
$\approx 0.36\%,$

which an investor probably would not worry about.

21. (a) The pictographs are misleading since the figure for 1999 appears to be more than twice as large as the one for 1990. One would guess the assets had more than doubled.

24. (a) $(17.5 + 8.2) \div$
$(17.5 + 8.2 + 17.4 + 6 + 1.5 + 1.2)$
$\approx 49.6\%$

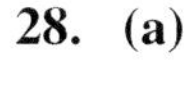
28. (a)

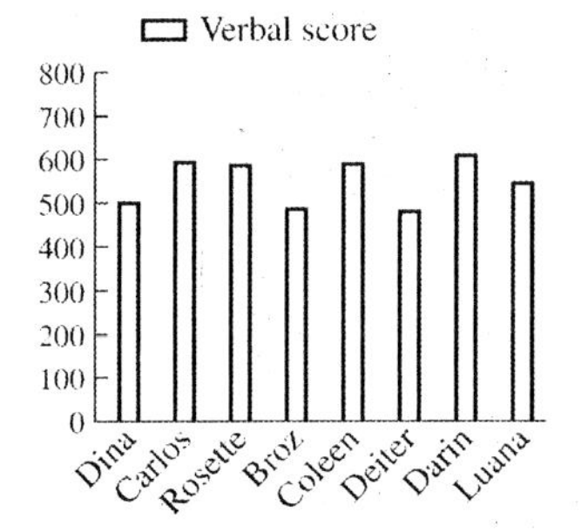

29. (a)

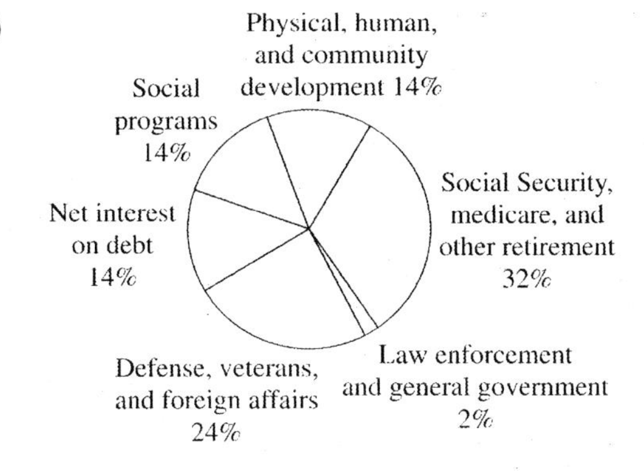

30.

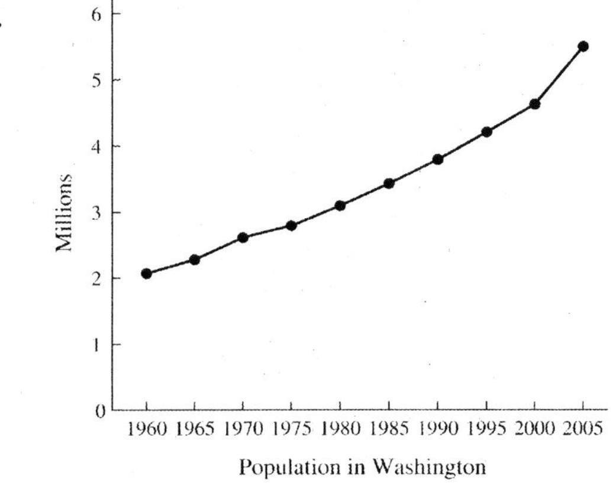

Problem Set 13.2

1. mean: $\overline{x} = \dfrac{\sum_{i=1}^{n} x_i}{n} = \dfrac{\sum_{i=1}^{16} x_i}{16}$

$= \dfrac{18 + 27 + \cdots + 27 + 30}{16} = \dfrac{329}{16}$

$= 20.5625 \doteq 20.6$

median: arrange the values in order from smallest to largest

14 15 17 17 18 18 18 19 19 21 22 23 24 27 27 30

The median is the average of the two middle values. $\hat{x} = \dfrac{19+19}{2} = 19$

mode: The mode is the value that occurs most often. In this case, the mode is 18.

3. (a) Listing the values from Problem 2 in order: 64, 67, 67, 69, 69, 69, 70, 70 70, 71, 71, 73, 74, 77, 77, 78, 79, 79, 79, 79, 80, 80, 81, 81, 81, 85, 86

Q_L is the median for the first 13 values: $Q_L = 70$. $\hat{x}$ is the median of the entire set: $\hat{x} = 77$. Q_U is the median for the last 13 values: $Q_U = 80$.

(b) The 5-number summary is: the smallest value – $Q_L - \hat{x} - Q_U$ – the largest value; i.e., 64 – 70 – 77 – 80 – 86.

6. (a) $\overline{x} = \dfrac{\sum_{i=1}^{16} x_i}{16} = \dfrac{329}{16} = 20.5625 \approx 20.56$

(b) $s = \sqrt{\dfrac{\sum_{i=1}^{16}(\bar{x} - x_i)^2}{16}} = \sqrt{\dfrac{(20.5625 - 18)^2 + (20.5625 - 27)^2 + \cdots + (20.5625 - 30)^2}{16}}$

$\approx \sqrt{\dfrac{6.5664 + 41.4414 + \cdots + 89.0664}{16}} \approx \sqrt{\dfrac{315.9375}{16}} \approx 4.44$

(c) To be within one standard deviation of the mean, a data value must be in the interval $(20.56 - 4.44, 20.56 + 4.44) = (16.12, 25)$.

There are 11 out of 16 values between 16.12 and 25. $\dfrac{11}{16} \approx 0.69 = 69\%$

(d) To be within two standard deviations of the mean, a data value must be in the interval $(20.56 - (2) \cdot (4.44), 20.56 + (2) \cdot (4.44)) = (11.68, 29.44)$

There are 15 out of 16 values between 11.68 and 29.44. $\dfrac{15}{16} \approx 0.94 = 94\%$

(e) To be within three standard deviations of the mean, a data value must be in the interval $(20.56 - (3) \cdot (4.44), 20.56 + (3) \cdot (4.44)) = (7.24, 33.88)$.

All 16 values are between 7.24 and 33.88: $\dfrac{16}{16} = 1 = 100\%$

12. (a) Except possibly for round off error, the sum of the deviations equals zero for every data set.

(b) Compute the sum in part (a) and point out that the negative terms balance out the positive terms so that the sum is zero.

13. (a) First calculate the mean of the data values. Do this by finding the sum of all the data values, and then dividing the sum by the total number of data values. Thus, the mean is

$$\frac{22.2 + 23.5 + 22.5 + 22.6 + 23.0 + 22.8 + 22.4 + 22.2 + 23.0 + 23.3 + 23.9 + 22.7}{12} \approx 22.8.$$

Therefore, the sum of the absolute values of the differences of the data values from the mean is $|22.2 - 22.8| + |23.5 - 22.8| + \cdots + |22.7 - 22.8| \approx 4.9$.

(b) We compute the mean or average value of the sum of the absolute values by dividing the result of part (a) by the number of data values, or 12. By doing so, we obtain $\dfrac{4.9}{12} \approx 0.41$. This number grows with greater variability, so could be used as a measure of variability.

(c) Compliment Leona on a good idea that avoids the canceling out of the effects of the various terms and note that her approach has an effect that is similar to the effect produced by squaring the terms in computing the standard deviation since squares are always positive. Here the standard deviation is 0.53, roughly the same as the 0.4 of part (b). The real reason for using the standard deviation is that it possesses useful theoretical properties that are too involved to discuss here while, at the same time, giving a reasonable measure of variability.

14. (a) Instead of finding the median for the number of students, Joseph found the median for the number of books reported. When calculating the mode, Joseph looked at the most number of books possible to report, not the most number of students who had read, 8. The mode is 1, and the median is 2.5.

(b) Explain the difference between the label (number of books reported) and the outcome (number of students who read that many books).

20. **(a)** mean: $\overline{x} = \dfrac{\sum_{i=1}^{n} x_i}{n} = \dfrac{28+34+41+19+17+23}{6} = \dfrac{162}{6} = 27$

standard deviation $s = \sqrt{\dfrac{1+49+196+64+100+16}{6}} = \sqrt{\dfrac{426}{6}} \approx 8.4$

(b) mean: $\overline{x} = \dfrac{33+39+46+24+22+28}{6} = \dfrac{192}{6} = 32$

standard deviation: $s = \sqrt{\dfrac{1+49+196+64+100+16}{6}} = \sqrt{\dfrac{426}{6}} \approx 8.4$

23. **(a)** $\overline{x} = \dfrac{1\cdot 12 + 1\cdot 18 + 5\cdot 22 + 1\cdot 24 + 3\cdot 26 + 2\cdot 28 + 4\cdot 30 + 3\cdot 32 + 1\cdot 36 + 2\cdot 38 + 2\cdot 42 + 1\cdot 60 + 3\cdot 100}{29}\cdot 1000$

$= \dfrac{1070}{29}\cdot 1000 \doteq 36,900$

To find the median, find the 15th ordered value: $\hat{x} = 30,000$. The mode is 22,000.

24. Answers will vary.

(a) Consider the condition: mean = median
Suppose $\overline{x} = \hat{x} = 10$. Now choose 2 values below $\overline{x}$, say 5 and 7. But we must also choose 2 values above $\overline{x}$ so that $\dfrac{\sum x_i}{n} = \overline{x}$, say 13 and 15. The condition mean = median has been met with the data: 5, 7, 10, 13, 15.
Consider the condition:
mean = median < mode. Using the data above, there must be 2 equal values larger than 10 to meet this condition. If 13 and 15 are replaced with their average of 14, then mean = median < mode for the data: 5, 7, 10, 14, 14.
For these values, $\overline{x} = 10, \hat{x} = 10$, and the mode is 14.

25. **(a)** If

$$s = \sqrt{\frac{\sum_{i=1}^{n}(\overline{x} - x_i)^2}{n}} = \sqrt{\frac{(\overline{x} - x_1)^2 + (\overline{x} - x_2)^2 + \cdots + (\overline{x} - x_n)^2}{n}} = 0$$

then $\sum_{i=1}^{n}(\overline{x} - x_i)^2 = 0$ and $(\overline{x} - x_i)^2 = 0$ for every x_i. That means that $\overline{x} = x_i$ for every x_i. Therefore, all the data values are equal.

27. **(a)** Given $n = 10$ and $\overline{x} = 3$, then $\sum x = n\cdot\overline{x} = 10\cdot 3 = 30$. Since $\sum x = 30$ for 10 values of ones, twos, and threes, the 10 values must all be 3s.

28. **(a)** $\overline{x}_A = \dfrac{27+38+25+29+41}{5} = \dfrac{160}{5} = 32$

(c) $\overline{x}_C = \dfrac{160+32+32}{7} = 32$

31. The total of data values in A is 30 · 45 = 1350. The total of data values in B is 40 · 65 = 2600. For the combined data,
$\overline{x} = \dfrac{1350+2600}{30+40} \approx 56.4$.

36. **(a)** No, since the number of individuals in each category varies widely. As a trivial example, suppose that one individual has no high school diploma and two have high school diplomas but no higher degree. It is tempting to think that the average median income for these two categories is ($16,000 + $24,000)/2 = $20,000. However, this is not so. Since one individual is in the first category and two are in the second, the average should be ($16,000 + $24,000 + $24,000)/3 = $21,333.33.

Problem Set 13.3

1. (a) All freshmen in U.S. colleges and universities in 2010

(c) All people in the U.S.

2. Yes. Many poorer people cannot afford telephones and some people have only cell phones; many people are irritated by telephone surveys and sales pitches, and so on. These factors could certainly bias a sample.

5. (a) This is surely a poor sampling procedure. The sample is clearly not random. The selection of the colleges or universities could easily reflect biases of the investigators. The choices of the faculty to be included in the study almost surely also reflects the bias of the administrators of the chosen schools.

(b) Presumably the population is all college and university faculty. But the opinions of faculty at large research universities are surely vastly different from those of their colleagues at small liberal arts colleges. Indeed, there are almost surely four distinct populations here.

11.

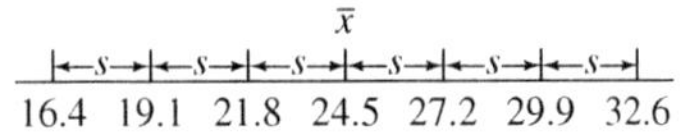

(a) $\overline{x} \pm \sigma$; The limit 2.7 units less than 24.5 is 21.8. The limit 2.7 units more than 24.5 is 27.2.

14. (a) $\overline{x} \approx 20.2$, $s \approx 2$. The z scores, $z = \dfrac{x - \overline{x}}{s}$, corresponding to the data in the order listed are –1.6, 0.9, 0.4, –0.6, 1.4, –0.6.

15. (a) $-1.6 + 0.9 + 0.4 + (-0.6) + 1.4 + (-0.6) = -0.1$

16. (a) Since five of the six data values are less than or equal to 22, $\dfrac{5}{6} = 0.83\overline{3}$, or 22 is the $83\frac{1}{3}$ percentile.

17. (a) A z-score of –1.75 corresponds to an area of 0.0401 = 0.04 = 4% to the nearest hundredth.

18. (a) Subtract the area to the left of $z = -1.75$ from that of $z = 1.75$. 0.96 – 0.04 = 0.92 = 92% to the nearest hundredth

23. Yes, since all sides of the die are equally likely to come up, all sequences of 0s and 1s are equally likely to appear.

27. (a) Since the z score indicates the location of a data point in a data set, in this case they are likely to be the same as in problem 14(a).

(b) The z scores, $z = \dfrac{x - \overline{x}}{s}$, are the same: –1.6, 0.9, 0.4, –0.6, 1.4, –0.6

28. Consider the data set {1, 2, 3}. Then $\overline{x} = \dfrac{1+2+3}{3}$ and the z scores are $\dfrac{1-\overline{x}}{s}$, $\dfrac{2-\overline{x}}{s}$, and $\dfrac{3-\overline{x}}{s}$. Therefore, the sum is

$$\frac{1-\overline{x}}{s} + \frac{2-\overline{x}}{s} + \frac{3-\overline{x}}{s} = \frac{1+2+3-3\left(\frac{1+2+3}{3}\right)}{s} = \frac{\frac{3(1+2+3)}{3} - 3\left(\frac{1+2+3}{3}\right)}{s} = 0$$

and the same computation would be true for $x_1, x_2, \cdots, x_n$.

29. (a) Probably twice what they were in problem 14(a).

(b) The z-scores, $z = \dfrac{x - \overline{x}}{s}$, are the same: –1.6, 0.9, 0.4, –0.6, 1.4, –0.6

(c) The data are all twice as big, but s is as well. Since the z scores are scaled by s, it is not surprising that they remain the same.

Chapter 13 Review Exercises

1. (a)

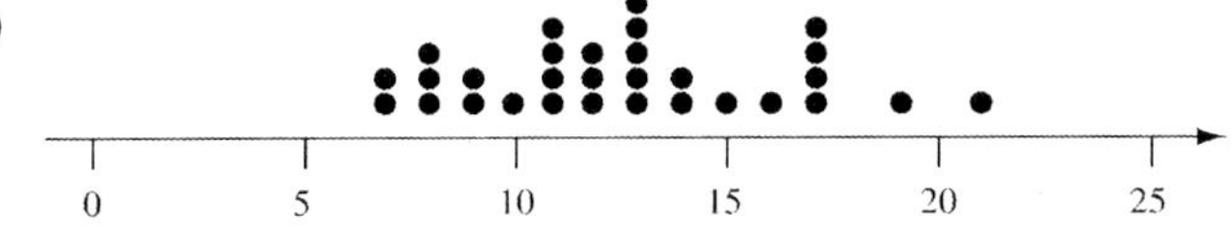

(b) About 13 hours per week

2.

0	7 7 8 8 8 9 9
1	0 1 1 1 1 2 2 2 3 3 3 3 3 4 4 5 6 7 7 7 7 9
2	1

3. Use a bar for each number of hours.

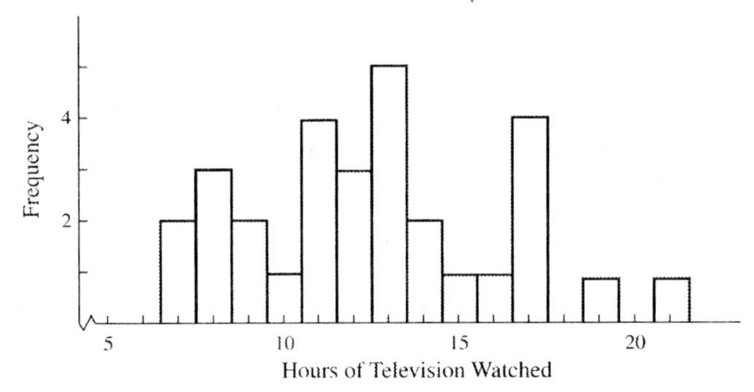

4. Put Mrs. Karnes class on the left of the stems and Ms. Stevens class on the right of the stems.

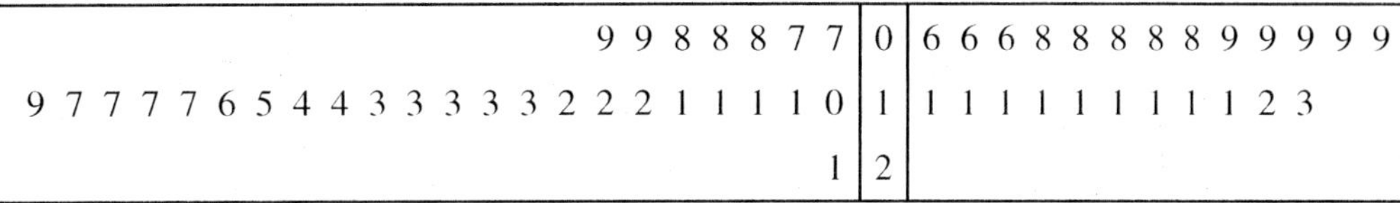

Mrs. Karnes		Ms. Stevens
9 9 8 8 8 7 7	0	6 6 6 8 8 8 8 8 9 9 9 9 9
9 7 7 7 7 6 5 4 4 3 3 3 3 3 2 2 2 1 1 1 1 0	1	1 1 1 1 1 1 1 1 1 2 3
1	2	

5. (a)

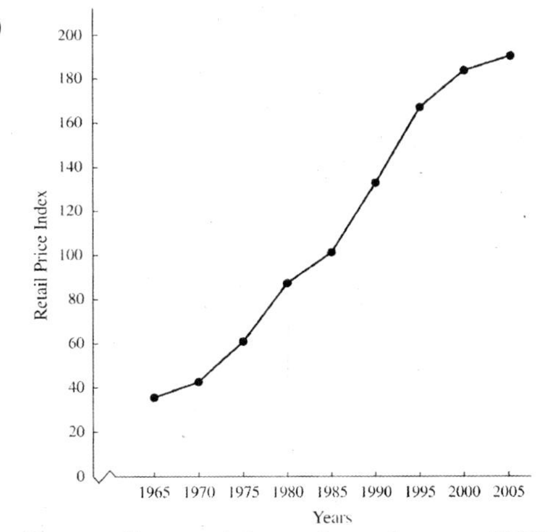

(b) Draw a line straight up from the year 1972 on the horizontal axis. Now draw a horizontal line from the point on the line graph above 1972 to the vertical axis. The retail price index for farm products in 1972 was about 48.

(c) Assuming the retail prices index continues the same upward trend as that from 2000 to 2005, the retail price index for farm products in 2010 will be about 200.

6. Many examples exist. One example is 1, 2, 3, 4, 90 with a mean of $\frac{1+2+3+4+90}{5} = 20$.

7. Find the central angle for each section of the budget.
Administration: 12% of 360° = 43.2°
New construction: 36% of 360° = 129.6°
Repairs: 48% of 360° = 172.8°
Miscellaneous: 4% of 360° = 14.4°

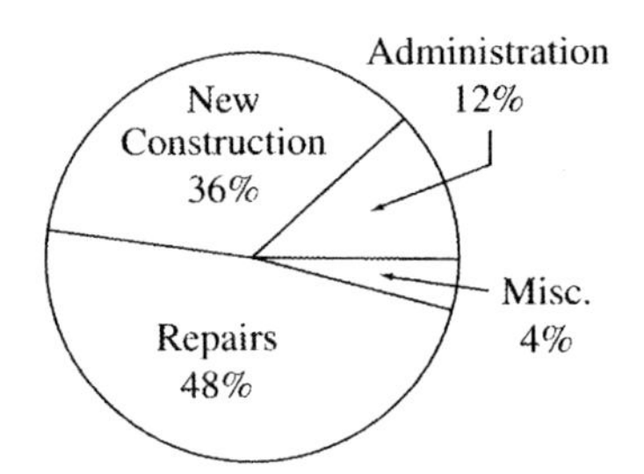

8. **(a)** The volume of the larger box is about double that of the smaller box. However the length of a side of the larger box is less than double the length of a side of the smaller box, suggesting that the change was less than doubling.

(b) The smaller cube has side length of 18 mm and volume of 5832 cubic mm. The larger cube has side length of 22.5 mm and volume of about 11,500 cubic mm. The volume of the larger cube is about twice that of the smaller cube, which defends the pictograph.

9. How is a "medical doctor" defined? Does this include all specialists? osteopaths? naturopaths? chiropractors? acupuncturists? How was the sampling done to determine the stated average?

10. $\overline{x} = \dfrac{378}{30} = 12.6$

The median is the average of the 15th and 16th ordered values.

$\hat{x} = \dfrac{12+13}{2} = 12.5$

The mode is 13.

$$s = \sqrt{\frac{(12.6-17)^2 + (12.6-8)^2 + \cdots + (12.6-12)^2}{30}}$$

$$= \sqrt{\frac{381.2}{30}} \doteq 3.6$$

11. **(a)** Q_L is the median of the first 15 ordered values, the 8th value.

$Q_L = 10$

Q_U is the median of the last 15 ordered values, the 23rd value.

$Q_U = 15$

(b) 5-number summary;
minimum $-Q_L - \hat{x} - Q_U -$ maximum
For Problem 1 data: 7 – 10 – 12.5 – 15 – 21

(c) $\text{IQR} = Q_U - Q_L = 15 - 10 = 5$

$Q_L - 1.5\ \text{IQR} = 10 - 1.5(5) = 2.5$

$Q_U + 1.5\ \text{IQR} = 15 + 1.5(5) = 22.5$

Since there are no data points less than 2.5 or greater than 22.5, there are no outliers.

(d) Q_L is the median for the first 12 ordered values, the average of the 6th and 7th values.

$Q_L = \dfrac{8+8}{2} = 8$

Median is the average of the 12th and 13th ordered values.

$\hat{x} = \dfrac{9+9}{2} = 9$

Q_U is the median for the last 12 ordered values, the average of the 18th and 19th values.

$Q_U = \dfrac{11+11}{2} = 11$

(e) 5-number summary for the Problem 4 data: 6 – 8 – 9 – 11 – 13

(f) $\text{IQR} = Q_U - Q_L = 11 - 8 = 3$

$Q_L - 1.5\ \text{IQR} = 8 - 1.5(3) = 3.5$

$Q_U + 1.5\ \text{IQR} = 11 + 1.5(3) = 15.5$

Since there are no data points less than 3.5 or greater than 15.5, there are no outliers.

(g)

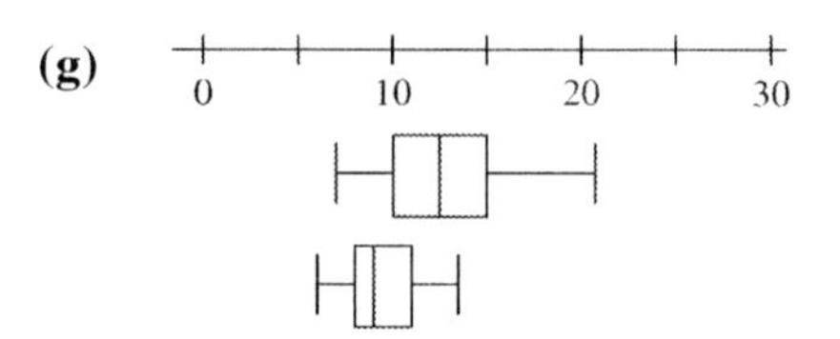

12. **(a)** $\bar{x} = \dfrac{1\cdot 24 + 4\cdot 25 + 7\cdot 26 + 8\cdot 27 + 6\cdot 28 + 5\cdot 29 + 5\cdot 30 + 3\cdot 31 + 1\cdot 33}{1+4+7+8+6+5+5+3+1} = \dfrac{1111}{40} \approx 27.8$

$s = \sqrt{\dfrac{1\cdot(27.8-24)^2 + 4\cdot(27.8-25)^2 + 7\cdot(27.8-26)^2 + \cdots + 1\cdot(27.8-33)^2}{40}} \approx \sqrt{\dfrac{163}{40}} \approx 2.0$

(b) $\bar{x} = \dfrac{1\cdot 24 + 3\cdot 25 + 8\cdot 26 + 12\cdot 27 + 9\cdot 28 + 4\cdot 29 + 2\cdot 30 + 1\cdot 33}{1+3+8+12+9+4+2+1} = \dfrac{1092}{40} = 27.3$

$s = \sqrt{\dfrac{1\cdot(27.3-24)^2 + 3\cdot(27.3-25)^2 + 8\cdot(27.3-26)^2 + \cdots + 1\cdot(27.3-33)^2}{40}} = \sqrt{\dfrac{104.4}{40}} \approx 1.6$

(c) The means are about the same for the two sets of data, but the standard deviation is smaller for the second histogram since the data is less spread out from the mean.

13. There are $21 \cdot 77 = 1617$ points for the 21 students. So there are $1617 + 69 + 62 + 91 = 1839$ points for all 24 students. Thus, the average is $1839 \div 24 \approx 76.6$.

14. There are $27 \cdot 75 = 2025$ points for the second period students and $30 \cdot 78 = 2340$ points for the fourth period students. Thus, the average for all students is $\dfrac{2340 + 2025}{57} \approx 76.6$.

15. No. The sample only represents the population of students at State University, not university students nationwide.

16. You might want to limit your sample to persons 20 years old and older who want to work. Alternatively, you might want to define several populations and determine figures for each: persons 20 years old and older who want to work, teenagers who want full-time employment, teenagers who want part-time employment, adults 20 years old and older who want part-time employment, and so on.

17. Voluntary responses to mailed questionnaires tend to come primarily from those who feel strongly (either positively or negatively) about an issue or who represent narrow special interest groups. They are rarely representative of the population as a whole.

18. $\bar{x} \doteq 9.3$ and $s \doteq 3$. The z scores, $z = \dfrac{x - \bar{x}}{s}$, in order are –0.8, –0.1, –1.1, 0.9, 1.9, –0.8, –0.1.

19. Since 12 is the 6th largest number in the data set, and $\dfrac{6}{7} = 0.8571$ to the nearest ten-thousandth, 12 is the 85.71 percentile to the nearest hundredth.

20. From Table 13.4, we see that 0.9505 corresponds to 1.65. therefore, 1.65 is essentially at the 95th percentile.

21. From Table 13.4, the entry for –0.9 is 0.1841 and the entry for 0.9 is 0.8159. Therefore, the desired percentage is 81.59% – 18.41% = 63.18%.

Chapter 13 Test

1. $(0.40) \cdot (360^\circ) = 144^\circ$

2.

40 50 60 70 80 90 100

3. **(a)** $\bar{x} = \dfrac{42 + 93 + 75 + \cdots + 57}{30} = \dfrac{2356}{30}$

≈ 78.5

(b) The median is the average of the 15th and 16th ordered values.

$\hat{x} = \dfrac{81 + 81}{2} = 81$

(c) mode = 87, the value that occurs most often

(d) $s \approx \sqrt{\dfrac{(78.5-42)^2 + (78.5-86)^2 + \cdots + (78.5-57)^2}{30}}$

$\approx \sqrt{\dfrac{5217.5}{30}} \approx 13.2$

4. The mean is $\frac{2356}{30} = 78.5\overline{3} \approx 78.53$. Thus,

$$s = \sqrt{\frac{(78.53-42)^2 + (78.53-93)^2 + \cdots + (78.53-57)^2}{30}}$$

≈ 13.2

The different z scores for 42, 86, and 80 from problem 3 as part of the entire data set are

$\frac{42-78.53}{13.2} \approx -2.8$

$\frac{86-78.53}{13.2} \approx 0.6$

$\frac{80-78.53}{13.2} \approx 0.1$

5.

Stem	Leaves
4	1 2
5	7
6	3 6
7	0 4 5 6 7 8 8
8	0 0 1 1 4 5 6 6 7 7 7 8 8 9
9	0 2 3 5

6. **(a)** Find Q_L = median of first 15 values = 75

Q_U = median of last 15 values = 87

The five-number summary is smallest value– $Q_L - \hat{x} - Q_u$ –largest value.

The 5-number summary for the data in Problem 1: 41 – 75 – 81 – 87 – 95

40 50 60 70 80 90 100

(b) 41 and 42 are outliers because they are less than $Q_L - 1.5 \cdot \text{IQR} = 57$.

7. If the sample is not chosen at random it is quite likely to reflect bias—bias of the sampler, bias reflecting the group from which the sample was actually chosen (views of teamsters, or AARP members), and so on.

8. A random sample is one chosen in such a way that every subset of size r in the population has an equal chance of being included.

9. Since 17 of the 30 scores in the data set are less than or equal to 84, and $\frac{17}{30} = 0.5\overline{6} \approx 57\%$, 84 is at the 57th percentile.

10. Consider each test is worth 100 points. Then all 5 tests would be worth 500 points. To average 80% on all five tests, Nanda must earn a total of 0.8(500) = 400 points. Nanda already has 77 + 79 + 72 = 228 points. Therefore, she must score at least 400 – 228 = 172 points on the remaining two exams.

Chapter 14 Probability

Problem Set 14.1

1. (a) There are 20 students in the class. Only one is less than five feet (60 inches) tall, so $P_e = \frac{1}{20}$ or 0.05 or 5%

3. (b) Individual probabilities will vary but the sum should always be 1. In our case, we have,
$P_e(1) + P_e(2) + P_e(3) + P_e(4) + P_e(5)$
$= 0.12 + 0.22 + 0.22 + 0.22 + 0.22$
$= 1.00.$

6. (a) $S = \begin{Bmatrix} H1, H2, H3, H4, H5, H6, \\ T1, T2, T3, T4, T5, T6 \end{Bmatrix}$

8. (a) $P_e(A) = 0$ means that event A never happened in the simulation. This is choice iv (or possibly ii).

(c) $P_e(C) = -0.5$ means that event C happened a negative number of times which is not possible. This is choice (ii).

(e) $P_e(E) = 1.7$ means that event E happened more than 100% of the time which is not possible. This is choice (ii).

11. (a) Answers will vary. For the 20 spins, the arrow ended up in Region A 12 times, Region B 6 times, and Region C only twice. Therefore, $P_e(\text{A}) = \frac{12}{20} = 0.6$,
$P_e(\text{B}) = \frac{6}{20} = 0.3$, $P_e(\text{C}) = \frac{2}{20} = 0.1$.

(b) For 20 spins, the arrow should end up in Region A $\left(\frac{240}{360}\right) \cdot 20 \approx 13$ times. This is close to the observed 12 times. The line on the thumb should end up in Region B $\left(\frac{90}{360}\right) \cdot 20 = 5$ times. This is close to the observed 6 times. The line on the thumb should end up in Region C $\left(\frac{30}{360}\right) \cdot 20 \approx 2$ times. That is exactly the observed result.

14. Answers will vary. Using the answer to 13(a), the first 6 occurred on the fourth roll twice. Therefore,

$P_e(\text{first 6 on fourth roll}) = \frac{2}{10} = 0.2$.

23. Answers will vary. When we did the experiment, the following cards were selected:

Hearts	Diamonds	Spades	Clubs
6	5	K	8
9	2	8	7
Q	Q	J	5
2	J	10	9
	2	5	
	6		
	J		

(a) Empirical probability of a red card $= \frac{\text{diamonds + hearts}}{20}$

$P_e(R) = \frac{4+7}{20} = \frac{11}{20} = 0.55$

(b) Empirical probability of a face card $= \frac{\text{J+Q+K}}{20}$

$P_e(F) = \frac{3+2+1}{20} = \frac{6}{20} = 0.30$

(c) Empirical probability of a red card or a face card

$= \frac{\text{hearts + diamonds + spades(J or Q or K) + clubs(J or Q or K)}}{20}$

$P_e(R \text{ or } F) = \frac{4+7+2+0}{20} = \frac{13}{20} = 0.65$

(d) Empirical probability of a red card and a face card

$= \frac{\text{hearts(J or Q or K) + diamonds(J or Q or K)}}{20}$

$P_e(R \text{ and } F) = \frac{3+1}{20} = \frac{4}{20} = 0.20$

(e) $P_e(R) + P_e(F) - P_e(R \text{ and } F)$
$= 0.55 + 0.30 - 0.20 = 0.65$

(f) The empirical probability of a red card or a face card can be found using the expression in (e), that is,
$P_e(R \text{ or } F) = P_e(R) + P_e(F) - P_e(R \text{ and } F)$.
In words, sum the empirical probability of a red card with the empirical probability of a face card then subtract the empirical probability of a red face card. This is unlike the result suggested by Problem 22.

24. Answers will vary. When we did the experiment we obtained these outcomes:
H5 H3 H2 H3 H4 T3 T2 T2 H2 T1 H2 H2 T4 T5 H1 H5 T1 H6 H4 T1

(a) $P_e(\text{H}) = \dfrac{12}{20} = 0.6$

(b) $P_e(5) = \dfrac{3}{20} = 0.15$

(c) $P_e(\text{H and } 5) = \dfrac{2}{20} = 0.1$

(d) $P_e(\text{H}) \cdot P_e(5) = (0.6) \cdot (0.15) = 0.09$

(e) Yes. Since the events H and 5 are independent, the number of simultaneous occurrences of H and 5 should be about $P_e(\text{H}) \cdot (\text{the number of occurrences of 5})$.
But then
$P_e(H \text{ and } 5)$
$\approx \dfrac{P_e(\text{H}) \cdot (\text{the number of occurrences of 5})}{20}$
$= P_e(\text{H}) \cdot P_e(5).$

30. Assuming that the blood types of spouses are independent events:

(a) $(0.45) \times (0.45) = 0.2025$ or about 20% of all couples.

31. (a) $\dfrac{9806}{10{,}000} = 0.9806$

33. Answers for (a), (b), and (c) will vary. The following line graphs are representative. The point with coordinate (x, y) shows x heads occurred y times, for $x = 0, 1, 2..., 10$.

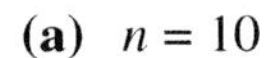

(a) $n = 10$

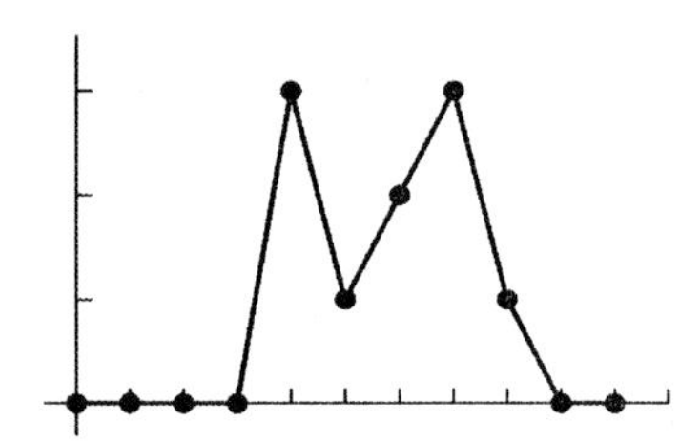

Problem Set 14.2

1. There are 3 ways to roll a 4:
$4 = 1 + 3 = 2 + 2 = 3 + 1$.
There are 5 ways to roll a 6:
$6 = 1 + 5 = 2 + 4 = 3 + 3 = 4 + 2 = 5 + 1$.
Therefore, there are $3 + 5 = 8$ ways to roll a 4 or 6.

3. The 12 possible outcomes are:
H1 H2 H3 H4 H5 H6 T1 T2 T3 T4 T5 T6

(a) There are 3 ways to get a head and an even number: H2 H4 H6

5. Draw a Venn Diagram with 2 overlapping circles. Label one circle "E" for English and the other circle "J" for Japanese. Since 11 of the students speak only English, the nonoverlapping part of "E" contains 11 members. 9 of the students speak only Japanese so the nonoverlapping part of "J" contains 9 members. That accounts for $11 + 9 = 20$ students out of 24 students in Mr. Walcott's class, so 4 students must be able to speak both English and Japanese (the overlapping part of "E" and "J").

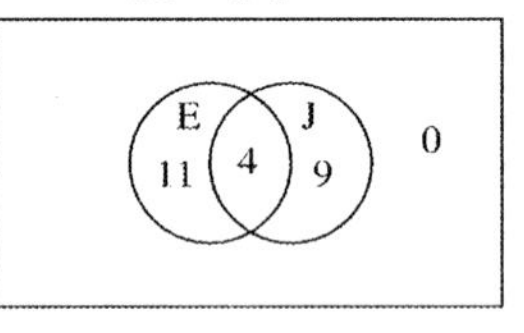

7. Let R be the set of red face cards.
{jack of diamonds, queen of diamonds, king of diamonds, jack of hearts, queen of hearts, king of hearts)
Let A be the set of black aces.
{ace of spades, ace of clubs}
Notice that $R \cap A = \varnothing$ since there are no elements that are common to both set R and set A. Since sets R and A are mutually exclusive, use the second principle of counting:
$n(R \text{ or } A) = n(R) + n(A) = 6 + 2 = 8$.
There are 8 ways to select a red face card or a black ace from an ordinary deck of playing cards.

8. (a) Since there are 6 numbers to choose from, the first digit has 6 possibilities. After the first digit is selected, the number cannot be repeated so there are 5 numbers to choose from for the second digit. The third digit has 4 remaining possibilities and the fourth digit has 3 numbers to choose from. Use the third principle of counting to find:

n(four-digit natural numbers with no repeat digits)

$= n(\text{1st digit}) \cdot n(\text{2nd digit} \mid \text{1st digit})$
$\cdot n(\text{3rd digit} \mid \text{1st and 2nd digits})$
$\cdot n(\text{4th digit} \mid \text{1st and 2nd and 3rd digits})$
$= 6 \cdot 5 \cdot 4 \cdot 3 = 360$

There are 360 four-digit natural numbers using the digits 1, 2, 3, 4, 5, or 6 at most once.

(b) 3 of the 6 digits are odd so there are 3 possibilities for the first digit. Once the first digit is chosen, there are 5 possibilities for the second digit, 4 possibilities for the third digit, and 3 possibilities for the fourth digit. Use the same reasoning presented in (a) to find that there are $3 \cdot 5 \cdot 4 \cdot 3 = 180$ four digit natural numbers using the digits 1, 2, 3, 4, 5, or 6 at most once and beginning with an odd digit.

(c) 3 of the 6 digits are odd and one of the three digits must appear in the first position. That leaves 2 of the 3 odd digits for the last position. Once the first and last digits are chosen, there are 4 possibilities for the second digit and 3 possibilities for the third digit. Use the same reasoning presented in (a) to find that there are $3 \cdot 2 \cdot 4 \cdot 3 = 72$ four digit natural numbers using the digits 1, 2, 3, 4, 5, or 6 at most once and beginning and ending with an odd digit.

10. To draw the possibility tree, begin by showing the 3 possible letters for the first letter and label these 3 limbs "a", "b", and "c". Now there are two possible letters remaining for the second letter so draw 2 limbs from each of the first letters and label each limb with the remaining possible letters. There is only one choice for the third letter after the first two have been selected so draw a limb and label the remaining letter. The possibility tree is shown in the next column.

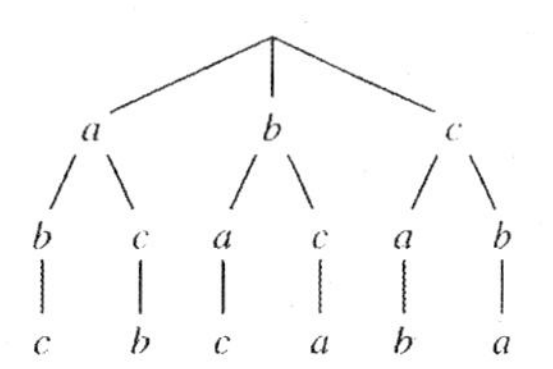

Following the limbs of the possibility tree, the code words are: *abc*, *acb*, *bac*, *bca*, *cab*, *cba*.

14. The digits 1, 3, 5, 7, and 9 are odd and the digits 0, 2, 4, 6, and 8 are even.

(a) There are 5 choices for the first odd digit, 5 choices of the second odd digit, 5 choices for the third odd digit, 5 choices for the fourth even digit, and 5 choices for the last even digit. Therefore, there are $5 \cdot 5 \cdot 5 \cdot 5 \cdot 5 = 5^5 = 3125$ 5-digit numbers with the first three digits odd and the last two digits even when repetition of digits is allowed.

16. (a)

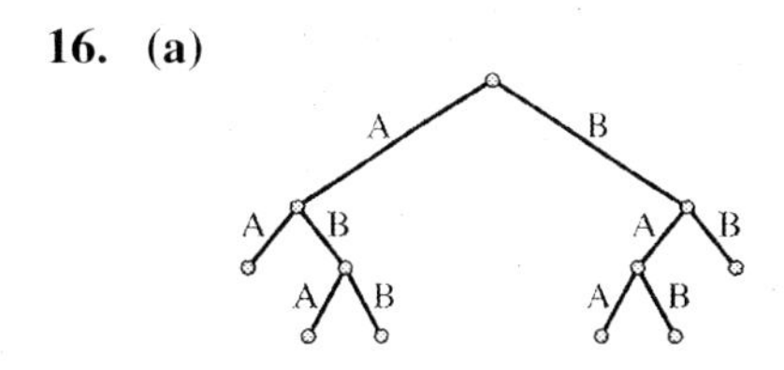

22. The possible scores and the number of ways to obtain each score are listed in the table below.

score (sum)	2	3	4	5	6	7
# of ways	1	2	3	4	5	6

score (sum)	8	9	10	11	12
# of ways	5	4	3	2	1

(a) The scores 2, 4, 6, 8, 10, and 12 are even and can be obtained $1 + 3 + 5 + 5 + 3 + 1 = 18$ ways.

23. (a) There are four kings and 40 non-face cards, so there are $4 \cdot 40 = 160$ ways to draw the cards so that the first card is a king and the second card is not a face card.

24. (a) There are ten types of straights, namely 5-high, 6-high, 7-high, ..., king-high, and ace-high. Each type can be filled in $4^5 = 1024$ ways, since there are four choices for each of the five cards. Then, there are $10 \cdot 1024 = 10,240$ possible straights.

27. **(a)** Each person's birthday can be in one of twelve months, so there are $12^4 = 20,736$ ways the months in which they celebrate their birthdays can occur.

29. **(a)** There are $9 \cdot 10 \cdot 10 \cdot 10 \cdot 10 = 90,000$ different five-digit numbers.

31. **(a)** There are 64 choices for the red checker and 63 choices for the black, so there are $64 \cdot 63 = 4032$ ways to place a red checker and a black checker on an 8 × 8 checkerboard.

35. **(a)** 2^4, or 16 words.

Problem Set 14.3

1. **(a)** combination. $C(10,5) = \dfrac{10 \cdot 9 \cdot 8 \cdot 7 \cdot 6}{5 \cdot 4 \cdot 3 \cdot 2 \cdot 1} = 252$

(b) permutation. $P(10,4) = 10 \cdot 9 \cdot 8 \cdot 7 = 5040$

4. **(a)** $P(6, 6) = 6! = 720$

5. **(a)** The club members can line up in $14! \approx 90$ billion ways.

7. **(a)** $7! = 7 \cdot 6 \cdot 5 \cdot 4 \cdot 3 \cdot 2 \cdot 1 = 5040$

(c) $9! \div 7! = \dfrac{9 \cdot 8 \cdot 7 \cdot 6 \cdot 5 \cdot 4 \cdot 3 \cdot 2 \cdot 1}{7 \cdot 6 \cdot 5 \cdot 4 \cdot 3 \cdot 2 \cdot 1} = 9 \cdot 8 = 72$

(e) $7 \cdot 7! = 7 \cdot 7 \cdot 6 \cdot 5 \cdot 4 \cdot 3 \cdot 2 \cdot 1 = 7 \cdot 5040 = 35,280$

8. **(a)** $P(13,8) = 13(13-1)(13-2)(13-3)(13-4)(13-5)(13-6)(13-7)$
$= 13 \cdot 12 \cdot 11 \cdot 10 \cdot 9 \cdot 8 \cdot 7 \cdot 6 = 51,891,840$

(c) $P(15, 2) = 15(15 - 1) = 15 \cdot 14 = 210$

(e) $C(15,15) = \dfrac{15(15-1)(15-2)\cdots(15-13)(15-14)}{15!} = \dfrac{15 \cdot 14 \cdot 13 \cdot 12 \cdot 11 \cdot 10 \cdot 9 \cdot 8 \cdot 7 \cdot 6 \cdot 5 \cdot 4 \cdot 3 \cdot 2 \cdot 1}{15 \cdot 14 \cdot 13 \cdot 12 \cdot 11 \cdot 10 \cdot 9 \cdot 8 \cdot 7 \cdot 6 \cdot 5 \cdot 4 \cdot 3 \cdot 2 \cdot 1} = 1$

10. **(a)** Since order doesn't matter, the answer is the number of 2-combinations chosen from the 10 + 15 balls in the bag. Therefore, there are $C(10 + 15, 2) = 300$ ways to choose 2 of the 25 balls.

(b) There are three mutually exclusive cases, depending on whether both balls are red, both are blue, or there is one of each color. There are $C(10, 2) = 45$ ways to choose 2 of the 10 red balls and there are $C(15, 2) = 105$ ways to choose 2 of the blue balls. Finally, there are $10 \cdot 15 = 150$ ways to choose a red ball and a blue ball. Altogether, this gives $C(10, 2) + C(15, 2) + 10 \cdot 15 = 300$ ways to choose 2 of the balls.

(c) Consider a bag with m red and n blue balls and ask how many ways two of the $m + n$ balls can be selected. One answer is $C(m + n, 2)$, but considering the cases where both balls are red, both are blue, or one is red and the other is blue give the answer $C(m + 2) + C(n + 2) + m \cdot n$. Equating the two answers proves the formula.

13. **(a)** There are 4 letters with two T's and two O's. The number of permutations of 4 things with 2 things alike and 2 other things alike is given by:

$\dfrac{4!}{2!2!} = \dfrac{4 \cdot 3 \cdot 2 \cdot 1}{2 \cdot 1 \cdot 2 \cdot 1} = \dfrac{4 \cdot 3}{2 \cdot 1} = 6$.

There are 6 different arrangements of the letters in TOOT.

22. (a) For every arrangement under consideration b immediately follows a so the sequence ab can be thought of as a single symbol. That is, how many arrangements of ab, c, d, and e are there? There are 4 choices for the first position, 3 choices for the second position, 2 choices for the third position, and one choice left for the last position. There are $4 \cdot 3 \cdot 2 \cdot 1 = 4! = 24$ of the 120 arrangements of a, b, c, d, and e where b immediately follows a.

(b) ab and ba are the two sequences where a and b are adjacent. Use the same reasoning as part (a) to find that there are 4! arrangements of ab, c, d, and e and 4! arrangements of ba, c, d, and e for a total of $4! + 4! = 24 + 24 = 48$ of the 120 arrangements of a, b, c, d, and e where a and b are adjacent.

(c) Consider all the cases where a precedes e.

(i) If a is first, e could be second, third, fourth, or fifth. This can happen in

$$1 \cdot 1 \cdot 3 \cdot 2 \cdot 1 = 6$$
$$1 \cdot 3 \cdot 1 \cdot 2 \cdot 1 = 6$$
$$1 \cdot 3 \cdot 2 \cdot 1 \cdot 1 = 6$$
$$1 \cdot 3 \cdot 2 \cdot 1 \cdot 1 = 6$$

24 ways.

(ii) If a is second, e could be third, fourth, or fifth. This can happen in

$$3 \cdot 1 \cdot 1 \cdot 2 \cdot 1 = 6$$
$$3 \cdot 1 \cdot 2 \cdot 1 \cdot 1 = 6$$
$$3 \cdot 1 \cdot 2 \cdot 1 \cdot 1 = 6$$

18 ways.

(iii) If a is third, e could be fourth or fifth. This can happen in

$$3 \cdot 2 \cdot 1 \cdot 1 \cdot 1 = 6$$
$$3 \cdot 2 \cdot 1 \cdot 1 \cdot 1 = 6$$

12 ways.

(iv) If a is fourth, e could be fifth. This can happen in $3 \cdot 2 \cdot 1 \cdot 1 \cdot 1 = 6$ ways. Combining the 4 cases above, there are $24 + 18 + 12 + 6 = 60$ ways a can precede e. Alternatively, by symmetry, in half the $5! = 120$ permutations of a, b, c, d, and e, a precedes e and in half e precedes a. Therefore, the desired number is

$$\frac{120}{2} = 60 \text{ as above.}$$

27. (a) The products of the terms in odd corners and in the even corners is the same wherever the hexagon is positioned.

(b)
$$\begin{aligned} & C(n-1, r-1)C(n, r+1)C(n+1, r) \\ &= \frac{(n-1)!}{(r-1)!(n-r)!}\frac{n!}{(r+1)!(n-r-1)!} \cdot \frac{(n+1)!}{r!(n+1-r)!} \\ &= \frac{(n-1)!}{r!(n-r-1)!}\frac{n!}{(r-1)!(n+1-r)!} \cdot \frac{(n+1)!}{(r+1)!(n-r)!} \\ &= C(n-1, r)C(n, r-1)C(n+1, r+1) \end{aligned}$$

28. (a) $C(8, 5)$

(b) $C(7, 5)$, since there are 7 non-red marbles from which to choose 5 marbles.

29. (a) First choose the members of the sextet in $C(11, 6)$ ways, and then choose the soloist from among the six already chosen.

30. (a) Choose eight of the fifteen members in $C(15, 8) = 6435$ ways, and the choose the two co-captains from the eight delegates in $C(8, 2)=28$ ways. Thus, there are $6435 \cdot 28 = 180{,}180$ ways to choose the delegation and co-captains.

31. (a)
$$C(6+4-1, 6) = C(9, 6) = C(9, 3) = \frac{9 \cdot 8 \cdot 7}{3 \cdot 2 \cdot 1} = 84$$

(d) The number of combinations with repetition is equivalent to determining the number of ways n ice cream cones can be ordered from k flavors. That is, it is the number of lists on n check marks and $k - 1$ bars that separate one flavor from the next. Therefore, n objects can be chosen from k types, with repetition allowed, in $C(n+k-1, n)$ ways.

32. (a) First notice that there are seven choices for the number of spots for each half-domino. Thus, there are

$C(7, 2) = \frac{7 \cdot 6}{2} = 21$ dominoes with a different number of spots on each half and seven more dominoes with the same number of each half, giving 21 + 7 = 28 in all. Alternatively, this can be viewed as a combination with repetition problem (see problem 31) where two objects are selected from seven types (number of spots) with it possible to select the same type twice. By the formula of problem 31(d), there are

$C(2+7-1, 2) = C(8, 2) = \frac{8 \cdot 7}{2} = 28$

dominoes is a double-six set.

33. (a) There are 16 trains of length 5.

(b) There are 1 one-car, 4 two-car, 6 three-car, 4 four-car, and 1 five-car trains of length 5.

36. (a) There are three ways. There are several ways to reason this. If the players are A, B, C, and D, there are three partners for player A, with the remaining two players forming the opposing team. Alternatively, choose two of the four players in $C(4, 2) = 6$ ways, and divide by 2 since it doesn't matter which team is chosen first.

(b) There are 15 ways. Choose the player sitting out in five ways and divide up the remaining four players in three ways as described in part (a). Alternatively, choose a team of two out of the five players in $C(5, 2) = 10$ ways and choose one of the remaining three players to sit out in three ways. This gives $\frac{10 \cdot 3}{2} = 15$ ways to set up the doubles match, where the division by 2 accounts for the fact that it doesn't matter which team is chosen first.

38. (a) 15 blocks. From 2nd Avenue to 8th Avenue requires six blocks of walking along avenues. Walking from B Street to K Street requires nine more blocks walking along streets.

(b) 5005. Each block walked is along either an avenue or a street, where six avenue blocks and nine street blocks must be walked. Each walk can be described as a permutation of six As and 9Ss. For example, the sequence SSAAASSSASASSAS indicates that Joe walks along a street, then a street, then an avenue, and so on. He can choose any six of the fifteen blocks as avenues with the remaining nine blocks walked along streets. Thus, Joe can reach his destination in $C(15, 6) = 5005$ ways.

39. (a) 362,880 **(c)** 2520

(e) 35

40. (a) $10! = 3{,}628{,}800$

(c) $C(13, 8) = 1287$

Problem Set 14.4

1. (a) There are two possible outcomes for each coin: Head (H) or Tail (T). The list of all outcomes for the experiment in order of penny, nickel, dime, and quarter are:

HHHH	THHH	TTHH	TTTH	TTTT
	HTHH	THTH	TTHT	
	HHTH	THHT	THTT	
	HHHT	HTTH	HTTT	
		HTHT		
		HHTT		

(c) There are 16 possible outcomes for the experiment and six of the outcomes have two heads and two tails (center column of outcomes from (a)). Therefore,

$P(\text{2 heads and 2 tails}) = \frac{6}{16} = \frac{3}{8} = 0.375$.

4. S = {3 red Fords, 4 white Fords, 2 black Fords, 6 red Hondas, 2 white Hondas, 5 black Hondas}
There are 22 equally likely cars to randomly choose from.

(a) Of all of the 22 cars available, four are white Fords. Therefore,

$P(\text{white Ford}) - \frac{4}{22} \approx 0.18$.

(c) There are 6 white cars (4 white Fords and 2 white Hondas) from the 22 available cars.

Therefore, $P(\text{white}) = \frac{6}{22} \approx 0.27$.

5. $S = \{B1, B2, B3, B4, B5, W1, W2, W3, W4, W5, W6, W7\}$
There are 12 equally likely balls to randomly choose from.

(a) Of the 12 balls to select, four are numbered 1 or 2 (B1, W1, B2, W2).
$P(1 \text{ or } 2) = P(1) + P(2)$
$= \frac{2}{12} + \frac{2}{12} = \frac{4}{12} \approx 0.33$

6. Let BR = brown, BL = blue, br = brunette, bl = blond.
$S = \{7BRbr, 2BLbr, 8BLbl, 3BRbl\}$. There are 20 equally likely children to randomly select.

(a) There are 7 brown-eyed brunettes out of the 20 children in the class. Therefore,
$P(BRbr) = \frac{7}{20} = 0.35$

7. $S = \{00, 01, 02, \cdots, 98, 99\}$. There are 100 equally likely 2-digit numbers possible.

(a) The set of numbers greater than 80 is $\{81, 82, \cdots, 98, 99\}$ and there are 19 such numbers. Therefore,
$P(\text{a number greater than } 80) = \frac{19}{100} = 0.19$.

(c) The set of numbers that are a multiple of 3 is $\{3, 6, 9, \cdots, 96, 99\}$ and there are $\frac{99}{3} = 33$ such numbers. Therefore,
$P(\text{a number is a multiple of } 3) = \frac{33}{100}$
$= 0.33$

(e) The set of even numbers less than 50 is $\{00, 02, \cdots, 46, 48\}$ and there are $\frac{50}{2} = 25$ such numbers. Therefore,
$P(\text{an even number less than } 50) = \frac{25}{100}$
$= 0.25$.

8. Assume the deck of cards is shuffled between deals so that the only known cards are the two dealt face up. That leaves 50 unknown cards.

(a) In order to win, the third card must be a six. There are four sixes left in the deck.
Therefore, $P(\text{win}) = \frac{4}{50} = 0.08$

(c) In order to win, the third card must be between the two cards dealt face, in this case, between a nine and a nine. There are no cards in the deck between two nines,
therefore, $P(\text{win}) = \frac{0}{50} = 0$.

10. **(a)** $\frac{3}{8} \cdot \frac{2}{7} = \frac{6}{56}$

12. **(a)**

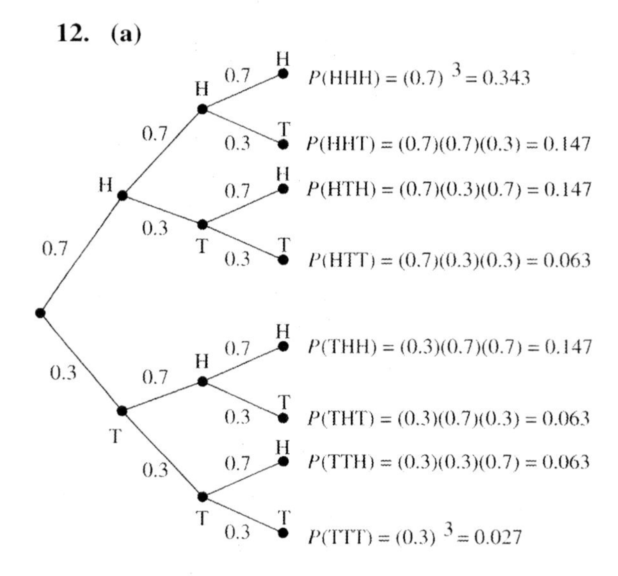

14. Three cards are randomly dealt from a deck of 52 cards in $C(52, 3)$ ways. 3 hearts can be randomly dealt in $C(13, 3)$ ways. Therefore,

$P(3 \text{ hearts}) = \frac{\text{number of ways for 3 hearts}}{\text{number of ways for 3 cards}}$

$= \frac{C(13, 3)}{C(52, 3)} = \frac{\frac{13 \cdot 12 \cdot 11}{3 \cdot 2 \cdot 1}}{\frac{52 \cdot 51 \cdot 50}{3 \cdot 2 \cdot 1}}$

$= \frac{13 \cdot 12 \cdot 11}{52 \cdot 51 \cdot 50} = \frac{1716}{132{,}600} \approx 0.013$

15. The sections 1–10 have equal areas so the central angle for each sector is $\frac{360°}{10} = 36°$.
Sections a–e have equal areas so the central angle for each sector is $\frac{360°}{5} = 72°$. The probability for each of the following solutions are determined by ratios of angular measures of appropriate regions.

(a) Three of the 10 regions are shaded so
$P(\text{shaded}) = \frac{3 \cdot 36°}{360°} = \frac{108°}{360°} = 0.3$.

(c) Regions 10 and 6 represent two of the 10 regions so

$P(\text{region 10 or region 6})$
$= P(\text{region 10}) + P(\text{region 6})$
$= \frac{36°}{360°} + \frac{36°}{360°} = \frac{72°}{360°} = 0.2$

(e) Region 8 is one of three shaded areas so

$P(\text{region 8} | \text{shaded area}) = \frac{36°}{3 \cdot 36°} = \frac{36°}{108°}$
$= \frac{1}{3} \approx 0.33$

(g) The vowels a and e cover 4 regions: 7, 8, 9, and 10. The odd regions are 1, 3, 5, 7, and 9. Regions 7 and 9 are odd and contain a vowel. Therefore,

$P(\text{vowel} \cup \text{odd-numbered region})$
$= P(\text{vowel}) + P(\text{odd-numbered region})$
$- P(\text{vowel and odd-numbered region})$
$= \frac{4 \cdot 36°}{360°} + \frac{5 \cdot 36°}{360°} - \frac{2 \cdot 36°}{360°}$
$= \frac{144° + 180° - 72°}{360°} = \frac{252°}{360°} = 0.7$

16. Notice the dart board measures 5 units wide and 5 units tall. To determine the area for region 5, notice that it is a square 1 unit by 1 unit so region 5 covers 1 sq. unit. To determine the areas for regions 1 and 3, find the area of the outside square and subtract the inside square area. The area for region 1 is $5^2 - 3^2 = 16$ sq. units. The area for region 3 is $3^2 - 1^2 = 8$ sq. units. The probabilities found below are ratios of regional areas.

(a) Region 1 takes up 16 of the 25 sq. units so $P(1) = \frac{16}{25} = 0.64$.

(c) Region 5 takes up 1 of the 25 sq. units so

$P(5) = \frac{1}{25} = 0.04$.

17. Area of a circle $= \pi r^2$

Area of the dart board $= \pi(2)^2 = 4\pi$

Area of the inner circle $= \pi(1)^2 = \pi$

Area of each of the regions 1 through 10

$= \frac{4\pi - \pi}{10} = \frac{3\pi}{10}$

Area of each of the regions a through e $= \frac{\pi}{5}$

Since the dart hits the board at random locations, the probability of hitting any particular region is the ratio of the area of the region and the area of the dart board.

(a) The area of region b is $\frac{\pi}{5}$ and the area of the dart board is 4π. Therefore,

$P(\text{region b}) = \frac{\frac{\pi}{5}}{4\pi} = \frac{\pi}{5} \cdot \frac{1}{4\pi} = \frac{1}{20} = 0.05$.

(b) The area of region b is $\frac{\pi}{5}$ and the area of the inner circle is π. Therefore,

$P(\text{not region b} | \text{inner circle}) = \frac{\pi - \frac{\pi}{5}}{\pi}$

$= \frac{\frac{4\pi}{5}}{\pi} = \frac{4\pi}{5} \cdot \frac{1}{\pi} = \frac{4}{5} = 0.8$.

18. **(a)** There are 5 ways to roll a six so there must be $36 - 5 = 31$ ways to not roll a six. If 6 represents the event of rolling a score of six, then the odds in favor of 6 are $n(6)$ to $n(\overline{6})$ or 5 to 31 written 5:31.

20. Since A and C are mutually exclusive,

$P(A \text{ or } C) = P(A) + P(C) = \frac{1}{2} + \frac{1}{6} = \frac{4}{6} = \frac{2}{3}$.

Therefore, the odds in favor of A or C are given by $\frac{P(A \text{ or } C)}{1 - P(A \text{ or } C)} = \frac{\frac{2}{3}}{1 - \frac{2}{3}} = \frac{2}{3} \cdot \frac{3}{1} = \frac{2}{1}$ or 2:1.

22. To determine the probability for each roll, count the number of ways to roll the score and divide by 36.

roll	2	3	4	5	6	7	8	9	10	11	12
probability	$\frac{1}{36}$	$\frac{2}{36}$	$\frac{3}{36}$	$\frac{4}{36}$	$\frac{5}{36}$	$\frac{6}{36}$	$\frac{5}{36}$	$\frac{4}{36}$	$\frac{3}{36}$	$\frac{2}{36}$	$\frac{1}{36}$

Expected win

$$= 4\left(\frac{1}{36}\right)+6\left(\frac{2}{36}\right)+8\left(\frac{3}{36}\right)+10\left(\frac{4}{36}\right)+20\left(\frac{5}{36}\right)+40\left(\frac{6}{36}\right)+20\left(\frac{5}{36}\right)+10\left(\frac{4}{36}\right)+8\left(\frac{3}{36}\right)+6\left(\frac{2}{36}\right)+4\left(\frac{1}{36}\right)$$

$$= \frac{600}{36} \approx 16.67$$

You can expect to win about \$16.67 per roll on this game.

28. Mischa is justifiably suspicious of Jorge's reasoning. There are three equally likely red spots that can be seen, and in just one of the cases is the hidden spot black. The correct probability is $\frac{1}{3}$.

29. **(a)** There are $C(6, 2)$ ways to choose 2 white balls and $C(6 + 8, 2) = C(14, 2)$ ways to choose any 2 balls from the urn. Therefore,

$$P(2\text{ white}) = \frac{n(\text{ways to draw 2 white balls})}{n(S)}$$

$$= \frac{C(6,\ 2)}{C(14,\ 2)} = \frac{\frac{6\cdot 5}{2\cdot 1}}{\frac{14\cdot 13}{2\cdot 1}} = \frac{6\cdot 5}{14\cdot 13}$$

$$= \frac{30}{182} \approx 0.16.$$

30. **(a)** There are $C(8, 4)$ ways to draw 4 red balls and $C(8 + 5 + 6, 4) = C(19, 4)$ ways to draw any 4 balls from the urn. Therefore,

$$P(\text{all four red}) = \frac{n(\text{ways to draw 4 red balls})}{n(S)}$$

$$= \frac{C(8,\ 4)}{C(19,\ 4)} = \frac{\frac{8\cdot 7\cdot 6\cdot 5}{4\cdot 3\cdot 2\cdot 1}}{\frac{19\cdot 18\cdot 17\cdot 16}{4\cdot 3\cdot 2\cdot 1}}$$

$$= \frac{8\cdot 7\cdot 6\cdot 5}{19\cdot 18\cdot 17\cdot 16} = \frac{1680}{93{,}024}$$

$$\approx 0.02.$$

31. There are $P(26, 5) = 26 \cdot 25 \cdot 24 \cdot 23 \cdot 22 = 7{,}893{,}600$ 5-letter code words without repetition of letters.

(a) If a code word begins with the letter *a*, then there are 25 choices for the second letter, 24 choices for the third letter, 23 choices for the fourth letter, and 22 choices for the last letter. There are $P(25, 4) = 25 \cdot 24 \cdot 23 \cdot 22 = 303{,}600$ ways to arrange the last 4 letters.

Therefore,

$$P(\text{a code word begins with } a) = \frac{P(25,\ 4)}{P(26,\ 5)} = \frac{303{,}600}{7{,}893{,}600} \approx 0.04$$

34. There are

$$C(52,\ 5) = \frac{52\cdot 51\cdot 50\cdot 49\cdot 48}{5\cdot 4\cdot 3\cdot 2\cdot 1} = \frac{311{,}875{,}200}{120} = 2{,}598{,}960$$

possible 5-card hands.

(a) There are

$$C(4,\ 2)\cdot C(48,\ 3) = \frac{4\cdot 3}{2\cdot 1}\cdot\frac{48\cdot 47\cdot 46}{3\cdot 2\cdot 1} = 103{,}776$$

possible 5-card hands with exactly two aces. Therefore,

$$P(\text{a five card hand has exactly two aces}) = \frac{C(4,\ 2)\cdot C(48,\ 3)}{C(52,\ 5)} = \frac{103{,}776}{2{,}598{,}960} \approx 0.040$$

35. One method of solving this problem is given in the back of the textbook. There is another method. There are

$6\cdot 6\cdot 6\cdot 6\cdot 6\cdot 6\cdot 6 = 6^7 = 279{,}936$ possible outcomes for tossing seven dice. If each number 1–6 is to appear at least once, then one of the numbers 1 through 6 must appear twice. For each such number, there are

$\frac{7!}{2!} = \frac{7\cdot 6\cdot 5\cdot 4\cdot 3\cdot 2\cdot 1}{2\cdot 1} = 2520$ possible

outcomes with the chosen number appearing twice and each other number appearing once. There are six possible numbers that can appear twice. Therefore,

$$P(\text{every number appears}) = \frac{6\cdot 2520}{279{,}936} = \frac{15{,}120}{279{,}936} \approx 0.05$$

36. (a) Since one of the nine keys is the correct one, $p_1 = \frac{1}{9}$. The first key tried is incorrect, with probability $\frac{8}{9}$, and the second key is correct with probability $\frac{1}{8}$. So, by conditional probabilities, $p_2 = \frac{8}{9} \cdot \frac{1}{8} = \frac{1}{9}$. Similarly, $p_3 = \frac{8}{9} \cdot \frac{7}{8} \cdot \frac{1}{7} = \frac{1}{9}$, and so on.

(b) The expected number of keys to be tried is

$$e_9 = 1 \cdot \frac{1}{9} + 2 \cdot \frac{1}{9} + \cdots + 9 \cdot \frac{1}{9}$$
$$= \frac{1}{9}(1 + 2 + \cdots + 9) = \frac{1}{9} \cdot \frac{9 \cdot 10}{2} = 5,$$

where the formula for the triangular numbers $1 + 2 + \cdots + n = \frac{n(n+1)}{2}$ is used in the case $n = 9$.

37. (a) A triangle with side lengths x, $y - x$, and $1 - y$ must satisfy
$x < y - x + 1 - y = 1 - x$,
$1 - y < x + y - x = y$, $y - x < x + 1 = y$.
These algebraically simplify to the inequalities $x < \frac{1}{2}$, $y > \frac{1}{2}$, $y > x + \frac{1}{2}$.

(b)

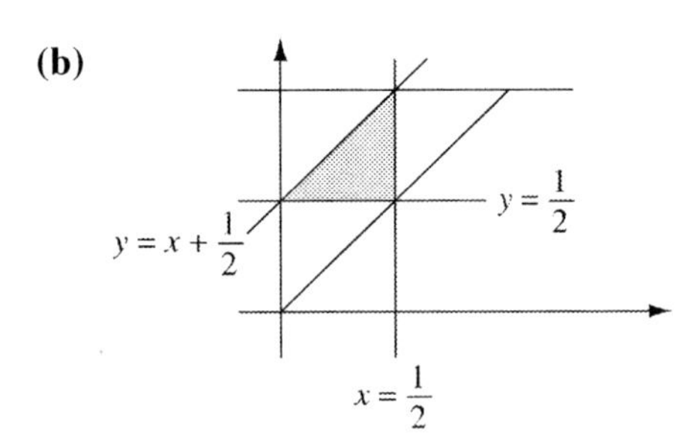

(c) The shaded triangle of part (b) is a right triangle with each leg of length $\frac{1}{2}$, so its area is $\frac{1}{8}$. The sample space is a triangle of area $\frac{1}{2}$, so the ratio of the areas is $\frac{1}{8} \div \frac{1}{2} = \frac{1}{4}$. By geometric probability, this is the probability of forming a triangle by a random choice of the points along the given segment.

38. (a) There are five patterns: four 2-loops (that is the small loops formed by two strings), two 2-loops and a 4-loop, one 2-loop and a 6-loop, two 4-loops, one 8-loop

(c) There are $C(8, 2) = \frac{8 \cdot 7}{2} = 28$ ways to choose the pair of 8 strings to be tied. Four of these pairs give a small loop. Thus, the first knot yields a small 2-loop with probability $\frac{4}{28} = \frac{1}{7}$. In Example 14.33, it was shown that the next three knots each give small loops with probability $\frac{1}{15}$. The probability of 4 small loops is therefore $\left(\frac{1}{7}\right) \cdot \left(\frac{1}{15}\right) = \frac{1}{105}$.

40. (a) $P(W_3) = \frac{7 \cdot 6 \cdot 5}{7 \cdot 7 \cdot 7} = \frac{30}{49}$

$P(\overline{W}_3) = 1 - P(W_3) = 1 - \frac{30}{49} = \frac{19}{49}$

42. (a) $\frac{1}{C(49,6)} = \frac{1}{13{,}983{,}816}$, or about 1 in nearly 14 million

43. (a) There are $C(80, 20)$ ways to draw 20 numbers from the barrel of 80. There are $C(8, 5)$ ways to choose the 5 numbers to match those marked as spots, and $C(72, 15)$ to choose the other 15 numbers from the 72 numbers not marked as spots. This gives the probability:
$\frac{C(8,5) \cdot C(72,15)}{C(80,20)} \approx 0.0183$

44. B. Because Cake B has a $\frac{1}{4}$ chance, Cake C has a $\frac{1}{5}$ chance, and Cakes A and D have a $\frac{1}{6}$ chance. It is the number of equal-sized pieces that determines the probabilities, not the size or shape of the cake.

Chapter 14 Review Exercises

1. Answers will vary. When we did the experiment we obtained the following:

4H and 0T	3H and 1T	2H and 2T	1H and 3T	0H and 4T
	卌	卌 ||||	卌 |	

The empirical probability of obtaining three heads and one tail for our experiment is:

$P_e(\text{3H and 1T}) = \dfrac{5}{20} = 0.25$.

2. Answers will vary. When we did the experiment we obtained the following:

2	3	4	5	6	7	8	9	10	11	12
|	|	||	||	|||	|||	||||	|	||	|	

(a) The empirical probability of obtaining a 3 or a 4 for our experiment is:

$P_e(3 \text{ or } 4) = P_e(3) + P_e(4) = \dfrac{1}{20} + \dfrac{2}{20} = \dfrac{3}{20} = 0.15$.

(b) The empirical probability of obtaining a score of at least 5 for our experiment is:

$$P_e(\text{at least } 5) = P_e(5) + P_e(6) + P_e(7) + P_e(8) + P_e(9) + P_e(10) + P_e(11) + P_e(12)$$
$$= \frac{2}{20} + \frac{3}{20} + \frac{3}{20} + \frac{4}{20} + \frac{1}{20} + \frac{2}{20} + \frac{1}{20} + \frac{0}{20} = \frac{16}{20} = 0.80.$$

3. Answers will vary. We use the results from our experiment shown in problem 2. Given that a score from 5 through 9 occurred, the empirical probability of obtaining a score from 5 through 7 for our experiment is: $P_e\left(5 \text{ or } 6 \text{ or } 7 \mid 5 \text{ or } 6 \text{ or } 7 \text{ or } 8 \text{ or } 9\right) = \dfrac{n(5 \text{ or } 6 \text{ or } 7)}{n(5 \text{ or } 6 \text{ or } 7 \text{ or } 8 \text{ or } 9)} = \dfrac{8}{13} \approx 0.6$.

4. Answers will vary. In our study we obtained the following:

Chocolate	Other
卌 ||	卌 卌 |||

The empirical probability that chocolate is the favorite flavor of ice cream for our survey is:

$P_e(\text{chocolate}) = \dfrac{7}{20} = 0.35$.

5. Answers will vary. When we did the experiment we obtained the following:

Number of tacks landing point up	5	4	3	2	1	0
Number of trials	||	||||	卌 |	卌 |	|	|

(a) The empirical probability that precisely three of the tacks land point up for our experiment is: $P_e(3) = \dfrac{6}{20} = 0.30$.

(b) The empirical probability that two or three of the tacks land point up for our experiment is:

$P_e(2 \text{ or } 3) = P_e(2) + P_e(3) = \dfrac{6}{20} + \dfrac{6}{20} = \dfrac{12}{20} = 0.60$.

6. Answers will vary. When we did the experiment we obtained the following:

Number of rolls to get a 5 or 6	1	2	3	4	5	6	7
Number of trials			II	IIII	III		I

(a) If we choose the average or mean number of rolls for our estimate of the number of rolls required to obtain a 5 or a 6, we obtain $\overline{x} = \frac{3+3+4+4+4+4+5+5+5+7}{10} = 4.4$. Since the number of rolls is a whole number, we estimate that it will take about 4 rolls.

(b) The empirical probability that it takes precisely five rolls to obtain a 5 or 6 for the first time for our experiment is: $P_e(\text{5 rolls to obtain a 5 or 6}) = \frac{3}{10} = 0.30$.

7. Answers will vary. When we did the experiment we obtained the following:

	Heart	Nonheart
Ace	I	II
Non-ace	~~IIII~~ III	~~IIII~~ IIII

(a) The empirical probability of drawing an ace or a heart for our experiment is:

$P_e(\text{ace or heart})$
$= P_e(\text{ace}) + P_e(\text{heart}) - P_e(\text{ace and heart})$
$= \frac{3}{20} + \frac{9}{20} - \frac{1}{20} = \frac{11}{20} = 0.55.$

(b) The empirical probability of drawing the ace of hearts for our experiment is:

$P_e(\text{ace and heart}) = \frac{1}{20} = 0.05$.

(c) Given that the card drawn was a heart, the empirical probability of drawing the ace for our experiment is:

$P_e(\text{ace} \mid \text{heart}) = \frac{1}{9} \approx 0.11$.

8. A match will occur with a probability of about 0.63. Thus, you will usually get around 15 or 16 matches in 25 trials.

9. (a) There are two choices, head (H) or tail (T), for each coin. The possible outcomes for the three coins are:

HHH HHT HTT TTT HTH
THT THH TTH

(b) Use the orderly list in (a) to see there are $\frac{3!}{2!1!} = 3$ ways to obtain two heads and one tail.

10. Draw a Venn Diagram with three overlapping circles and label one circle "F" for French, another circle "E" for English, and the last circle "G" for German. 17 students speak German, French, and English so label the section where all three circles overlap 17. 24 students speak both French and English so the entire overlap between circles F and E must sum to 24. 17 of these students have been recognized so the other overlapping section must be for 24 – 17 = 7 students. 27 students speak German and English with 17 already recognized so the other overlapping section of circles G and E must be for 27 – 17 = 10 students. A total of 38 students speak English so the remaining section of circle E must be for 38 – (17 + 7 + 10) = 4 students. The information provided recognizes the language categories for 38 of the 90 students in Ferry Hall. Since all of the 90 students belong somewhere inside the circles F, E, and G, the three remaining sections (French only, German only, and French and German only) must sum to 90 – 38 = 52.

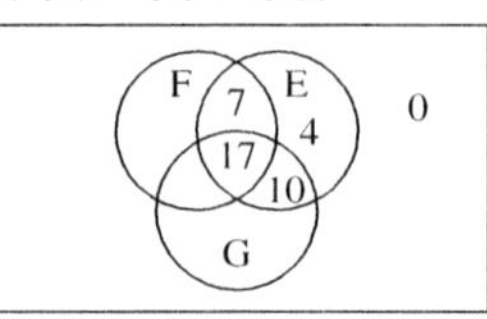

(a) Use the Venn Diagram above and notice that there are only four students who do not speak French or German. Therefore, 90 – 4 = 86 students speak German or French.

(b) Use the Venn Diagram above to find the overlapping section of circles F and E that does not overlap with circle G. There are 7 students who speak French and English but not German.

11. **(a)** If repetition of letters is not allowed, each successive letter after the first has one less possible choice. That is, there are $P(26, 5) = 26 \cdot 25 \cdot 24 \cdot 23 \cdot 22$ $= 7{,}893{,}600$ possible 5-letter words.

(b) There are 5 choices (a, e, i, o, u) for the first letter and 4 choices for the last letter. Removing these two letters leaves 24, 23, and 22 letters for the second, third, and fourth letters. Therefore, there are $5 \cdot 24 \cdot 23 \cdot 22 \cdot 4 = 242{,}880$ possible 5-letter code words that begin and end with vowels.

(c) The three letter sequence can appear in $C(3, 1) = \frac{3}{1} = 3$ sequence positions: *aef* _ _, _ *aef* _, or _ _ *aef*. There are $26 - 3 = 23$ letters remaining for the first blank and 22 letters remaining for the second blank. Therefore, there are $3 \cdot 23 \cdot 22 = 1518$ possible 5-letter code words which contain the sequence *aef*.

12. **(a)** There are two ways to use a single symbol, four ways to use two symbols, and eight ways to use three symbols, so $2 + 4 + 8 = 14$ letters can be formed if at most three symbols are allowed.

(b) There are 16 ways to use four symbols, so $2 + 4 + 8 + 16 = 30$ letters can be formed using up to four symbols. Therefore, four symbols are sufficient to code each letter.

13. The $10^4 = 10{,}000$ combinations can all be tried in $10{,}000 \div 60 = 166.666\ldots$ minutes, which is a bit under 2 hours and 47 minutes.

14. **(a)** There are five digits that still look like a digit when upside down: 0, 1, 6, 8, and 9. Therefore, there are $5^5 = 3125$ ZIP numbers that still appear to be ZIP numbers when read upside down.

(b) For a number to be detour prone, the middle digit must be unchanged when read upside down, so it is a 0, 1, or 8 on reversible, non-detour prone numbers such as 99166. Since these reversible, non-detour prone numbers have first and second digits that are the same upside down as the fourth and fifth digits, there are $5 \cdot 5 \cdot 3 = 75$ ZIP numbers that are unchanged when read upside down. This means that $3125 - 75 = 3050$ ZIP numbers are detour prone.

15.

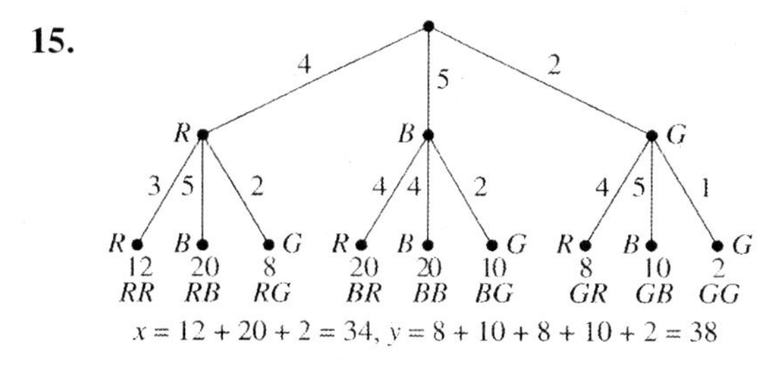

$x = 12 + 20 + 2 = 34$, $y = 8 + 10 + 8 + 10 + 2 = 38$

16. **(a)** There are

$$C(15, 9) = \frac{15 \cdot 14 \cdot 13 \cdot 12 \cdot 11 \cdot 10 \cdot 9 \cdot 8 \cdot 7}{9 \cdot 8 \cdot 7 \cdot 6 \cdot 5 \cdot 4 \cdot 3 \cdot 2 \cdot 1} = \frac{1{,}816{,}214{,}400}{362{,}880} = 5005$$

ways to choose nine players from among 15 players without regard to the position of the player.

(b) There are $C(2, 1) = \frac{2}{1} = 2$ ways to select a pitcher and $C(3, 1) = \frac{3}{1} = 3$ ways to select a catcher. Once the pitcher and catcher have been selected, there are 13 players remaining to fill the other seven positions. The rest of the team can be selected in

$$C(13, 7) = \frac{13 \cdot 12 \cdot 11 \cdot 10 \cdot 9 \cdot 8 \cdot 7}{7 \cdot 6 \cdot 5 \cdot 4 \cdot 3 \cdot 2 \cdot 1} = \frac{8{,}648{,}640}{5040} = 1716 \text{ ways.}$$

Therefore, the team can be selected in $2 \cdot 3 \cdot 1716 = 10{,}296$ ways.

17. **(a)** There are 13 clubs in an ordinary deck of playing cards so there are $C(13, 2) = \frac{13 \cdot 12}{2 \cdot 1} = \frac{156}{2} = 78$ ways to select two clubs.

(b) The face cards are jacks, queens, and kings and there are 12 all told in an ordinary deck of playing cards. Therefore, there are $C(12, 2) = \frac{12 \cdot 11}{2 \cdot 1} = \frac{132}{2} = 66$ ways to select two face cards.

(c) Use $n(A \text{ or } B) = n(A) + n(B) - n(A \cap B)$ where event A is to select two clubs and event B is to select two face cards. From (a), $n(A) = 78$ and from (b), $n(B) = 66$. There are 3 face cards that are clubs so there are $C(3, 2) = \frac{3 \cdot 2}{2 \cdot 1} = 3$ ways to choose two face cards that are clubs. Therefore, $n(A \text{ or } B) = 78 + 66 - 3 = 141$. There are 141 ways to select two clubs or two face cards from an ordinary deck of playing cards.

18. (a) STREETS has two S's, two T's, one R, and two E's. These letters can be arranged in $\frac{7!}{2!2!1!2!} = \frac{7 \cdot 6 \cdot 5 \cdot 4 \cdot 3 \cdot 2 \cdot 1}{2 \cdot 1 \cdot 2 \cdot 1 \cdot 1 \cdot 2 \cdot 1}$ $= \frac{5040}{8} = 630$ recognizably different orders.

(b) Now, STREETS has two S's, two T's, one R, and *one* EE. These letters can be arranged in

$$\frac{6!}{2!2!1!1!} = \frac{6 \cdot 5 \cdot 4 \cdot 3 \cdot 2 \cdot 1}{2 \cdot 1 \cdot 2 \cdot 1 \cdot 1 \cdot 1} = \frac{720}{4} = 180$$

recognizably different orders with two E's adjacent.

(c) Since one T starts the word, there are two S's, one T, one R, and two E's for the remaining positions in the word. These letters can be arranged in

$$\frac{6!}{2!1!1!2!} = \frac{6 \cdot 5 \cdot 4 \cdot 3 \cdot 2 \cdot 1}{2 \cdot 1 \cdot 1 \cdot 1 \cdot 2 \cdot 1} = \frac{720}{4} = 180$$

recognizably different orders.

19. (a) With no restriction, the children can line up in $5! = 120$ ways.

(b) If the two boys are side-by-side, there are $2 \cdot 4! = 48$ ways to line up.

(c) If no girls are side-by-side, there are $3! \cdot 2! = 12$ ways to line up.

20. (a)

The diagram shows the 6 points of intersection.

(b)

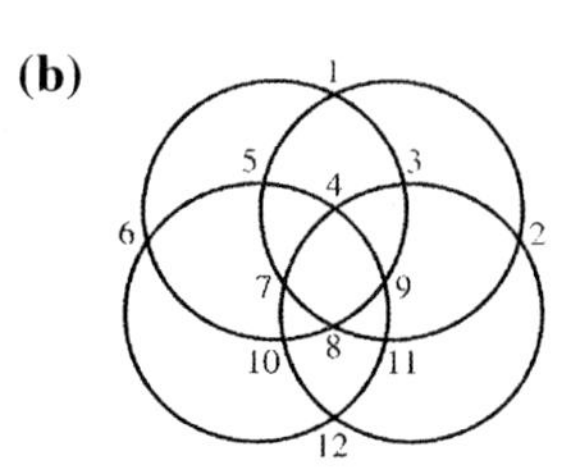

The diagram shows the 12 points of intersection.

(c) There are $C(n, 2) = \frac{n(n-1)}{2}$ ways to choose a pair of circles, and each pair creates up to two points of intersection. Altogether, the n circles can intersect in up to $n(n-1)$ points.

21. (a)

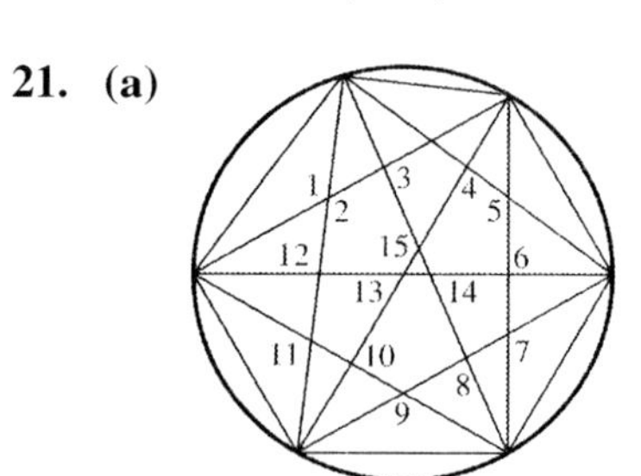

The diagram shows the 15 points of intersection.

(b) The location of the four endpoints of the intersecting chords determines the point of intersection.

(c) There are $C(n, 4)$ ways to choose a set of 4 of the n points, and each choice gives one point of intersection. Therefore, the number of points of intersection is $C(n, 4)$.

22. (a) There are two possible outcomes for each of the two coins, head (H) or tail (T), and six possible outcomes for the die (1, 2, 3, 4, 5, 6).

$$S = \begin{Bmatrix} \text{HH1} & \text{HH2} & \text{HH3} & \text{HH4} & \text{HH5} & \text{HH6} \\ \text{HT1} & \text{HT2} & \text{HT3} & \text{HT4} & \text{HT5} & \text{HT6} \\ \text{TH1} & \text{TH2} & \text{TH3} & \text{TH4} & \text{TH5} & \text{TH6} \\ \text{TT1} & \text{TT2} & \text{TT3} & \text{TT4} & \text{TT5} & \text{TT6} \end{Bmatrix}$$

(b) From (a), there are 24 equally likely outcomes and TT5 is one of those outcomes. Therefore,

$$P(\text{T}, \text{T}, 5) = \frac{1}{24} \approx 0.04.$$

23. Use the sample space in exercise 22(a) to note six outcomes that have tails for both coins. One of those six outcomes has a five for the die. Therefore, $P(5 \mid \text{T, T}) = \frac{1}{6} \approx 0.17$.

24.

Sum	2	3	4	5	6	7	8	9	10	11	12
Probability	$\frac{1}{36}$	$\frac{2}{36}$	$\frac{3}{36}$	$\frac{4}{36}$	$\frac{5}{36}$	$\frac{6}{36}$	$\frac{5}{36}$	$\frac{4}{36}$	$\frac{3}{36}$	$\frac{2}{36}$	$\frac{1}{36}$

Solution 1: To compute the probability of obtaining a sum of at most 11, add the probabilities associated with sums of 2 through 11.

$$P(\text{sum at most } 11) = \frac{1}{36}+\frac{2}{36}+\frac{3}{36}+\frac{4}{36}+\frac{5}{36}+\frac{6}{36}+\frac{5}{36}+\frac{4}{36}+\frac{3}{36}+\frac{2}{36} = \frac{35}{36} \approx 0.97$$

Solution 2: If you fail to obtain a sum of at most 11 you obtain a sum of 12.
Use $P(A) = 1 - P(\bar{A})$ where A is a sum of at most 11. Then,

$$P(\text{sum at most } 11) = 1 - P(\text{sum of } 12) = 1 - \frac{1}{36} = \frac{35}{36} \approx 0.97.$$

25. $S = \{5 \text{ white}, 6 \text{ red}, 4 \text{ black}\}$
There are

$$C(5+6+4,\ 2) = C(15,\ 2) = \frac{15 \cdot 14}{2 \cdot 1} = \frac{210}{2} = 105$$

ways to randomly select two balls.

(a) There are $C(5,\ 2) = \frac{5 \cdot 4}{2 \cdot 1} = \frac{20}{2} = 10$ ways to select two white balls,

$C(6,\ 2) = \frac{6 \cdot 5}{2 \cdot 1} = \frac{30}{2} = 15$ ways to select two red balls, and

$C(4,\ 2) = \frac{4 \cdot 3}{2 \cdot 1} = \frac{12}{2} = 6$ ways to select two black balls. Therefore,

$P(2 \text{ balls the same color})$

$$= \frac{C(5,\ 2) + C(6,\ 2) + C(4,\ 2)}{C(15,\ 2)} = \frac{10+15+6}{105} = \frac{31}{105} \approx 0.30$$

(b) From (a), there are $C(5, 2) = 10$ ways to select two white balls, thus

$$P(2 \text{ white}) = \frac{C(5,\ 2)}{C(15,\ 2)} = \frac{10}{105} \approx 0.10.$$

(c) From (a), there are $C(5, 2) + C(6, 2) + C(4, 2) = 10 + 15 + 6 = 31$ ways to select two balls the same color, and $C(5, 2) = 10$ ways to select two white balls. Therefore,

$P(2 \text{ white} \mid 2 \text{ same color})$

$$= \frac{C(5,\ 2)}{C(5,\ 2) + C(6,\ 2) + C(4,\ 2)} = \frac{10}{31} \approx 0.32.$$

(d) See below for possibility tree.

$$P(\text{same color}) = \frac{2}{21} + \frac{1}{7} + \frac{2}{35} = \frac{10+15+6}{105} = \frac{31}{105}$$

$$P(\text{WW}) = \frac{2}{21}$$

$P(\text{WW} \mid \text{same color})$

$$= \frac{P(\text{WW \& same color})}{P(\text{same color})} = \frac{2/21}{31/105} = \frac{10}{31}$$

25. (d)

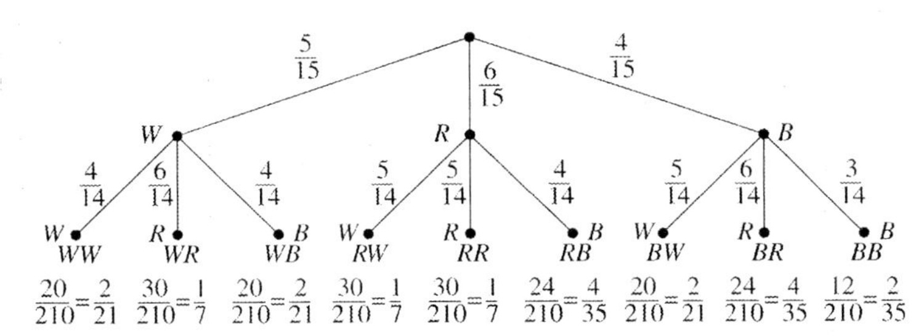

26. There are 12 equal size outer sections, each with central angle $\frac{360°}{12} = 30°$. There are 4 equal sized inner sections, each with central angle $\frac{360°}{4} = 90°$. The probabilities are found using the appropriate ratios of angular measures.

(a) Sections *b* and 8 do not overlap so $b \cap 8 = \varnothing$. Therefore, $P(\text{b and } 8) = 0$.

(b) Since b and 8 do not overlap, they are mutually exclusive. Sum the probability for each event to find:

$$P(\text{b or } 8) = P(\text{b}) + P(8) = \frac{90°}{360°} + \frac{30°}{360°} = \frac{120°}{360°} = \frac{1}{3} \approx 0.33.$$

(c) Given that the spinner lands in section 8, it cannot land in section *b*. Therefore, $P(\text{b}|8) = 0$.

(d) Section 2 is a proper subset of section *b*. Therefore,

$$P(\text{b and } 2) = P(2) = \frac{30°}{360°} \approx 0.08.$$

(e) Sections *b* and 2 are not mutually exclusive, therefore,

$$P(\text{b or } 2) = P(\text{b}) + P(2) - P(\text{b and } 2) = \frac{90°}{360°} + \frac{30°}{360°} - \frac{30°}{360°} = \frac{90°}{360°} = 0.25.$$

(f) Given that the spinner lands in section *b*, it can land on Sections 1, 2, or 3. Therefore, use (d) to find:

$$P(2|\text{b}) = \frac{P(2 \text{ and b})}{P(\text{b})} = \frac{\frac{30°}{360°}}{\frac{90°}{360°}} = \frac{30°}{90°} = \frac{1}{3} \approx 0.33.$$

27. The possible outcomes for tossing three coins are:

HHH	HHT	HTT	TTT
HTH	THT	THH	TTH

Each of these eight possible outcomes are equally likely.

(a) Use the list of outcomes above to note there are three ways to get two heads and one tail so there are five ways to not get two heads and one tail. The odds in favor of two heads and one tail are:

$n(\text{2H and 1T})$ to $n\left(\overline{\text{2H and 1T}}\right)$ or $3:5$.

(b) There is only one outcome of three heads so there are seven outcomes that are not three heads. The odds in favor of three heads are: $n(\text{HHH})$ to $n\left(\overline{\text{HHH}}\right)$ or $1:7$.

28. **(a)** Using the method of Example 10.35, the odds in favor of 4 are given by the ratio $\frac{P(A)}{1-P(A)} = \frac{0.85}{1-0.85} = \frac{0.85}{0.15} = \frac{17}{3}$ which is often written as 17:3.

(b) Use $n(S) = n(A) + n(\bar{A}) = 17 + 8 = 25$; then $P(A) = \frac{n(A)}{n(S)} = \frac{17}{25} = 0.68$.

29. **(a)** Expected value = \$0(0.15) + \$5(0.50) + \$10(0.25) + \$20(0.10)
= \$2.50 + \$2.50 + \$2.00 = \$7.00
The expected value of the game is \$7.00.

(b) You expect to collect \$7.00 per play but you must pay \$10.00 per play for an average loss of \$3.00 over the long run. If you want to make money, it is not wise to play. If your enjoyment during the play of the game is worth at least \$3.00 per play, then you can play for fun.

30. There are 8 equally likely possibilities for the 3 children, BBB, BBG, BGB, BGG, GBB, GBG, GGB, and GGG. However, the family has at least one girl, so there are 7 remaining equally likely possibilities. Out of those 7, only 3 have two boys. So the probability that the other two children are both boys is $\frac{3}{7}$.

Chapter 14 Test

1. **(a)** $7! = 7 \cdot 6 \cdot 5 \cdot 4 \cdot 3 \cdot 2 \cdot 1 = 5040$

(b) $\frac{9!}{6!} = \frac{9 \cdot 8 \cdot 7 \cdot 6!}{6!} = 9 \cdot 8 \cdot 7 = 504$

(c) $\dfrac{8!}{(8-8)!} = \dfrac{8!}{0!} = \dfrac{8\cdot7\cdot6\cdot5\cdot4\cdot3\cdot2\cdot1}{1}$
$= 40,320$

(d) $7 \cdot 6! = 7! = 5040$

(e) $P(8, 5) = 8 \cdot 7 \cdot 6 \cdot 5 \cdot 4 = 6720$

(f) $P(8, 8) = 8 \cdot 7 \cdot 6 \cdot 5 \cdot 4 \cdot 3 \cdot 2 \cdot 1 = 8!$
$= 40,320$

(g) $C(9, 3) = \dfrac{9\cdot8\cdot7}{3\cdot2\cdot1} = \dfrac{504}{6} = 84$

(h) $C(9, 9) = \dfrac{9\cdot8\cdot7\cdot6\cdot5\cdot4\cdot3\cdot2\cdot1}{9\cdot8\cdot7\cdot6\cdot5\cdot4\cdot3\cdot2\cdot1} = \dfrac{9!}{9!} = 1$

2. $S = \begin{Bmatrix} Y_1, Y_2, Y_3, Y_4, Y_5, B_1, B_2, B_3, B_4, \\ G_1, G_2, G_3, G_4, G_5, G_6, G_7, G_8 \end{Bmatrix}$

(a) There are
$C(8, 5) = \dfrac{8\cdot7\cdot6\cdot5\cdot4}{5\cdot4\cdot3\cdot2\cdot1} = \dfrac{6720}{120} = 56$ ways
to select five green marbles.

(b) There are
$C(5, 5)\cdot C(8, 5) = \dfrac{5\cdot4\cdot3\cdot2\cdot1}{5\cdot4\cdot3\cdot2\cdot1}\cdot\dfrac{8\cdot7\cdot6\cdot5\cdot4}{5\cdot4\cdot3\cdot2\cdot1}$
$= \dfrac{6720}{120} = 56$
ways to select five yellow and five of the eight green marbles.

(c) From (a), there are 56 ways to select five green marbles. There is only
$C(5, 5) = \dfrac{5\cdot4\cdot3\cdot2\cdot1}{5\cdot4\cdot3\cdot2\cdot1} = \dfrac{5!}{5!} = 1$ way to
select five yellow marbles. Therefore, there are 56 + 1 = 57 ways to choose either all five of the five yellow or five of the eight green marbles.

3. Since we are given that 3 of the balls selected are green, we must select 2 more of the 14 balls in $C(14, 2)$ ways. But we can select 2 yellow balls in $C(5, 2)$ ways. Thus,
$P(2 \text{ yellow} \mid 3 \text{ green})$

$= \dfrac{C(5, 2)}{C(14\cdot2)} = \dfrac{\frac{5\cdot4}{1\cdot2}}{\frac{14\cdot13}{1\cdot2}} \approx 0.11 .$

4. Prepare a card as shown and ask a number of people to choose number. Calculate the empirical probability of choosing 3 as the number of times 3 is chosen divided by the number of people questioned.

5. The odds in favor of E are:
$\dfrac{P(E)}{1-P(E)} = \dfrac{0.35}{1-0.35} = \dfrac{0.35}{0.65} = \dfrac{7}{13}$ or 7:13 .

6. (a) There are seven choices for each letter so
$7\cdot7\cdot7\cdot7 = 7^4 = 2401$ 4-letter code words are possible when repetition of letters is allowed.

(b) There are $P(7, 4) = 7 \cdot 6 \cdot 5 \cdot 4 = 840$ 4-letter code words when repetition of letters is not allowed.

7. (a) There are two vowels available to begin the word so the first letter has two possible choices. That leaves $P(6, 3) = 6 \cdot 5 \cdot 4 = 120$ ways to choose the remaining three letters. Therefore, there are $2 \cdot 120 = 240$ 4-letter words beginning with a vowel when repetition of letters is not allowed.

(b) Think of *cd* as a single unit. There are 3 ways to position *cd* in a four letter code word and the other two positions can be filled in $5 \cdot 4 = 20$ ways. Therefore, there are $3 \cdot 20 = 60$ four letter code words with *c* followed by *d*. Similarly, there are 60 such words with *d* followed by *c*. Therefore, there are 60 + 60 = 120 four letter code words with *c* and *d* adjacent.

8. This would be an empirical probability obtained by keeping records for a large number of trials of treating strep throat with penicillin.

9. Draw a Venn Diagram of three overlapping circles. Label one circle "F" for French, another circle "G" for German, and the remaining circle "C" for Chinese. One student studies all three languages so write "1" in the section where all three circles overlap. Two students study French and Chinese so write 2 – 1 = "1" in the empty section of the overlapping part of circles F and C. Three students study German and Chinese so write 3 – 1 = "2" in the empty section of the overlapping part of circles G and C. Twelve students study German and French so write 12 – 1 = "11" in the empty section of the overlapping part of circles G and F.

(*continued on next page*)

(*continued*)

17 students study Chinese so write 17 – (1 + 1 + 2) = "13" in the empty section of circle C. 29 students study German so write 29 – (11 + 1 + 2) = "15" in the empty section of circle G. 27 students study French so write 27 – (11 + 1 + 1) = "14" in the empty section of circle F. All of Mrs. Spangler's calculus students study at least one foreign language so there are no students outside the circles in the Venn Diagram.

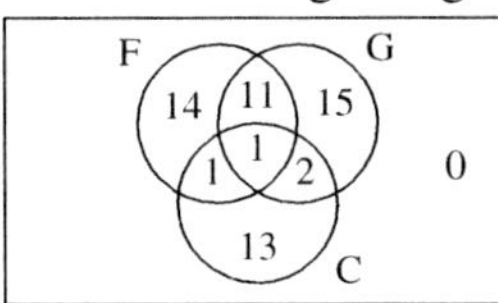

(a) To find the number of students in Mrs. Spangler's class, add the numbers in the circles on the Venn Diagram:
14 + 11 + 15 + 1 + 1 + 2 + 13 = 57

(b) Find the nonoverlapping section of circle C. 13 students only study Chinese.

(c) Find the overlapping section of circles F and G that does not include circle C. 11 students study French and German but not Chinese.

10. There are five yellow marbles and 4 + 8 = 12 nonyellow marbles. The odds in favor of selecting a yellow marble is: $n(\text{yellow})$ to $n\left(\overline{\text{yellow}}\right)$ or 5:12.

11.

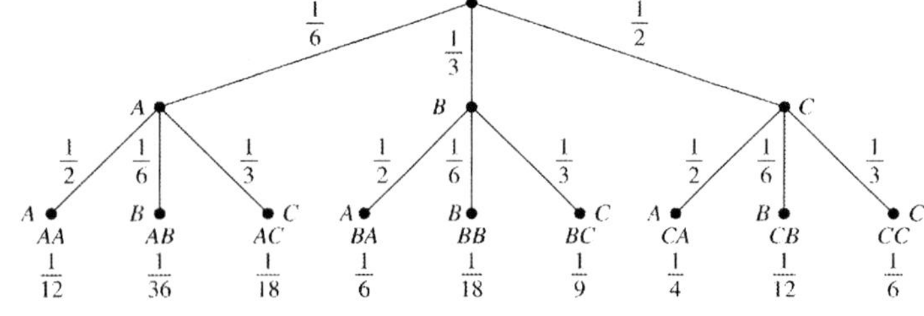

(a) $P(AA \cup BB \cup CC)$
$= P(AA) + P(BB) + P(CC)$
$= \frac{1}{12} + \frac{1}{18} + \frac{1}{6} = \frac{11}{36}$

(b) $P(AA \cup AB \cup AC \cup BA \cup CA) = P(AA) + P(AB) + P(AC) + P(BA) + P(CA)$
$= \frac{1}{12} + \frac{1}{36} + \frac{1}{18} + \frac{1}{6} + \frac{1}{4} = \frac{3+1+2+6+9}{36} = \frac{21}{36} = \frac{7}{12}$

or

$P(A \text{ on red cube} \cup A \text{ on blue cube})$
$= P(A \text{ on red cube}) + (A \text{ on blue cube}) - P(A \text{ on red cube and } A \text{ on blue cube})$
$= \frac{1}{6} + \frac{1}{2} - \frac{1}{12} = \frac{21}{36} = \frac{7}{12}$

Appendix B Getting the Most Out of Your Calculator

Problem Set B.1

1. **(a)** 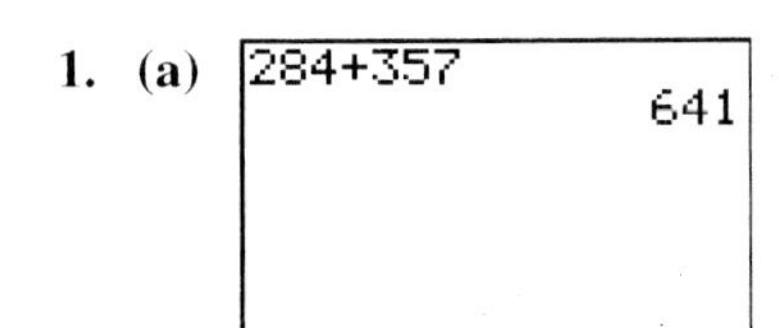

(c) 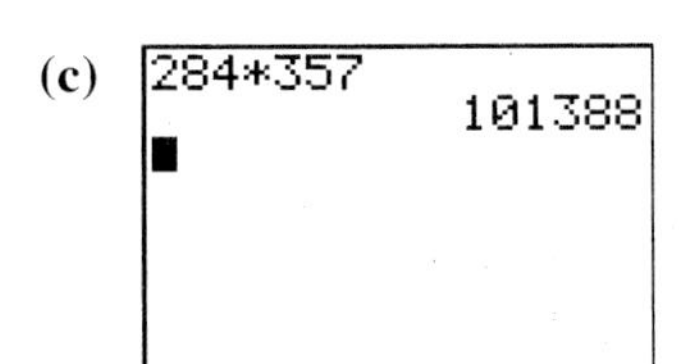

(e) 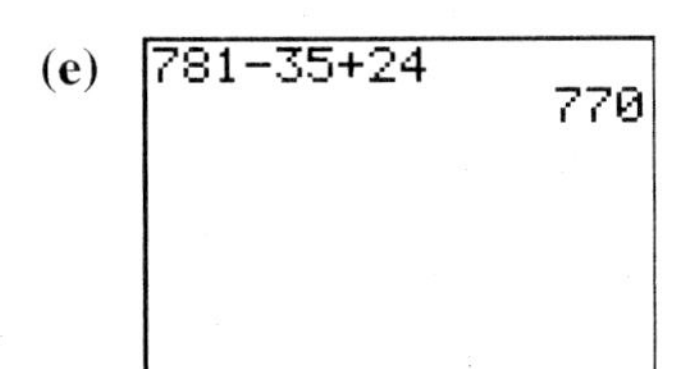

(g) 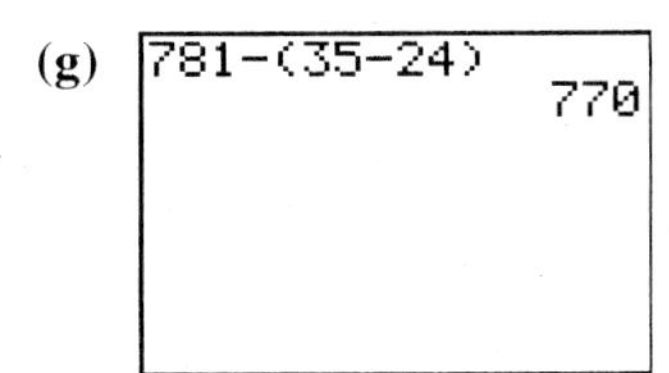

(i) 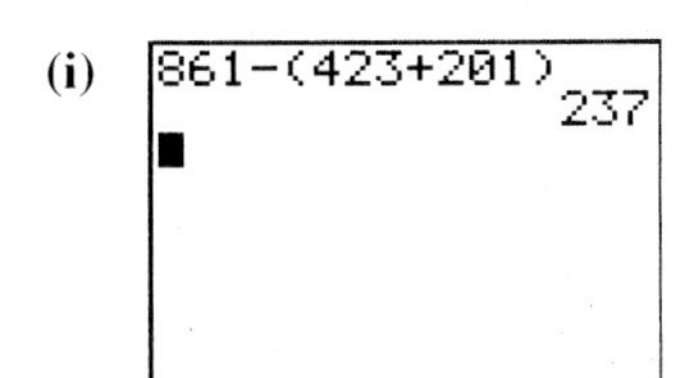

2. **(a)** 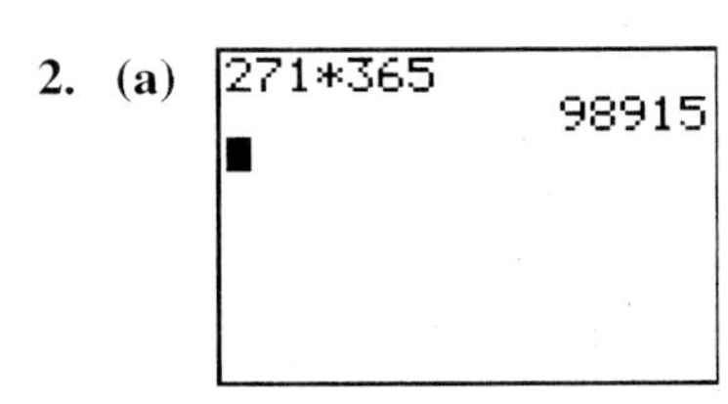

(c) 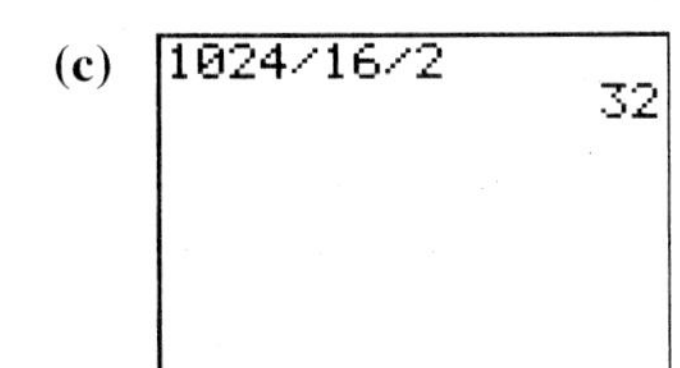

3. **(a)** 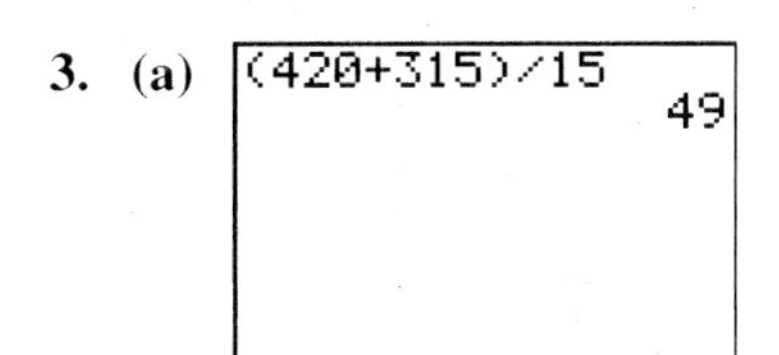

(c) 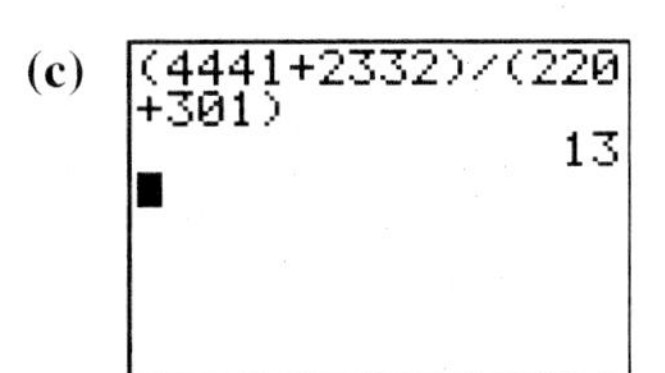

4. **(a)** 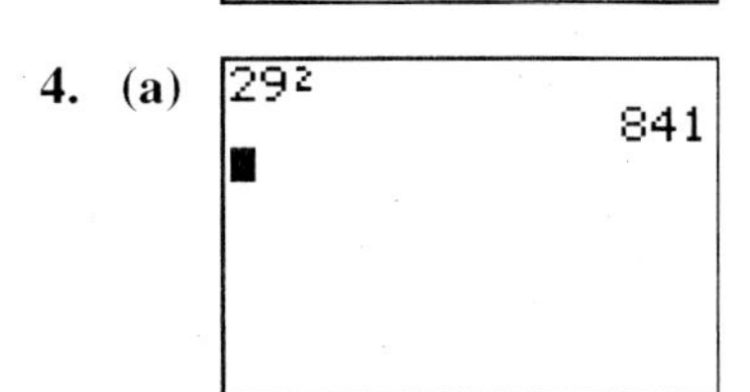

(c) 

5. **(a)** 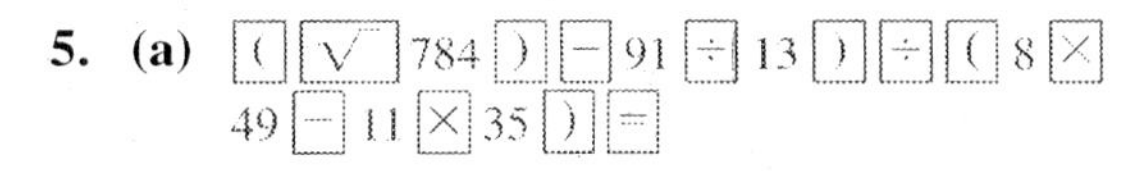

7. **(a)** $1831-(17\times 28)+34$

13. **(a)** The [=] key needs to be pressed 14 times.
5 [M+] [+] 3 [=] [M+] [=] [M+] [=] . . .
[=] [M+] [MR], 390

15. We need both a sum of the data and a sum of the squared data. This can be accomplished by summing each new datum x_i into the memory before its value is squared. Use the following keystrokes:

4 [×] [(] 6.59 [M+] [x^2] [+] 13.04 [M+] [x^2] [+] 4.89 [M+] [x^2] [+] 7.48 [M+] [x^2] [)] [=] 8

The following additional keystrokes compute the standard deviation:

[−] [MR] [x^2] [=] [√] [÷] 4 [=] 3.055

(*continued on next page*)

(*continued*)

$$\bar{x} = \frac{6.59 + 13.04 + 4.89 + 7.48}{4} = 8$$

$$s = \frac{1}{4}\sqrt{\begin{array}{l} 4\left(6.59^2 + 13.04^2 + 4.89^2 + 7.48^2\right) \\ \quad - (6.59 + 13.04 + 4.89 + 7.48)^2 \end{array}}$$
$$\approx 3.055$$

16. a. The algorithm below generates the Lucas numbers. We show, at each step, the entry, the value of x in the display, and the value M in the memory. The Lucas numbers are highlighted.

Entry	**1**	[M+]	**3**	[+]
x	**1**	1	3	**3**
M	0	1	1	1
Entry	[M+]	[MR]	[+]	[M+]
x	3	**4**	**7**	7
M	4	4	4	11
Entry	[MR]	[+]	[M+]	[MR]
x	**11**	**18**	18	**29**
M	11	11	29	29
Entry	[+]	...		
x	**47**	...		
M	29	...		

17. (c) The coefficients are the successive Fibonacci numbers $\tau^7 - 13\tau = 8$, $\tau^8 - 21\tau = 13$, and $\tau^9 - 34\tau = 21$. In general, we expect that $\tau^n - F_n\tau = F_{n-1}$.

(e) Notice that 3, 4, 7, 11, and 18 are Lucas numbers, so we surmise that the next three equations are $\left\{\tau^7\right\} = 29$, $\left\{\tau^8\right\} = 47$, and $\left\{\tau^9\right\} = 76$. This can be verified as follows:

$\tau^7 \approx 29.034$, so $\left\{\tau^7\right\} = 29$.

$\tau^8 \approx 46.979$, so $\left\{\tau^8\right\} = 47$.

$\tau^9 \approx 76.013$, so $\left\{\tau^9\right\} = 76$.

This can be generalized as $\left\{\tau^n\right\} = L_n$, $n \geq 2$.

Appendix C A Brief Guide to *The Geometer's Sketchpad*

Problem Set C.1

1.

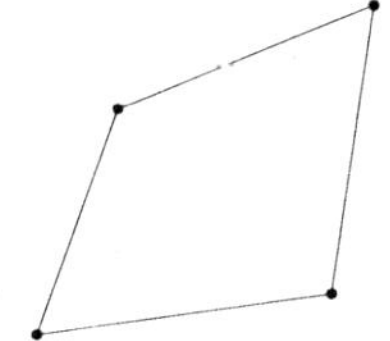

(a) AB = 3.27 cm
BC = 3.25 cm
CD = 3.27 cm
DA = 3.25 cm

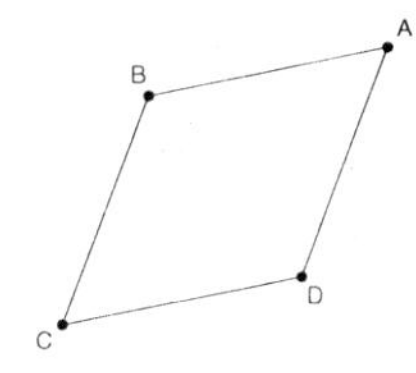

The rhombus is a drawing. It has not been constructed to have the property of having sides of equal length. It has only been drawn to look as if it has sides of equal length.

2. (a) Construct the midpoint M of the segment $\overline{AB}$. Construct the perpendicular bisector to the segment through M. Construct any point $C \neq M$ on the perpendicular bisector. Triangle ABC is isosceles.

(c) Construct any point M on the perpendicular bisector of segment $\overline{AB}$. Construct a line parallel to $\overline{AB}$ through point M. Construct any circle centered at M. If the circle intersects the parallel line at points C and D, then quadrilateral $ABCD$ is an isosceles trapezoid.

(e) Construct the segment $\overline{AC}$ and the perpendicular bisector of the segment. Construct a point B on the perpendicular bisector. Construct a circle centered at A through B. If the circle intersects the perpendicular bisector at D, then $ABCD$ is a rhombus with opposite vertices at A and C.

4.

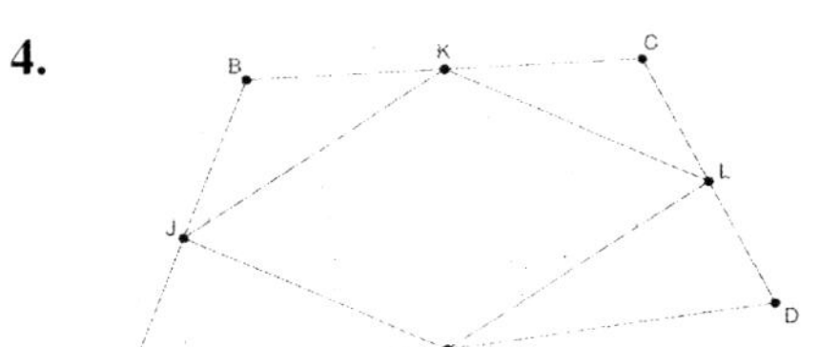

(a)

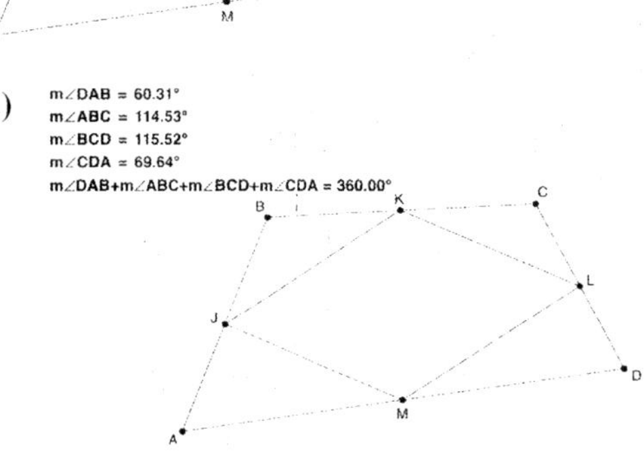

The sum of the angle measures is 360°.

5. (a)

m∠ ABC = 90.00°

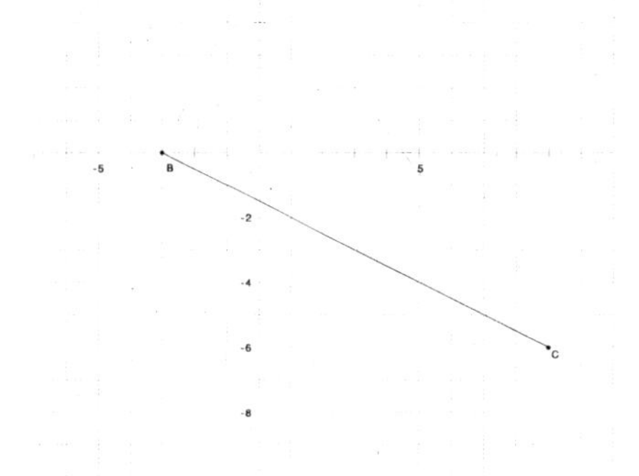

There are several methods to show that triangle CDE is a right triangle with the right angle at vertex D. *Method One*: Use **Length** to measure the lengths CD, DE, and CE, and use **Calculate**... to show that $CD^2 + DE^2 = CE^2$. Triangle CDE is a right triangle by the converse of the Pythagorean theorem.
Method Two: Measure the slopes of $\overline{CD}$ and $\overline{DE}$, and show that the product of the slopes is -1. This shows that the segments are perpendicular to one another.
Method Three: Measure angle CDE and see that it is 90°.Since one of the angles of the triangle is 90°, it is a right triangle.

6. (a)

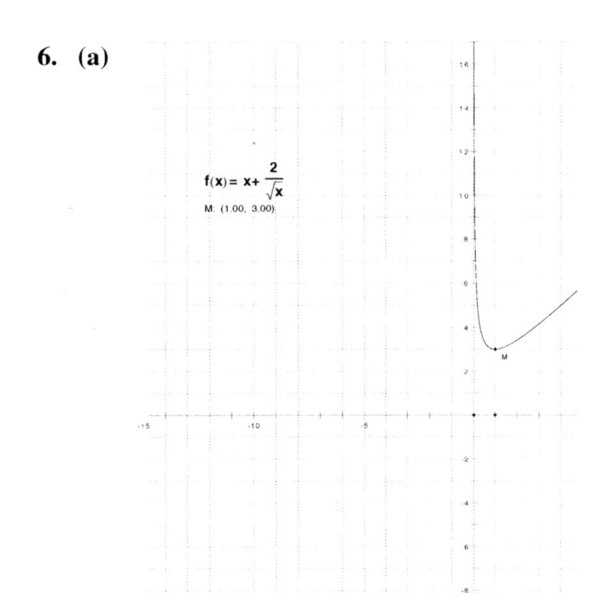

The coordinates of the minimum value of the function are (1, 3).